▪ SCIENCE INCARNATE ▪

■ SCIENCE INCARNATE ■

Historical Embodiments of Natural Knowledge

EDITED BY

CHRISTOPHER LAWRENCE

AND

STEVEN SHAPIN

THE UNIVERSITY OF CHICAGO PRESS

Chicago & London

CHRISTOPHER LAWRENCE is reader at the Wellcome
Institute for the History of Medicine, London.
STEVEN SHAPIN is professor in the Department of Sociology
at the University of California, San Diego.

THE UNIVERSITY OF CHICAGO PRESS, CHICAGO 60637
THE UNIVERSITY OF CHICAGO PRESS, LTD., LONDON
© 1998 by The University of Chicago
All rights reserved. Published 1998
Printed in the United States of America

07 06 05 04 03 02 01 00 99 98   1 2 3 4 5

ISBN: 0-226-47012-1 (cloth)
ISBN: 0-226-47014-8 (paper)

Library of Congress Cataloging-in-Publication Data

Science incarnate : historical embodiments of natural knowledge /
   edited by Christopher Lawrence and Steven Shapin.
      p.   cm.
   Includes bibliographical references and index.
   ISBN 0-226-47012-1 (cloth : alk. paper).—ISBN 0-226-47014-8
(pbk. : alk. paper)
      1. Science—Social aspects.   2. Science—Philosophy.   3. Mind and
body.   I. Lawrence, Christopher, 1947– .   II. Shapin, Steven.
Q175.5.S3645   1998
303.48′3—dc21   97-28762
                                                            CIP

⊗ The paper used in this publication meets the minimum requirements of the American
National Standard for Information Sciences—Permanence of Paper for Printed Library
Materials, ANSI Z39.48-1992.

CONTENTS

ACKNOWLEDGMENTS

EARLY VERSIONS OF some of the chapters in this volume were delivered as papers to a one-day conference entitled "The Body of Knowledge" held 28 March 1994 at the Wellcome Institute for the History of Medicine in London. We thank the Institute for its support in organizing that conference and during the editing process.

Part of the editing work was done while Shapin was a Fellow at the Center for Advanced Study in the Behavioral Sciences, Stanford, California. He is grateful for the center's support and for financial support provided by The Andrew W. Mellon Foundation.

INTRODUCTION
The Body of Knowledge

STEVEN SHAPIN AND CHRISTOPHER LAWRENCE

■ *Why It Is Mysterious, Funny, and Important* ■

HOW IS ONE TO THINK about the relationship between the body and the body of knowledge? In dominant modernist sensibilities one can take that question only so far before one runs up against either a big mystery or a bad joke. In the seventeenth century Descartes argued that the world contained two qualitatively distinct sorts of stuff—mental and material—mysteriously juxtaposed in the brain's pineal gland. That juxtaposition offered an avenue for reciprocal influence. The passions, considered as physiological states of the body, might influence mental states, and the passions, in turn, might be brought under rational control. Indeed, this book seeks to recover the extraordinarily rich repertoires we once possessed for speaking about the bodily circumstances that either assisted or handicapped the processes by which genuine knowledge was to be attained. It was not so very long ago that our official culture abundantly testified to the significance for knowledge-making processes of intellectuals' humors, complexions, constitutions, dietetics, regimens, habitations, and habits of living. The way we lived, that is to say, was once understood to be intimately connected to the way we think.

Yet even in these past cultures such notions as "knowledge itself," "free-floating concepts," "disembodied ideas," "Truth"—whatever the corporeal conditions of their production, or even of their validation—shuffled off their mortal coils. Descartes's scheme was one of many ancient and early modern frameworks for understanding the reciprocal influence of bodily and mental states, but it was not considered that the corporeal circumstances that bore upon the *processes* of thinking were attached to the *products* of rightly conducted rational thought.[1] The mystery of disembodiment occurred somewhere between corporeal process and intellectual product. The disembodiment of "knowledge itself" was understood to make that knowledge what it was and to give it value. In dominant Western cultural traditions knowledge itself was to be identified

---

1. See, e.g., Porter, "History of the Body," esp. 206–13; idem, *"Barely Touching"*; Rousseau and Porter, "Introduction: Toward a Natural History of Mind and Body."

1

neither with its ink-on-paper and leather-bound inscriptions nor with the embodied human processes by which ideas were generated, represented, communicated, sustained, and justified.[2]

Accordingly, intellectuals have enshrined their products, and the objects of
their interpretations, in the transcendental and disembodied domain that knowledge itself was understood to inhabit. Such versions of where knowledge was located did not need to be justified; in much academic discourse they have been
taken for granted. In dominant sensibilities, therefore, the response to the mystery inscribed in the relationship between embodied knowledge-making and disembodied knowledge itself was just not to talk about it. Indeed, not talking about
it has been a way of ensuring its continued mysterious status.

If big mystery resides in the ineffable point of contact between the material
and the authentically intellectual, the bad joke that accompanies it has classic
form: the iconoclastic juxtaposition of high and low, lèse-majesté, the carnivalesque.[3] What has truth to do with the belly? (Ever heard the one about the
Wittgenstein cookbook? Or the Oppenheimer abs-toner?) What difference does
it make to knowledge itself whether Einstein rode a bicycle, whether Russell was
randy, or whether Darwin was flatulent? To bring body and knowledge into contact in these ways is occasionally taken as funny, sometimes as enraging, more
often just as pointless.[4]

Yet the claimed pointlessness of such juxtaposition presents itself as one of
the central problems faced by much modern intellectual biography and related
genres. What *is* the relationship between embodied lives and disembodied knowledge? How are we enjoined or permitted to talk about that relationship, and
with what consequences for our culture's self-understanding? Give no account
of what thinkers looked like, how and whom they loved, the quotidian rhythms
and textures of their lives, and you produce accounts that are widely judged to
be flat, hollow, lifeless, and ultimately unreadable. Give rich and detailed accounts of physiognomy, passions, habits, and regimens, and you will invite from
academic readers (at least) puzzled inquiries about what all this can possibly
have to do with "knowledge itself."[5]

2. Of course, in official stories, embodiment—like situatedness—is not only tolerated but insisted upon when ideas of *low value* (folklore, common sense, the products of practical reasoning) or
erroneous ideas are at issue. For an introduction to parallel topics concerning local and situated knowledge, see Ophir and Shapin, "The Place of Knowledge"; Shapin, "Placing the View from Nowhere."
    3. Bakhtin, *Rabelais and His World.*
    4. One must, however, recognize a partial and local exemption for stories connecting gonads
and knowledge, approved as legitimate and interesting wherever Freudian writ still runs.
    5. This point is touched on by a number of contributions to Shortland and Yeo, eds, *Telling
Lives in Science,* as well as in the introduction to that book, esp. 7–11. We emphasize "academic read-

In 1880 T. H. Huxley wrote about a Babylonian philosopher:

> Happily Zadig is in the position of a great many other philosophers. What he was like when he was in the flesh, indeed whether he existed at all, are matters of no great consequence. What we care about in a light is that it shows the way, not whether it is lamp or candle, tallow or wax.[6]

And that is a sentiment that modern accounts of intellectual life are obliged to confront, and against which they are often hard put to find a reply. How *could* such things matter—why *should* they matter—to knowledge itself?

The chapters in this book do not even try to offer a coherent and systematic response to this question. Rather, they are exercises in cultural history of ideas (especially of scientific ideas) that attempt inter alia to exemplify how one might go about countering sentiments such as Huxley's, how one might write about the history of scientific (and related) ideas through a history of their embodied forms and vicissitudes, and how one might do so legitimately, constructively, and interestingly. The object in so doing need not be exclusionary or denunciatory, and for our authors it is neither. First, to say (as one certainly may) that no one has ever seen a disembodied idea is not to say that such things do not exist but only that they are *theoretical* in character. (No one has ever seen a physical force, or a neutrino, or a social interest either.) When our contributors draw attention to what may be learned by looking at the embodied presentations of knowledge, they do not necessarily mean to say that disembodied ideas do not exist or that common speech about such things ought to be banned (though several contributors do indeed suggest that there is no *obligation* at all that we refer to genuine knowledge as disembodied). Rather, they aim to provide a novel way of understanding the mundane procedures by which such theoretical entities are brought into existence, made credible, and sustained. Historians' routine concern with concrete materials is thus encouraged and extended, and the aim is to tell mundane stories about things traditionally taken as supramundane.

Notions like "knowledge itself" are not so much rejected as respecified. Here, our authors suggest, are other types of stories one can tell about the making, maintenance, and modification of knowledge. Are these stories superior to, should they take the place of, more traditional tales about truth-making? Our authors are by no means united in their willingness to make claims of this kind, nor in their interest in programmatic assertions of anything like that scope. Yet

---

ers" here, since it is by no means obvious that the troubles of speaking of the links between "the personal" and "the intellectual" present themselves quite so forcefully to lay readers.

6. Huxley, "Method of Zadig," 2.

they all evidently presume that rich and detailed stories about the embodied processes of knowledge-making and knowledge-portraying are legitimate ways of getting at aspects of what might plausibly be meant by "knowledge itself." Bodily practices that visibly and publicly portray the status, identity, and worth of knowledge help *create* the notion of what knowledge is. So our authors take themselves to be talking *about* knowledge, and not merely about some trivial circumstances *attending* knowledge. That is their challenge to both modern and pre-modern sensibilities. Past cultures had elaborate vocabularies for talking about the bodily processes of knowledge-making, including resources for the embodied portrayal of disembodied knowing. By recovering these cultural vocabularies and their uses, our contributors implicitly table the question "What else is necessary to the notion of 'knowledge itself'?"

Second, it has undeniably been a persistent feature of learned culture over a great span of time and place that the worth of knowledge has been linked to its stipulated elevation above the mundane and the corporeal. The pragmatist John Dewey, for example, noted and condemned

> the age-long association of knowing and thinking with immaterial and spiritual principles, and of the arts, of all practical activity in doing and making, with matter. For work is done with the body . . . and is directed upon material things. The disrepute which has attended the thought of material things in comparison with immaterial thought has been transferred to everything associated with practice.[7]

Accordingly, it has been, and continues to be, widely assumed that bringing knowledge (or, rather, conceptions of knowledge) "back to earth" can be motivated only by a desire to denigrate or can have the effect only of devaluing the knowledge concerned. Yet it is that very assumption that some of our authors suggest may be turned into a topic of historical inquiry. In these essays our authors mean neither to denigrate nor to celebrate; they want to understand.

■ *The Embodiment of Knowledge as Academic Topic* ■

WHY IS IT that Truth and the body are so pervasively set in opposition? How is that opposition manifested and enforced in various settings? Are there circumstances in which it can be rejected, modified, set aside, or even inverted? How did past cultures speak about body and knowledge, body and knowledge-making? Is it in fact an assumption of our culture as a whole or is it specifically linked with

7. Dewey, *The Quest for Certainty,* 5.

the sentiments and interests of the priesthood and allied learned classes?[8] Has the strength of the linkage between knowledge and disembodiment changed over time? Steven Shapin's chapter (chapter 1) explicitly raises a number of these questions, and the other chapters offer concrete materials for thinking about them. Collectively, we mean to put such questions about the body of knowledge on the agenda of cultural and intellectual history: to initiate historical inquiry, not to assert the robust or exact findings of a mature research program. Yet despite these modest and deflationary disclaimers, we do not wish to escape a legitimate obligation to give some account of the intellectual currents to which these chapters respond, the agendas to which they might make a contribution, and the present-day circumstances that prompt us to think about the body of knowledge in these ways.

Frank Parkin's brilliant 1980s satire of the modern British university was aptly titled *The Mind and Body Shop*. The title referred to the academy's rough division into two camps: those dedicated to the explication of body (the natural sciences and their engineering associates) and those whose objects of study are the productions of mind (the humanities and some—not all—of the human and social sciences). It is a division of labor that has existed in basically stable form for centuries and that maintains such peace, order, and civility as obtain between the faculties and the disciplines. Yet, even as Parkin wrote, this dualistic arrangement was breaking down, a change accelerating though the 1980s and early 1990s and arguably reaching its furthest extent in current arrangements in the universities of the American West Coast.

Some departments of philosophy—responding partly to Anglo-American currents of "naturalistic epistemology" and partly to the siren calls of cognitive science and neurophysiology—increasingly took seriously the reduction of mind to matter that was advertised in nineteenth-century materialist utopias. Where the noumenal meets the neural net, deans and vice-chancellors might be promised always-welcome departmental downsizing from *The Mind and Body Shop* to just another cost-effective body shop. At the same time, those academic projects fundamentally shaped by Nietzschean and post-Nietzschean philosophical forces, which overlap only very slightly with so-called "wet philosophy," turned

---

8. This is definitely not to say that formally learned cultures have not intermittently produced pragmatic or iconoclastic *rejections* of the value of disembodied knowledge or even of its existence. Of course they have done so, and the current spread of academic interest in embodiment and culture may index a major shift in late modern learned sentiment. However, by qualifying the object of discussion as "learned" culture, we mean to hold open the possibility—scarcely investigated—that lay and civic cultures have pervasively seen less point in identifying the worth of knowledge with its disembodiment or transcendence.

their attention to the body as part of a developing twentieth-century reaction against philosophical rationalisms and idealisms. "In retrospect," as one modern social theorist has said, "the problem of the body in society will be seen to have dominated the development of Western philosophy throughout this century."[9]

Nietzsche's antirationalist philosophy was extended and transformed by twentieth-century Continental phenomenology and existentialism, both genres broadly asserting the primacy of everyday lived experience in the world, of the emotions, and of the embodied self against what was seen as the defining and disfiguring mark of modernism, the allegedly absolute and desiccated Cartesian mind-body dualism.[10] These phenomenological currents probably exerted a substantial—though largely diffuse and indirect—influence on work done from the 1970s by sociologists of scientific knowledge and historians concerned with the mundane details of scientific and technological practice. Against modernist traditions that centered attention on how scientific concepts *represent* the world, some interest began to be deflected to the study of science and technology as embodied ways of living in the world, of shaping both the world and the knowing self. If one wants to interpret scientific experience of the world, one has to understand, for example, the "embodiment relations" by which instruments materially extend human perception. The subject of scientific experience is, so to speak, an instrument-body hybrid.[11] Accordingly, as Michael Polanyi wrote, just as we can conceive of scientific instruments as extensions of the body, so we can regard scientists' bodies as part of an instrumental exploration of the world:

> We use instruments as an extension of our hands and they may serve also as an extension of our sense. We assimilate them to our body by pouring ourselves into them. And we must realize then also that our own body has a special place in the universe: we never attend to our body as an object in itself. Our body is always in use as the basic instrument of our intellectual and practical control over our surroundings. . . . Every time we assimilate a tool to our body our identity undergoes some change; our person expands into new modes of being. . . . Our whole articulate equipment turns out to

9. Turner, "Inner Self as Outward Appearance"; see also idem, *The Body and Society*; Schilling, *The Body and Social Theory*; Featherstone, Hepworth, and Turner, *The Body: Social Process and Cultural Theory*; and Rorty, *Philosophy and the Mirror of Nature*, e.g., 7, 32 (for dissolution of the mind-body problem via neopragmatist philosophy).

10. We have, however, already noted that the historical (as opposed to the mythical) Descartes wrote extensively about the *dependence* of mental and bodily states; and see Dear's chapter (chapter 2) in this volume.

11. See, especially, the précis of Heidegger's, Husserl's, and Merleau-Ponty's phenomenology in Ihde, *Technology and the Lifeworld*, chap. 3.

be merely a tool-box, a supremely effective instrument for deploying our inarticulate faculties. . . . Our body is the ultimate instrument of all our external knowledge.[12]

So far as contemporary academic concerns with body and self are concerned, none of Nietzsche's intellectual heirs has influenced as great a range of projects as Michel Foucault. From *The Birth of the Clinic* (1963) and *Discipline and Punish* (1975) to his last work on *The History of Sexuality* (1976–84), Foucault treated techniques of bodily management, presentation, and control as integral to the constitution of social power and cultural meaning.[13] The body expresses power and has power inscribed upon it. The body is politically as well as biologically vulnerable: "power relations," Foucault wrote, "have an immediate hold upon it; they invest it, mark it, train it, torture it, force it to carry out tasks, to perform ceremonies, to emit signs."[14] Prisons, asylums, clinics, and schools are places where power is performed by disciplining the body.

By the early 1990s there was scarcely any significantly audible academic voice concerned with "the projects of the self" that did not assert the necessity of attending to its embodiment. "Our experience of the human body," writes the moral philosopher John Casey, "is necessary to our sense of ourselves as persons; and our sense of ourselves as persons is necessary to our perception and understanding of the human body."[15] Contemporary historians and sociologists are, no doubt, arguing in good faith and with good reason when they persistently condemn academic neglect of the body side of the mind-body relationship and attribute such neglect to the perduring idealism of our culture,[16] yet current fashion in these parts of the academic world would suggest that they are, at least locally, flogging a very sick horse.

12. Polanyi, *The Study of Man,* 25, 30–31; idem, *The Tacit Dimension,* 15; see also idem, *Personal Knowledge,* 55–63, 174–76. Historians and sociologists of science have explored and extended Polanyi's stress on the importance of "tacit knowledge" in a number of detailed studies of the embodiment of scientific work; see, among many examples, Collins, "The TEA Set"; Schaffer, "Astronomers Mark Time"; idem, "Self-Evidence" (and idem, chapter 3 in this volume); Lawrence, "Incommunicable Knowledge"; MacKenzie and Spinardi, "Tacit Knowledge, Weapons Design"; Shapin and Schaffer, *Leviathan and the Air-Pump,* chap. 6; Shapin, *A Social History of Truth,* chap. 8; Olesko, "Tacit Knowledge and School Formation"; Pinch, Collins, and Carbone, "Inside Knowledge."

13. Driver, "Bodies in Space: Foucault's Account of Disciplinary Power."

14. Foucault, *Discipline and Punish,* 25; see also Leder, *The Absent Body,* esp. chap. 3; Lock, "Cultivating the Body," 144–48.

15. Casey, *Pagan Virtues,* 29; see also, e.g., Carrithers, Collins, and Lukes, *The Category of the Person;* Csordas, *Embodiment and Experience;* Sennett, *The Fall of Public Man,* esp. chap. 8; idem, *Flesh and Stone.*

16. E.g., Porter, *"Barely Touching,"* 45.

■ *Embodiment as Historians' Topic* ■

SOME CULTURAL AND intellectual historians, of course, were responsive to the systematic preoccupations of philosophers and sociocultural theorists. Yet it would be wrong to point to a very coherent set of inspirations for the extent of historians' recent concern with the body. In the case of medical historians, traditional engagement with ideas *about* the human body was importantly supplemented by a focus on the lived experience *of* the body, as the discipline responded to wider historical impulses to write "history from below," or, specifically, "from the patient's point of view."[17] Those historians sensitive to anthropological resources—and these included historians of ideas as well as social historians—could draw upon long-standing Durkheimian sensibilities, including dispositions to interpret the human body as a pattern for members' social and cosmological ideas and, more generally, to regard the body as a culturally resonant expressive resource.[18] "The human body," says Mary Douglas, "is always treated as an image of society and . . . there can be no natural way of considering the body that does not involve at the same time a social dimension." What we think about bodily boundaries, and transactions across them, is embedded in thought about, and practices organized around, social boundaries and social transactions.[19] And within broadly functionalist anthropological traditions culture might be conceived, in Malinowski's words, as "an instrumental enhancement of human anatomy," referring "directly or indirectly to the satisfaction of a bodily need." Human biological functions provided culture with both its instrumental purpose and its metaphorical base.[20]

For feminist historians the body has been a topic of more than academic interest. The Western cultural tradition that preferentially defined women in terms of their bodies and bodily functions presented itself at once as a topic of historical research and (as is shown here by Alison Winter's chapter on Ada Lovelace [chapter 6]) as a continuing obstacle to women's full presence in intellectual dis-

17. See, among very many examples, Porter, *Patients and Practitioners;* idem, "The Patient's View"; Rosenberg, "The Therapeutic Revolution"; Fissell, "Disappearance of the Patient's Narrative."

18. E.g., Durkheim, *Elementary Forms of the Religious Life;* Mauss, "Techniques of the Body"; Hertz, "The Pre-eminence of the Right Hand"; Douglas, *Natural Symbols;* Lock, "Cultivating the Body"; Kantorowicz, *The King's Two Bodies;* Le Goff, "Head or Heart?"; Camporesi, *The Incorruptible Flesh;* idem, *The Anatomy of the Senses;* Benthall and Polhemus, *The Body as a Medium of Expression;* Shapin, "Homo phrenologicus"; Gilman, *The Jew's Body,* esp. chaps. 2, 7; and, for specific appropriations of such sensibilities in the history of medicine, see, e.g., Sawday, *The Body Emblazoned;* Pouchelle, *The Body and Surgery;* Lawrence, "Nervous System and Society."

19. Douglas, *Natural Symbols,* 70. For anthropological criticism of dispositions to treat bodily practice as secondary to linguistic practice, and therefore to reduce bodily gestures and routines to semantics, see Jackson, "Knowledge of the Body," esp. 328–30.

20. Malinowski, *Scientific Theory of Culture,* 171; see also Sahlins, *Culture and Practical Reason,* chap. 2.

course. Women's nature was pervasively said to be far more strongly shaped by their bodily circumstances than was men's. Men were intellectual, active, and externally oriented; women nurturing, reproductive, and internally oriented. Women's place in the privacy of the home and men's in the institutions of public life naturally flowed from their respective constitutions. Women's bodies themselves were said to be differently constituted than men's—colder in temperature, more friable, more labile, more potent in shaping (and limiting) mind and its products. Feminist historians and social scientists have sought both to remedy traditional silence about the body and to revalue its place in our understanding of how identity and culture are made. In this, they join many other currents directing attention to the body, while the continuing overlap between academic topic and political predicament gives feminist work on the body a generally sharper critical bite than it elsewhere has.[21]

For cultural and social historians, and especially for early modernists, Norbert Elias's *The Civilizing Process* has probably been the decisive inspiration in recent engagements with the body. Elias offered a way of understanding aspects of European modernization through accounts of how and why bodily control and inhibition developed. The minutiae of posture, gesture, and the management of bodily excretions—matters previously considered too vulgar or too trivial to merit historians' notice—now stood revealed as complex cultural codes that both signaled and constituted a new social order, with its characteristic hierarchies of virtue, its important boundaries between public and private spaces, its regimens of body management appropriate to different sorts of spaces, and its regulative sensitivities to what was shameful and what decent. If you wanted to interpret changing locations of cultural authority or changing conceptions of the self in early modern society, Elias's arguments about body management were increasingly treated as central. Consequential accounts of macrosociological change, of modernization, could and should be told—it was now reckoned—through stories about changes in how and where it was deemed fit to feed and to fart.[22]

---

21. For selective entry to the now-vast feminist literature on the body and culture, see, e.g., Outram, *The Body and the French Revolution;* idem, "Fat, Gorilla and Misogyny"; Jordanova, *Sexual Visions;* Schiebinger, *The Mind Has No Sex?* For vigorous current disputes over the historical construal of sex and sex-difference, see, e.g., Laqueur, *Making Sex;* Cadden, *The Meaning of Sex Difference;* and Park and Nye, "Destiny Is Anatomy." For widely influential studies of ancient and medieval appreciations of the gendered body and its religious significance, see, e.g., Bynum, *Holy Feast and Holy Fast;* idem, "The Female Body and Religious Practice"; idem, "Why All the Fuss about the Body?" (for a critical review of the literature); Brown, *The Body and Society;* and Foucault, *The History of Sexuality.*

22. See, again among very many examples, Greenblatt, *Renaissance Self-Fashioning;* Biagioli, *Galileo, Courtier;* Bryson, "The Rhetoric of Status"; Becker, *Civility and Society;* Goldgar, *Impolite*

■ *How the Body Signals Knowledge* ■

ELIAS'S FOCUS on the culturally shaped and culture-constituting body is evident in a number of our chapters, and not just those concerned with early modern topics. Yet the claims to originality of some of the work contained here consist in bringing sensibilities such as Elias's to bear on questions having to do with the identity, status, and value of *knowledge,* questions that have traditionally belonged to philosophers, and, especially, to epistemologists. The body is indeed a culturally embedded, and culture-constituting, signaling system, and one of the things the body can signal is the possession and reliable representation of *truth.*

Our authors' concerns are undeniably diverse, and it would do them a collective disservice to claim to discern a single interpretative goal to which they all aspire. It is, nevertheless, clear that a number of the chapters interpretatively link knowledge to body by way of the historical *identities* of knowledge-makers. Across a great range of cultures, questions about the status and worth of knowledge have been partially dealt with by the bodily presentations of those who produce and report upon that knowledge. Why should I accept that what you say is true? Why should I accept that what you represent as knowledge is genuine? Because you can see—or accept on others' trustworthy reports—that I am the *kind of person* who lives for truth. Put another way, truth may be conceived as a personal *performance,* an individual act that uses culturally given materials for its point and value.

The bodily identity of the truth-seeker undoubtedly varies from culture to culture and from time to time within a culture. Shapin's survey (chapter 1), however, puts the topic of *asceticism* and its relationship to truth firmly on the agenda. If proper knowledge is understood to derive from sacred sources and to be transcendent in its nature and scope, then one powerful resource available so to portray knowledge-claims is through a presentation of the *knower* as otherworldly and disengaged, that is, the embodied public display of a disembodied mind, of a mind which was not *there* in its own body because it was understood to be somewhere else, in the domain where genuine knowledge was to be had. There were many bodily techniques by which such disengagement could be displayed: physiognomic, postural and gestural, sumptuary, situational, dietetic. In turn, there were many cultural resources available that could advise how the

Learning; Bremmer and Roodenburg, *A Cultural History of Gesture;* Gent and Llewellyn, *Renaissance Bodies;* Shapin, *A Social History of Truth,* esp. chaps. 3–4; and Chartier, *Cultural History,* esp. 71–94. In addition, some cultural historians have explored the rich resources offered by Erving Goffman's symbolic interactionist sociology (e.g., his *Presentation of Self*) for appreciating the body as a culturally embedded signaling system.

body should be regulated so as to put it in a fit condition for the reception and production of knowledge. Shapin argues that, at least until fairly recently, Western culture has absolutely understood the *point* of a relationship that now appears substantially as a joke: the relationship between the belly and knowledge.

Rob Iliffe's chapter on Isaac Newton's care of, and presentation of, his body (chapter 4) describes the humoral theory that linked the melancholic temperament to the scholarly role. Iliffe makes the fragility of the balance between philosophical genius and madness into a cultural-historical as well as a biographical topic. Newton's identity as the greatest of natural philosophers, and that of the knowledge he produced, was assembled from preexisting cultural resources, including those that allowed the genuineness of knowledge to be read off the special constitution of the philosopher's body. Similarly, Simon Schaffer (chapter 3) traces an early modern cultural nexus that linked the authenticity of experiential testimony to the credibility of philosophical knowledge by way of philosophers' exceptional bodily constitutions, including their special sensory capacities. Adam, it was believed, "needed no Spectacles," and English Restoration culture debated whether and how the postlapsarian experimental natural philosopher might achieve sensory regeneration. The question was not so much whether the experimental philosopher was constituted "as other men," but in precisely which ways he was *not.* The instrumental use of the philosopher's body was closely linked to contested traditions of magical, religious, and symbolic action.

Peter Dear's chapter (chapter 2) attempts to make historical sense of specifically Cartesian relationships between early modern *norms* of bodily comportment and criteria of intelligibility in natural philosophy. He shows how mechanized natural philosophy—and its version of an intelligible nature— "constrained (or expressed) the structure of human behavior so intimately connected to it." Broadly following Elias, Dear argues that the intelligibility of Cartesian accounts of the passions and the mechanical human body was shaped by substantial realities of early modern social life: "Social life, formalized through manners, meant letting people see each other as automata under the control of reason; so automata is what they became."

Christopher Lawrence is mainly interested in long-standing traditions of bodily presentation and their uses by physicians and surgeons in specific historical contexts to legitimate their roles. In these connections the mind-body dualism takes a particularly concrete and vivid form. Lawrence's richly illustrated chapter (chapter 5) displays the bodily techniques by which the archetypal roles of scholarly physicians and manually working surgeons were made visible and justified. Yet Lawrence also draws attention to the complexity and problematic nature of these bodily presentations, particularly in relation to contests over the proprietorship of medical practice.

A number of our contributors remark explicitly on the significance of gender in the bodily presentations of an almost wholly male scientific and philosophical culture from antiquity through the early modern. Alison Winter's chapter (chapter 6), however, is this volume's most direct engagement with the gendered intellectual body. Her account of the aspiring Victorian mathematical analyst Ada Lovelace describes how contemporary understandings of female bodies disqualified women from the highest intellectual activities. She also assesses those existing cultural roles for which women's constitutions were reckoned specially to fit them, and points out some relationships between femininity, invalidism, and intellectual work. Winter details how a very singular intellectual woman negotiated her way through the difficult obstacle course constituted by Victorian culture for *all* women with intellectual ambitions. Finally, Winter relates how, in the end, Lovelace's solution to gender/knowledge issues was the reflexive transformation of her personal mind-body relationship into a topic of research.

Ada Lovelace was an intellectual invalid, but the greatest, and most famous, Victorian scientific invalid of all was Charles Darwin. Janet Browne (chapter 7) does not seek in any way to deny the fact that Darwin *was* ill for much of his life: indeed, she contributes to ongoing historical discussions of just what it was that ailed him. However, she wants mainly to describe how Darwin's illness was portrayed and understood, and with what consequences for his personal authority and for the authority of his scientific work. (Taking the topic of portrayal quite literally, Browne reproduces, and traces the iconographical significance of, many examples of Darwin's contemporary portraiture.) Darwin's "sick body" can, Browne shows, be considered "as one further professional resource in a rich repertoire of resources" that worked publicly to establish his identity and his intellectual bona fides. Darwin's illness worked inter alia as a buffer distancing himself from scientific controversy and social friction. Disease symbolically underwrote disinterestedness.

None of our contributors seeks to argue against either the pertinence or the legitimacy of what might be called a *realist* sensibility toward the bodily presentation of knowledge. For all any of us know, an ascetic way of life might just be either the physiological outcome of certain sorts of intellectual work or a disposing factor in intellectual impulses. Again, for all we know, philosophizing and vegetarianism might be causally linked in some currently plausible (or future possible) physiological scheme, and some study might establish that physicians are just statistically thinner than surgeons. But whatever the answers to such questions might be, our contributors mean to show that a *symbolic* perspective is formally independent of a realist sensibility and that the former is both possible and historically important. "The way that philosophers really are"—and this is, of course, a highly varying and contested notion—is never a sufficient explanation of "the way they are presented as being" or of "the way they are believed to be" in any given culture. Cultural phenomena can never be sufficiently accounted for by physiological explanations. The *credibility* of accounts of "how philosophers really are" always involves cultural interpretation.

Our last chapter is no exception to these dicta, yet its sensibility perhaps comes closer to the realist sensibility than the others. Andrew Warwick (chapter 8) writes about the physical culture that accompanied heroic preparations for the mathematical Tripos examinations in Victorian Cambridge. His topic is emergence and change in a particular version of the "mind-body" relationship, taken both as a cultural construct and as a practical regimen. Warwick understands the Tripos as a form in which the learning process was effectively *industrialized,* and he shows how Cambridge undergraduates sought to ward off potential breakdown by employing "regular physical exercise both to regulate the working day and in the belief that it preserved a robust constitution." The common distinction between "aesthetes" and "athletes," Warwick says, must be treated with caution in the Victorian period. And a specific version of a disciplined body of mathematical knowledge was understood to be sustained by a particular version of the physically disciplined student body.

■ *What Is Special about the Scientific Body?* ■

WE SHOULD, FINALLY, say something about the cultural territory occupied by our authors. In the main, and reflecting their own interests and competences, the materials treated here are scientific and medical in nature. The intrinsic importance of these materials needs no justification, while both the special circumstances attending the interpretation of such topics and the generalizability of findings derived from them require brief comment. First, it should be understood that in our late modern culture science counts as Truth, and how science is interpreted counts as a story about Truth. There is no present-day body of culture that competes with science in any significant way for the mantle of Truth. That alone makes science a massively special object of inquiry, for, as we have briefly suggested, the conditions affecting the interpretation of what counts as genuine knowledge and what as error, or knowledge of lesser value, clearly differ very significantly. That the idea of telling stories about the embodied production and validation of authentically scientific knowledge seems relatively original is testimony to the extent to which scientific Truth enjoys a special protection not extended to other, lower, forms of culture.

Yet this admitted specialness of science, and the distinctive conditions affecting inquiry about science, also prompt thoughts about the terms in which the historical and sociological interpretation of science may have wider interest and applicability. It is important to note that the evaluation of different forms of culture is historically variable. If natural science now defines true knowledge for late moderns, Christian religion did so for our cultural ancestors. Resources and repertoires used to identify and underwrite conceptions of disembodied religious knowledge and the religious knower were adapted (as the chapters by Shapin, Iliffe, and Schaffer argue) to perform similar tasks for early modern natural philosophy. The content of Truth significantly varied but the approved stories told about the nature of true knowledge and the authentic knower remained interestingly stable over a great span of time. Nor has science been the only form of culture to clothe its knowers in garments originally cut to fit the religious ascetic and to adapt to its purposes religious tropes of otherworldliness and self-denial. Romantic conceptions of artistic truth—whether the arts concerned are literary, poetic, painterly, plastic, or musical—have also made consequential stipulations about the body of the artist that bear pronounced family resemblances to those attaching to the priest-scientist. And Romanticism should not, in these connections, be identified too restrictively with its late eighteenth- or early nineteenth-century manifestations: its use of notions of mystic union between knower and known have ancient and medieval antecedents, while its individualism continues to color both lay and expert late modern understandings of the creative process in general.

Our contributors are historians, and their purposes here are directed almost wholly to the interpretation of past cultures. However, even as we track changing conceptions of the body of knowledge over time, so we find ourselves in a position to appreciate the contingency of present sensibilities. Our authors' work is meant to count as sober history, yet each of the chapters—and the volume as a whole—trades upon the shock value of speaking about scientific knowledge-making in relation to the body. That shock is an indication that our culture's official attitudes may have changed very significantly from what they were in the early modern period, or even, perhaps, from what they were in the late nineteenth century. Those past cultures, as we have already indicated, very much acknowledged the pertinence of speaking about the special bodily constitution, temperament, complexion, and dietetics of the Truth-seeker, and the notion that the philosopher was *differently constituted* than the ordinary person was not necessarily treated as at all funny.

What has (apparently) changed since then?[23] First, a humoral theory of the human constitution (and its implications for understanding mind-body relationships) has effectively disappeared from our official medical vocabulary, with the consequence that the dualism between bodily and mental states so widely (and erroneously) attributed to Cartesian modernism is now (at last) a pronounced feature of much of our official culture. The ancients, the seventeenth-century moderns, and the Victorians found it sensible to consider, to manage, and to speak about the philosopher's special dietetics and bodily regimens: in the main, we do not. Second, a sacred theory of knowledge has also apparently been eliminated, again at least from our official accounts. When what is the case about nature no longer counts as God's Truth, then the act of knowing is no longer seen as the imitation of otherworldly divinity. And, to the extent that intellectuals' knowledge is effectively seen as desacralized, the ancient distinction between the worldly conditions for achieving *expertise* and the otherworldly conditions for attaining a simulacrum of God's knowledge is collapsed. So one might say that all we now have is expertise: the expert is not understood to be imitating God, merely trained up in those worldly institutions in which expertise is on offer. The acquisition of expertise, as Warwick suggests, may indeed be understood to require bodily discipline, but the cultural appreciation of such discipline is perhaps wholly this-worldly. Finally, the professionalization and bureaucratization of intellectual life have thrown up a well-supported picture of the knowledge expert (no longer the Truth-seeker, and probably no longer even the "intellectual") as "doing a job like anyone else." Why should anyone doing an ordinary, but skilled, job of work be differently motivated, or differently constituted, from anybody else?

So historical inquiry into the body of knowledge has one sort of justification: the humor evidently generated by the very idea of such a project is itself a product of late modernity. And historians of ideas should be sensitive to the rich vocabularies cultures through the nineteenth century possessed for talking about the body of knowledge. Those scholars and theorists interested in describing late modern culture should take such an inquiry seriously, for it draws together sentiments about knowledge and the knower that lie at the heart of our present arrangements.

A final caveat: we should always be careful not to explain more than we know is the case. And that is why we have intermittently qualified our sketch of recent changes in conceptions of knowledge and the knower by making reference to "dominant" or "official" culture. Our very late twentieth-century "official" cul-

---

23. Some of these speculations are addressed at greater length in Shapin's chapter (chapter 1).

ture indeed lacks a physiologically sanctioned theory of humors, temperaments, and complexions; it has officially rejected a sacred conception of knowledge; and it officially offers a sociological and psychological portrait of expert knowers that insists upon their constitutional and moral equivalence with everyone else in their societies. It is, therefore, this official picture that makes inquiries such as ours appear funny. What we do not know is just how far the writ of this official version runs. We do not know whether the public authority of expert knowledge in late modern societies has wholly dispensed with appreciations of knowledge and the knower such as those described in this volume. There are impressionistic grounds for suggesting that such cultural changes have been more piecemeal and more localized than we currently appreciate. And, if this is the case, then these studies of the body of knowledge in past cultures may tell us more about our present selves than is superficially apparent.

■ REFERENCES ■

Bakhtin, Mikhail. *Rabelais and His World.* Trans. Hélène Iswolsky. Bloomington: Indiana University Press, 1984.

Barnes, Barry, and Steven Shapin, eds. *Natural Order: Historical Studies of Scientific Culture.* Beverly Hills, Calif.: Sage, 1979.

Becker, Marvin B. *Civility and Society in Western Europe, 1300–1600.* Bloomington: Indiana University Press, 1988.

Benthall, Jonathan, and Ted Polhemus, eds. *The Body as a Medium of Expression.* London: Allen Lane, 1975.

Biagioli, Mario. *Galileo, Courtier: The Practice of Science in the Culture of Absolutism.* Chicago: University of Chicago Press, 1993.

Bremmer, Jan, and Herman Roodenburg, eds. *A Cultural History of Gesture.* Cambridge: Polity Press, 1991.

Brown, Peter. *The Body and Society: Men, Women and Sexual Renunciation in Early Christianity.* London: Faber and Faber, 1989.

Bryson, Anna. "The Rhetoric of Status: Gesture, Demeanour and the Image of the Gentleman in Sixteenth- and Seventeenth-Century England." In Gent and Llewellyn, 136–53.

Bynum, Caroline Walker. "The Female Body and Religious Practice in the Later Middle Ages." In Feher et al., 1 : 161–219.

———. *Holy Feast and Holy Fast: The Religious Significance of Food to Medieval Women.* Berkeley and Los Angeles: University of California Press, 1987.

———. "Why All the Fuss about the Body? A Medievalist's Perspective." *Critical Inquiry* 22 (1995): 1–33.

Cadden, Joan. *The Meaning of Sex Difference in the Middle Ages.* Cambridge: Cambridge University Press, 1993.

Camporesi, Piero. *The Anatomy of the Senses: Natural Symbols in Medieval and Early Modern Italy.* Trans. Allan Cameron. Cambridge: Polity Press, 1994.

———. *The Incorruptible Flesh: Bodily Mutation and Mortification in Religion and Folklore.* Trans. Tania Croft-Murray and Helen Elsom. Cambridge: Cambridge University Press, 1988.

Carrithers, Michael, Steven Collins, and Steven Lukes, eds. *The Category of the Person: Anthropology, Philosophy, History.* Cambridge: Cambridge University Press, 1985.

Casey, John. *Pagan Virtues: An Essay in Ethics.* Oxford: Clarendon Press, 1990.

Chartier, Roger. *Cultural History: Between Practices and Representations.* Trans. Lydia G. Cochrane. Cambridge: Polity Press, 1988.

Collins, H. M. "The TEA Set: Tacit Knowledge and Scientific Networks." *Science Studies* 4 (1974): 165–86.

Csordas, Thomas J., ed. *Embodiment and Experience: The Existential Ground of Culture and Self.* Cambridge: Cambridge University Press, 1995.

Dewey, John. *The Quest for Certainty: A Study of the Relation of Knowledge and Action.* The Gifford Lectures 1929. New York: Minton, Balch, 1929.

Douglas, Mary. *Natural Symbols: Explorations in Cosmology.* London: Barrie and Rockliff, Cresset Press, 1970.

Driver, Felix. "Bodies in Space: Foucault's Account of Disciplinary Power." In *Reassessing Foucault: Power, Medicine, and the Body,* ed. Colin Jones and Roy Porter, 113–31. London: Routledge, 1994.

Durkheim, Emile. *The Elementary Forms of the Religious Life.* Trans. Joseph Ward Swain. 1915. Reprint, New York: Free Press, 1965.

Elias, Norbert. *The Civilizing Process.* Trans. Edmund Jephcott. 2 vols. Oxford: Blackwell, 1978, 1983.

Featherstone, Mike, Mike Hepworth, and Bryan S. Turner, eds. *The Body: Social Process and Cultural Theory.* London: Sage, 1991.

Feher, Michel, et al., eds. *Fragments for a History of the Human Body.* 3 vols. New York: Zone Books, 1989.

Fissell, Mary E. "The Disappearance of the Patient's Narrative and the Invention of Hospital Medicine." In *British Medicine in an Age of Reform,* ed. Roger French and Andrew Wear, 92–109. London: Routledge, 1991.

Foucault, Michel. *Discipline and Punish: The Birth of the Prison.* Trans. Alan Sheridan. New York: Vintage Books, 1979.

———. *The History of Sexuality.* Trans. Robert Hurley. 3 vols. (vol. 1 = *An Introduction;* vol. 2 = *The Use of Pleasure;* vol. 3 = *The Care of the Self*). New York: Vintage Books, 1988, 1990.

Frank, Arthur W. "Bringing Bodies Back In: A Decade Review." *Theory, Culture and Society* 7 (1990): 131–62.

Gent, Lucy, and Nigel Llewellyn, eds. *Renaissance Bodies: The Human Figure in English Culture c. 1540–1660.* London: Reaktion Books, 1990.

Gilman, Sander. *The Jew's Body.* London: Routledge, 1991.

Goffman, Erving. *The Presentation of Self in Everyday Life.* London: Allen Lane, 1969.

Goldgar, Anne. *Impolite Learning: Conduct and Community in the Republic of Letters, 1680–1750.* New Haven: Yale University Press, 1995.

Greenblatt, Stephen. *Renaissance Self-Fashioning: From More to Shakespeare.* Chicago: University of Chicago Press, 1980.

Harpham, Geoffrey Galt. *The Ascetic Imperative in Culture and Criticism.* Chicago: University of Chicago Press, 1987.

Hertz, Robert. "The Pre-eminence of the Right Hand: A Study in Religious Polarity [1909]." In *Death and The Right Hand,* trans. Rodney and Claudia Needham, 89–113. Glencoe, Ill.: The Free Press, 1960.

Huxley, Thomas Henry. "On the Method of Zadig." In *Collected Essays,* vol. 4, *Science and Hebrew Tradition,* 1–23. 1880. Reprint, New York: D. Appleton, 1900.

Ihde, Don. *Technology and the Lifeworld: From Garden to Earth.* Bloomington: Indiana University Press, 1990.

Jackson, Michael. "Knowledge of the Body." *Man,* n.s., 18 (1983): 327–45.

Jordanova, Ludmilla. *Sexual Visions: Images of Gender in Science and Medicine between the Eighteenth and Twentieth Centuries.* Madison: University of Wisconsin Press, 1989.

Kantorowicz, Ernst H. *The King's Two Bodies: A Study in Mediaeval Political Theology.* Princeton: Princeton University Press, 1957.

Laqueur, Thomas. *Making Sex: Body and Gender from the Greeks to Freud.* Cambridge: Harvard University Press, 1990.

Lawrence, Christopher J. "Incommunicable Knowledge: Science, Technology and the Clinical Art in Britain 1850–1914." *Journal of Contemporary History* 20 (1985): 503–20.

———. "The Nervous System and Society in the Scottish Enlightenment." In Barnes and Shapin, 19–40.

Leder, Drew. *The Absent Body.* Chicago: University of Chicago Press, 1990.

Le Goff, Jacques. "Head or Heart? The Political Use of Body Metaphors in the Middle Ages." In Feher et al., 3:13–26.

Lock, Margaret. "Cultivating the Body: Anthropology and Epistemologies of Bodily Practice and Knowledge." *Annual Review of Anthropology* 22 (1993): 133–55.

MacKenzie, Donald, and Graham Spinardi. "Tacit Knowledge, Weapons Design, and the Uninvention of Nuclear Weapons." *American Journal of Sociology* 101 (1995): 44–99.

Malinowski, Bronislaw. *A Scientific Theory of Culture, and Other Essays.* Chapel Hill: University of North Carolina Press, 1944.

Mauss, Marcel. "Techniques of the Body." *Economy and Society* 2 (1973): 70–88.

Olesko, Kathryn M. "Tacit Knowledge and School Formation," *Osiris,* n.s., 8 (1993): 16–29.

Ophir, Adi, and Steven Shapin. "The Place of Knowledge: A Methodological Survey." *Science in Context* 4 (1991): 3–21.

Outram, Dorinda. *The Body and the French Revolution: Sex, Class and Political Culture.* New Haven: Yale University Press, 1989.

———. "Fat, Gorilla and Misogyny: Women's History in Science." *British Journal for the History of Science* 24 (1991): 361–67.

Park, Katharine, and Robert A. Nye. "Destiny Is Anatomy" (essay review of Laqueur, *Making Sex*), *New Republic*, 18 February 1991, 53–57.

Pinch, Trevor J., H. M. Collins, and Larry Carbone. "Inside Knowledge: Second Order Measures of Skill." *Sociological Review* 44 (1996): 163–86.

Polanyi, Michael. *Personal Knowledge: Towards a Post-critical Philosophy.* Chicago: University of Chicago Press, 1958.

———. *The Study of Man: The Lindsay Memorial Lectures.* Chicago: University of Chicago Press, 1959.

———. *The Tacit Dimension.* Garden City, N.Y.: Doubleday, 1966.

Porter, Roy. "*Barely Touching:* A Social Perspective on Mind and Body." In Rousseau, 45–80.

———. "Bodies of Thought: Thoughts about the Body in Eighteenth-Century England." In *Interpretation and Cultural History*, ed. Joan H. Pittock and Andrew Wear, 82–108. Basingstoke: Macmillan, 1991.

———. "History of the Body." In *New Perspectives on Historical Writing*, ed. Peter Burke, 206–32. Cambridge: Polity Press, 1991.

———. "The Patient's View: Doing Medical History from Below." *Theory and Society* 14 (1985): 175–98.

———, ed. *Patients and Practitioners: Lay Perceptions of Medicine in Pre-industrial Society.* Cambridge: Cambridge University Press, 1985.

Pouchelle, Marie-Christine. *The Body and Surgery in the Middle Ages.* Trans. Rosemary Morris. Cambridge: Polity Press, 1990.

Rorty, Richard. *Philosophy and the Mirror of Nature.* Princeton: Princeton University Press, 1979.

Rosenberg, Charles E. "The Therapeutic Revolution: Medicine, Meaning, and Social Change in Nineteenth-Century America." *Perspectives in Biology and Medicine* 20 (1977): 485–506.

Rousseau, G. S., ed. *The Languages of Psyche: Mind and Body in Enlightenment Thought.* Clark Library Lectures 1985–86. Berkeley and Los Angeles: University of California Press, 1990.

Rousseau, G. S., and Roy Porter, 1990. "Introduction: Toward a Natural History of Mind and Body." In Rousseau, 3–44.

Sahlins, Marshall. *Culture and Practical Reason.* Chicago: University of Chicago Press, 1976.

Sawday, Jonathan. *The Body Emblazoned: Dissection and the Human Body in Renaissance Culture.* London: Routledge, 1995.

Schaffer, Simon. "Astronomers Mark Time: Discipline and the Personal Equation." *Science in Context* 2 (1988): 115–45.

———. "Self-Evidence." *Critical Inquiry* 18 (1992): 327–62.

Schiebinger, Londa. *The Mind Has No Sex? Women in the Origins of Modern Science.* Cambridge: Harvard University Press, 1989.

Schilling, Chris. *The Body and Social Theory.* Thousand Oaks, Calif.: Sage, 1993.

Sennett, Richard. *The Fall of Public Man.* Cambridge: Cambridge University Press, 1974.

———. *Flesh and Stone: The Body and the City in Western Civilization.* New York: W. W. Norton, 1994.

Shapin, Steven. "Homo phrenologicus: Anthropological Perspectives on an Historical Problem." In Barnes and Shapin, 41–71.

———. "Placing the View from Nowhere: Historical and Sociological Problems in the Location of Science." *Transactions: An International Journal of Geographical Research,* in press.

———. *A Social History of Truth: Civility and Science in Seventeenth-Century England.* Chicago: University of Chicago Press, 1994.

Shapin, Steven, and Simon Schaffer. *Leviathan and the Air-Pump: Hobbes, Boyle, and the Experimental Life.* Princeton: Princeton University Press, 1985.

Shortland, Michael, and Richard Yeo, eds. *Telling Lives in Science: Essays on Scientific Biography.* Cambridge: Cambridge University Press, 1996.

Turner, Bryan S. *The Body and Society: Explorations in Social Theory.* Oxford: Blackwell, 1984.

———. "Inner Self as Outward Appearance." *Times Higher Education Supplement,* 19 November 1993, 24–25.

ONE

THE PHILOSOPHER AND THE CHICKEN

On the Dietetics of Disembodied Knowledge

STEVEN SHAPIN

This work, though it deals only with eating and drinking, which are regarded in the eyes of our supernaturalistic mock-culture as the lowest acts, is of the greatest philosophic significance and importance. . . . How former philosophers have broken their heads over the question of the bond between body and soul! Now we know, on scientific grounds, what the masses know from long experience, that eating and drinking hold together body and soul, that the searched-for bond is nutrition.

—Ludwig Feuerbach, review of Jacob Moleschott's
Theory of Nutrition (1850)

■ *Introduction* ■

A STORY IS TOLD—and much repeated—about Sir Isaac Newton when he was living in London toward the end of his life:

> His intimate friend Dr. [William] Stukel[e]y, who had been deputy to Dr. [Edmond] Halley as secretary to the Royal Society, was one day shown into Sir Isaac's dining-room, where his dinner had been for some time served up. Dr. Stukel[e]y waited for a considerable time, and getting impatient, he removed the cover from a chicken, which he ate, replacing the bones under the cover. In a short time Sir Isaac entered the room, and after the usual compliments sat down to his dinner, but on taking off the cover, and seeing nothing but bones, he remarked, "How absent we philosophers are. I really thought that I had not dined." [1]

Here is another story, circulating among modern academic philosophers, and it is about another, and much later, Cambridge philosopher. In 1934 Ludwig Wittgenstein came to stay with his friend Maurice Drury at a cottage in rural Ireland, and, as Drury relates,

> Thinking my guests would be hungry after their long journey and night crossing, I had prepared a rather elaborate meal: roast chicken followed by

---

1. Brewster, *Life of Newton,* 341 n. For a representative twentieth-century retelling of the chicken story, see Grove Wilson, *The Human Side of Science,* 198.

suet pudding and treacle. Wittgenstein rather silent during the meal. When we had finished [Wittgenstein said], "Now let it be quite clear that while we are here we are not going to live in this style. We will have a plate of porridge for breakfast, vegetables from the garden for lunch, and a boiled egg in the evening." This was then our routine for the rest of his visit.[2]

In 1945 his American former student Norman Malcolm visited Wittgenstein in his Whewell Court rooms at Cambridge. Malcolm relates that Wittgenstein

> prepared supper for us. The pièce de résistance was powdered eggs. Wittgenstein asked whether I cared for them, and knowing how he valued sincerity, I told him that in truth they were dreadful. He did not like this reply. He muttered something to the effect that if they were good enough for him they were good enough for me. Later he related this incident to [Yorick] Smythies, and (according to Smythies) Wittgenstein took my distaste for powdered eggs as a sign that I had become a snob.[3]

That was wartime. Afterward, when Wittgenstein lived in Dublin, "he would go to Bewley's Café, in Grafton Street for his midday meal—always the same: an omelette and a cup of coffee." What especially pleased Wittgenstein was that he became so well known at the café that he did not have to utter a word to order his food: it just came. "'An excellent shop: there must be very good management behind this organization.'"[4]

My concern here is not to do with a late-Wittgensteinian solution to the problem of chicken-egg priority. Nor is it the moral of these—and a series of strikingly similar—stories that those who love wisdom do not love chicken: there is no reason to suppose that there is some special philosophic foulness that attaches to chicken. Rather, the point made by those telling these stories is publicly to say something of consequence about the special constitution of individuals who give themselves up wholly to the pursuit of truth. These are *stipulations* about the bodies of truth-seekers. The chicken is both real and figurative—made into symbolic capital for the quality of knowledge. It is, so to speak, epistemological chicken. And what these stories stipulate is that the truth-seeker is someone who attains truth by denying the demands of the stomach and, more generally, of the body. That is one way in which it is said that the individuals in question *are* truth-lovers—that is, *philosophers*—and one way available to philosophers to be recognized as such. And, if (as is likely) there is now a distinct

2. Drury, "Conversations with Wittgenstein," 125.
3. Malcolm, *Wittgenstein: A Memoir*, 40.
4. Drury, "Conversations with Wittgenstein," 156.

sense of the bizarre in discussing truth in relation to the stomach, it is that very oddness of association that is my topic of inquiry. Why is it that the belly is conceived to stand at the opposite pole to truth?[5]

These stories—and many like them—are unusually widely distributed and persistent over a broad sweep of Western culture. I find them fascinating and important, and I want to tell a few more of them as I go on. My fascination with these stories proceeds partly from a puzzle I sometimes encounter in conversation with academic colleagues in philosophy and in the history of ideas. They occasionally say that, in contrast with some social historians and sociologists of knowledge, their concern is with "disembodied knowledge," with knowledge *itself,* rather than with its embodied production, maintenance, and reproduction.[6] Such locutions are standard, well institutionalized in a range of academic practices, and rarely contested. Yet, to tell the truth, I have never seen a "disembodied idea," nor, I suspect, have those who say they study such things. What I and they have seen is embodied people *portraying* their disembodiment and that of the knowledge they produce or the documentary records of such portrayals. These portrayals are the topic in which I am interested here. How are they done? With what cultural materials are they accomplished? To what ends? I start with a prejudice: it is that the portrayal of our culture's most highly esteemed knowers and forms of knowledge as disembodied has been one of the major resources we have had for displaying the truth, objectivity, and potency of knowledge.[7] These stories, and the cultural practices they describe, constitute that portrayal. They are stories about the meager and the physiologically disciplined bodies of truth-lovers.

My particular interest has been with early modern natural philosophers and the stories attached to their bodies. And I will briefly rehash some familiar stories attaching to Robert Boyle, Henry More, Isaac Newton (again), and Henry Cavendish. But the stories are, indeed, *attached,* since they were associated with

5. Here I should say that stories about truth-lovers' stomachs are only one potential focus for thinking about disembodiment as a topic in practical epistemology. One could imagine an extended study divided into chapters: the face, the eyes, the loins, the skin, the hands, gesture, costume, the body in solitude. (See, for example, the topical organization of Onians, *Origins of European Thought about the Body.*) I concentrate here on the belly partly because of the strength of the opposition between it and the mind, while other chapters in the present volume range more widely over corporeal terrain.

6. This was the same intellectual subject that Nietzsche recognized and opposed: "a 'pure, will-less, painless, timeless subject of knowledge,'" "'knowledge-in-itself'": "What Is the Meaning of Ascetic Ideals?" in *Genealogy of Morals,* 717–93, quoting 744.

7. For the iconography of intellectuals, see, e.g., Zanker, *The Mask of Socrates;* Fletcher, "Iconographies of Thought"; and Janet Browne's contribution (chapter 7) to this volume. I have treated the related topic of solitude as an epistemological resource in Shapin, "'The Mind Is Its Own Place.'"

other truth-lovers in other, and much earlier, settings. So I want to get to Newton et al. by way of settings from which emerge our earliest knowledge of such
stories.[8] And at the end of this chapter I want to suggest that these stories
no longer attach to present-day truth-seekers in quite the same way. The career
of such stories, I speculate, tracks the development of modern conceptions of
knowledge and the knower in a perspicuous way.

■ *The Ascetic Ideal and Its Classical Tropes* ■

WE ARE DEALING HERE with a *trope,* one of very great antiquity and perva-
siveness, a trope that has been consequentially attached in a range of settings to
those who are said to be authentic lovers of truth.[9] Possibly the original of the
trope is found in Plato. In the *Phaedrus,* Socrates tells a charming story about the
origins of the race of cicadas. Once upon a time, before the Muses were called
into being, cicadas were human beings. And when the Muses were created,

> some of the people of those days were so thrilled with pleasure that they
> went on singing, and quite forgot to eat and drink until they actually died
> without noticing it. From them in due course sprang the race of cicadas, to
> which the Muses have granted the boon of needing no sustenance right
> from their birth, but of singing from the very first, without food or drink,
> until the day of their death.

And when the cicadas die they report to the Muses "how they severally are paid
honor among mankind, and by whom." These people—dancers, singers, historians, and the like—are the blessed of the Muses. But of these some are specially
blessed: "To the eldest, Calliope [Muse of epic poetry], and to her next sister,
Urania [Muse of astronomy], they tell of those who live a life of *philosophy* and
so do honor to the music of those twain whose theme is the heavens and all the
story of gods and men, and whose song is the noblest of them all." [10]

At the very end of his life, Socrates made clear the special affinity between
the cicada's way of life and that of the philosopher. Sentenced to death, Socrates
argued against those of his friends who would have him flee Athens and avoid
the hemlock. In the *Phaedo* Socrates brings Simmias round to the view that of all
men the philosopher is one who, rather than fearing death, should embrace it.
The argument proceeds by way of the role of the body, its desires and require-

---

8. See also chapters in this book by Peter Dear (chapter 2, on Descartes), Robert Iliffe (chapter 4,
on Newton), and Simon Schaffer (chapter 3, on English Restoration natural philosophers in general).
9. For the significance of similar tropes in non-Western as well as European cultures, see, e.g.,
Goody, *Cooking, Cuisine, and Class,* chap. 4.
10. Plato, *Phaedrus,* 259 b–e.

ments, in the philosopher's search for truth. Socrates: "Do we believe that there is such a thing as death?" Simmias: "Most certainly." Socrates: "Is it simply the release of the soul from the body? Is death nothing more or less than this, the separate condition of the body by itself when it is released from the soul, and the separate condition by itself of the soul when released from the body? Is death anything else than this?" Simmias: "No, just that." Socrates: "Well then, my boy, see whether you agree with me. . . . Do you think that it is right for a philosopher to concern himself with the so-called pleasures connected with food and drink?" Simmias: "Certainly not, Socrates."

Socrates went on to establish that the philosopher is a different sort of person from the ordinary run of humanity: he "frees his soul from association with the body, so far as is possible, to a greater extent than other men." And if the philosopher's disembodiment is the condition for his hope to attain truth during mortal life, so death, which is the final freeing of the soul from the constraints of the body, is not to be shunned but welcomed: "Surely the soul can best reflect when it is free of all distractions such as hearing or sight or pain or pleasure of any kind—that is, when it ignores the body and becomes as far as possible independent, avoiding all physical contacts and associations as much as it can, in its search for reality." In "despising the body and avoiding it, and endeavoring to become independent—the philosopher's soul is ahead of all the rest. . . . If we are ever to have pure knowledge of anything, we must get rid of the body and contemplate things by themselves with the soul by itself." In this way, the practice of philosophy during life was the imitation of death, since both philosophy and death act to free the soul from its bodily prison: "True philosophers make dying their profession." [11]

The ancient Greek association between the truth-lover's way of life and the denial of the body was widespread and mutatis mutandis persistent. Diogenes the Cynic was advertised as a philosopher who cared so little for fleshly and material rewards that, when asked by the great Alexander what thing he might desire of him, he requested only that Alexander should "stand out of my light." [12] Stoic philosophers, content with water and plain bread, able to miss their dinner without complaint or even without noticing, were celebrated for the simplicity of their diet. Epicurus, whose identification of pleasure as the goal of life was much misunderstood, was "thrilled with pleasure in the body, when I live on bread and water," and commanded a friend to "[s]end me some preserved

11. Plato, *Phaedo*, 64d–66d, 67e; cf. idem, *Gorgias*, 524–27. For treatment of the pervasive (but "very curious") association between death and philosophy, see Arendt, *Life of the Mind*, 1:79–81.
12. Diogenes Laërtius, *Lives of Eminent Philosophers*, 2:41; cf. Stanley, *History of Philosophy* (1687), 410.

cheese, that when I like I may have a feast."[13] The Greek seeker after truth was recurrently said to eat only enough to keep life going. To eat more than a bare minimum, or to yearn after delicacies, was to compromise the philosopher's ideal self-sufficiency. The condition for truth was an austere dietetics.

Pythagoras and his followers were famous for their abstemiousness. Legend had it that they routinely performed "an exercise of temperance": "There being prepared and set before them all sorts of delicate food, they looked upon it a good while, and after that their appetites were fully provoked by the sight thereof, they commanded it to be taken off, and given to the servants." Later commentators made much of Pythagorean vegetarianism and the prohibition against eating beans. Both animal flesh and beans produced noxious effluvia that corrupted the body and rendered it impure and unfit for intellectual activity.[14] Accordingly, a frugal diet was not only a display of dedication to knowledge and an emblem of a person who cared little for its pleasures and needs, it might also be understood as a *physiological condition* for putting the body in a fit posture for the intellectual and spiritual quest. (As Ludwig Feuerbach much later punned, "Der Mensch ist, was er iszt.")[15] Broadly Pythagorean sentiments persisted into the later Roman Empire, Plotinus and his pupil Porphyry arguing strenuously for abstemiousness and vegetarianism for all, but especially for those intending to live a philosophical life: "Abstinence from animal food . . . is not simply recommended to all men, but to philosophers." Porphyry's tract commending vegetarianism was written against a philosophical friend who took up flesh-eating on his conversion to Christianity. You are what you eat, and those who consumed flesh fed their animal natures while they poisoned their souls.[16]

■ *The Ascetic Ideal and Its Christian Tropes* ■

AFTER JESUS WANDERED in the desert for forty days and nights, "he hungered" (Matthew 4 : 1–2; Luke 4 : 1–2). Satan's first temptation was not power but food: "If thou art the Son of God, command that these stones become bread. But

13. Oates, *Stoic and Epicurean Philosophers*, 48 (for Epicurus); for Stoic dietetics, see Epictetus, *Discourses*, 434, 439, 443; Seneca, *Moral Essays*, 1 : 128, 151; and see also Nussbaum, *Therapy of Desire*, 112–14, and Brown, *Body and Society*, 27.

14. Stanley, *History of Philosophy*, 493, 506–7, 511, 518 (quoted passage), 564; see also Grmek, *Diseases in the Ancient Greek World*, chap. 9 ("The Harm in Broad Beans"); Dodds, *The Greeks and the Irrational*, 143–54; Camporesi, *The Magic Harvest*, 11, 15.

15. For an introduction to the origins of this pun, see Wartofsky, *Feuerbach*, 413–14, 451 n. 6. The project of giving an account of the "connection between what you eat and how you think" has not been wholly abandoned by modern medical science; see, e.g., Bourre, *Brainfood*, esp. chaps. 2, 8.

16. Iamblichus, *On the Pythagorean Life*, esp. 4–7, 14, 24, 43–44, 47–48 (on Pythagorean dietetics); Porphyry, *On Abstinence from Animal Food* (ca. A.D. 250), e.g., 54–56, 64, 99–100; see also Bynum, *Resurrection of the Body*, 33–41; Osborne, "Ancient Vegetarianism," 218–23.

he answered and said, It is written [quoting Deuteronomy 8:3] Man shall not live by bread alone" (Matthew 4:3–4; Luke 4:3–4); "Is not the life more than the food and the body more than the raiment?" (Matthew 6:25). When the disciples wondered that the Rabbi did not eat, "he said unto them, I have meat to eat that ye know not" (John 4:30–32). For the faithful, Jesus himself was "the bread of life: he that cometh to me shall not hunger, and he that believeth on me shall never thirst" (John 6:35). Paul lectured the Corinthians: "Meats for the belly, and the belly for meats; but God shall bring to nought both it and them" (I Corinthians 6:13).[17]

The early Christian idiom for expressing the relationship between the denial of bodily wants and the attainment of spiritual knowledge is probably more familiar than that of Greek Antiquity, and fine recent historical work has yielded new understandings of the ascetic culture produced by the Egyptian monks of the third and fourth centuries. Peter Brown, for example, has corrected dominant modern assumptions about the temptations Christian hermits and anchorites took themselves to the desert to confront and surmount. For Saint Anthony, the desert was "a zone of the non-human," and, for this reason, Brown writes,

> the most bitter struggle of the desert ascetic was presented not so much as a struggle with his sexuality as with his belly. It was his triumph in the struggle with hunger that released, in the popular imagination, the most majestic and the most haunting images of a new humanity. . . . The titillating whispers of the "demon of fornication," much though they appear to fascinate modern readers, seemed trivial compared with [the obsession with food].[18]

(In the Middle Ages, the skin disease erysipelas was known as Saint Anthony's blush, because, as one legend has it, the anchorite saint blushed every time he was obliged to eat.) The Desert Fathers regarded eating as a matter of both shame and spiritual danger: "The body prospers in the measure in which the soul is weak-

---

17. See also Bynum, *Holy Feast and Holy Fast*, 3. Forty days and forty nights was also the period of Elijah's fast: I Kings 19:8. For Judaic and early Christian conceptions of food as embodying God's knowledge, see Feeley-Harnik, *The Lord's Table*, 82–91.

18. Brown, *Body and Society*, 218–21. Interviewed about the first volume of his *History of Sexuality*, Michel Foucault "confessed" that "sex is boring," and that it was so for the Greeks and early Christians as well:

> [Sex] was not a great issue. Compare, for instance, what they say about the place of food and diet. I think it is very, very interesting to see the move, the very slow move, from the privileging of food which was overwhelming in Greece, to interest in sex. Food was still much more important during the early Christian days than sex. For instance, in the rules for monks, the problem was food, food, food. Then you can see a very slow shift during the Middle Ages when they were in a kind of equilibrium . . . and after the seventeenth century it was sex. (Foucault, "On the Genealogy of Ethics," 229)

ened and the soul prospers in the measure in which the body is weakened."[19] Another legend tells of a friend, concerned for the health of the hermit Abba Macarius, bringing him a bunch of grapes. Macarius was unwilling to indulge himself and sent them to another hermit, who then passed them on to still another, until at last they came back to Macarius, uneaten.[20] Here the religious life of the mind appears not just disembodied but specifically disemboweled.

The ascetics of Late Antiquity tended to conceive of the human body as an "autarkic" system. In ideal conditions, and, tellingly, before Adam's original sin—it was food, after all, that brought him down—the body was thought capable of running "on its own heat." It needed just enough food to maintain that heat. It was only "the twisted will of fallen men" that gorged the body with surplus food, and it was this dietary surfeit that produced the excess energy manifested in "physical appetite, in anger, and in the sexual urge." The passions, including that of sexuality, were thus in part epiphenomena of dietetics: food before sex. Brown writes that

> in reducing the intake to which he had become accustomed, the ascetic slowly remade his body. He turned it into an exactly calibrated instrument. Its drastic physical changes, after years of ascetic discipline, registered with satisfying precision the essential, preliminary stages of the long return of the human person, body and soul together, to an original, natural and uncorrupted state.

In Genesis (1:29) the Lord said that "I have given you every herb . . . and every tree . . . and to you it shall be for meat." From the early Christian era well into the eighteenth century and beyond it was debated whether Adam and Eve were vegetarians and whether they ate only raw foods; whether this was the natural diet of prelapsarian humans; whether the Fall from Grace altered the human constitution so that we now required flesh and cooked foods; and, importantly, whether fallen humans might restore their pure state, and their pristine and powerful intellectual capacities, by a pure and primitive diet.[21]

19. Abba Daniel (ca. 450), in Desert Fathers, *Sayings of the Desert Fathers*, 43–44; cf. Musurillo, "The Problem of Ascetical Fasting." Note the typical gesture here at what Max Weber ("Religious Rejections," 327) called "the Janus-face" of asceticism: the world and the flesh are denied, but in such a way as to attain mastery—if not of this world, then of a greater world.

20. Desert Fathers, *Lives of the Desert Fathers*, 109; Bynum, *Holy Feast and Holy Fast*, 38; Gould, *The Desert Fathers*, 143; Camporesi, *The Anatomy of the Senses*, esp. chap. 4.

21. Brown, *Body and Society*, 223; see also Grimm, "Fasting Women," 231–34. Ancient theories of "innate heat" and its relation to diet are treated in Mendelsohn, *Heat and Life*, chap. 2, and in Temkin, "Nutrition," 85–88. For continuing medical speculation on the natural dietetics of human beings before the Fall and in Antiquity, see, e.g., Cheyne, *Essay of Health and Long Life* (1724), 91–92; Mackenzie, *History of Health* (1760), 17–53; and Smith, *Sure Guide in Sickness and Health* (1776), 78–81. And for the causal influence of dietetics on the sexual appetite, see Bynum, *Holy Feast and*

In the early Christian era, Saint Augustine was perhaps the most influential voice advertising the disciplined body as the condition for spirituality. The Jews feared certain foods, while to the Christian all foods were equally clean or unclean: "It is not the impurity of food I fear but that of uncontrolled desire." God taught Augustine "to take food in the way I take medicines. But while I pass from the discomfort of need to the tranquillity of satisfaction, the very transition contains for me an insidious trap of uncontrolled desire." Moreover, the variety of fleshly pleasures offered by the variety of foods was a snare. Routine consumption of *the same* foods—for Augustine as for Wittgenstein—was a way of ensuring against the "tumult of the flesh" and "bringing the body into captivity."[22] That was the human condition: to be human was not only to err but to eat, and, in eating, people inevitably fed those animal wants that had the potential to corrupt the soul.[23] In this way, the Eucharist Host and Communion wine expressed not only particularly Christian worship but also the general human predicament, until such time as bread was replaced by the Bread of Heaven. After the Resurrection, there would be no need to eat in order to prevent decay.[24]

By contrast, Jewish traditions of asceticism, and ascetic warrants for knowledge, were relatively poorly developed. Immediately after the Old Testament's most eloquent commendation of decorum—"To every thing there is a season"—the aged Solomon wrote that there is nothing better for men "than to rejoice, and to do good so long as they live. And also that every man should eat and drink, and enjoy good in all his labour, is the gift of God. . . . Go thy way, eat thy bread with joy, and drink thy wine with a merry heart; for God hath already accepted thy works" (Ecclesiastes 3:1, 3:12–13, 9:7).[25] In the twelfth century, the Spanish-Jewish physician Moses Maimonides worried about the effects, both on pious Gentiles and on his own coreligionists, of the heroic asceticism of Christian "saintly ones." In fact, Maimonides said, such abstinence was best understood as

---

Holy Fast, 37; Camporesi, *The Anatomy of the Senses,* 67–69; and Rouselle, *Porneia,* 169–78. The dependence of lust on diet remained proverbial into the early modern period; see Erasmus's quotation (*Proverbs or Adages* [1569], 34v) of the adages "Without meate and drinke the lust of the body is colde"; "The beste way to tame carnall lust, is to kepe abstinence of meates and drinkes"; and "A licourouse [licentious] mouth, a licourouse taile."

22. Augustine, *Confessions,* 171, 204–7. Saint Gregory of Nyssa (d. 395) described taste as "the mother of all vice" (quoted in Bynum, *Holy Feast and Holy Fast,* 38). And Camporesi refers (*The Anatomy of the Senses,* 65; cf. 147) to a Christian "anti-cuisine," aiming at "an alienation of taste . . . a cuisine with a minus sign, a protest against the physiological game we are forced to play by the organic cycles of the flesh."

23. In the second century Porphyry (*On Abstinence from Animal Food,* 54) wrote specifically against taking a *variety* of foods, for such diversity only fed the "variety of pleasure . . . and in this respect resembles venereal enjoyments, and the drinking of foreign wines."

24. Bynum, *Resurrection of the Body,* 102; cf. 124–28, 148. For an anthropological interpretation of the Eucharist, see Feeley-Harnik, *The Lord's Table,* esp. 63–70.

25. Cf. Ecclesiastes 5:18, 8:15, 10:19, and Luke 12:19.

a periodic means of "restoring the health of their souls" and as a contingent re-action against "the immorality of the towns-people." The mistake of the igno-rant was to think that extreme abstinence was virtuous in itself, "that by this means man would approach nearer to God, as if He hated the human body, and desired its destruction. It never dawned on them, however, that these actions were bad and resulted in moral imperfection of the soul." Aristotelian modera-tion was identified as the dietetics of both spiritual and civic well-being.[26]

Nor is it the case that pagan and early Christian ethical and medical author-ity issued blanket recommendations of severe abstemiousness. From the pre-Socratics through the Hippocratic and Galenic corpus, and the writings of such Stoic philosophers as Epictetus and Seneca, health was seen to flow from observ-ing *moderation*—in exercise, in study, and in diet. Both gluttony and excessive fasting were explicitly identified as recipes for moral and physiological disaster. Let the body serve the rational mind, not the mind the body; in eating, let your aim be to "quench the desires of Nature, not to fill your belly"; "allow thy belly what thou shouldst, not what thou mayest"; "eat to live, not live to eat."[27] Such advice, as well as the physiological schema that justified it, proved remarkably stable over a great span of European history. Tweaked, tuned, and idiosyncrati-cally interpreted by individual writers, balance, stability, and moderation re-mained the dominant dietetic counsel from Antiquity to the modern period.[28] The lay wisdom of an early modern proverb had it that "[h]e that is ashamed to eat is ashamed to live." Yet the prudent "middle way" to which free civic actors

26. Maimonides, *Eight Chapters on Ethics,* 60–62; idem, *Medical Aphorisms,* 1:122; 2:41–46; cf. Bynum, *Holy Feast and Holy Fast,* 36 (for Patristic citation of Old Testament examples of holy fast-ing). In the seventeenth century Spinoza's dietetics substantially fell in with the dominant tradition of Jewish philosophical moderation:

> [I]t is the part of a wise man to refresh and recreate himself with moderate and pleasant food and drink. . . . For the human body is composed of very numerous parts, of diverse nature, which continually stand in need of fresh and varied nourishment, so that the whole body may be equally capable of performing all the actions, which follow from the necessity of its own nature; and, consequently, so that the mind may also be equally ca-pable of understanding many things simultaneously. This way of life, then, agrees best with our principles, and also with general practice. (Spinoza, "Ethics," 219–20; cf. 241)

For the comparatively restrained Jewish traditions of self-denial, see Solomon, "Asceticism." A dom-inant Gentile sentiment is inverted by the Yiddish proverb: "Az der mogn iz leydik iz der moyekh oykh leydik" (When the stomach is empty so is the brain) (cf. note 30 below).

27. E.g., Diogenes Laërtius, *Lives of Eminent Philosophers,* 1:165 (of Socrates: "He would say that the rest of the world lived to eat, while he himself ate to live"); Galen, *On the Passions of the Soul,* 49–51; Seneca, *Moral Essays,* 2:119, 137, 157; Epictetus, *Discourses,* 458–59.

28. Wesley Smith ("Development of Classical Dietetic Theory," 443–44) refers to such counsel, and the dietetic knowledge that underpinned it, as "the common property of the culture": "It is probably because the tradition belonged to everyone that it did not easily take the impress of a special point of view or group and persisted essentially unchanged through the centuries." In classical usage, "dietetics" included the study and regulation of food and drink, but the term more generally signified regimen or the management of ways of living, or, in medical terms of art, the "non-naturals."

were enjoined created at the same time a way of understanding, and celebrating, the special dietetic self-denial of truth-seekers.[29] The philosopher *was not as other men:* his discipline of the belly was recognized in the culture both as the condition of spirituality and as a badge by which authentic truth-lovers might be identified. A "lean and hungry look," like a specially ascetic way of life, might visibly mark not only the politically risky person—"He thinks too much: such men are dangerous"—but also the exceptionally virtuous and wise man.[30]

Heroic abstinence constituted a potential problem as well as a resource for the developing institutions of Christianity. By Late Antiquity a Church that had assumed substantial responsibilities of civic management was in a different position with respect to gestures of otherworldly disengagement from the one that had once stood on the political periphery. While the solitary ascetic continued symbolically to represent piety in its purest and highest form, such examples could not be effectively offered as a pattern for the ordinary conduct of the whole body of the faithful. The clerical hierarchy increasingly worried about the uncontrollability of individual gestures of heroic asceticism and about the potentially subversive alternative claims to religious authority that such gestures might represent. Orthodoxy was now in a position where its canons formally celebrated heroic asceticism while its institutions reserved the right to counsel a temperate course and to monitor the authenticity and interpretation of individual ascetic gestures. When the bishop lived in a mansion and kept a sumptuous table, personal acts of heroic asceticism might plausibly be treated as subversive critique. The temperate and highly ordered dietetics of monasticism was one way of managing the problem: the sixth-century monastic *Rule* of Saint Benedict, for example, provided for victuals (excluding "the flesh of quadrupeds" but including a ration of wine) whose nature and quantity were prudently adapted to the local climate as well as to individual brothers' work routines, constitutions, and momentary states of health.[31] Another was the careful surveillance of

29. For surveys of the dietetic literature of Early and Late Antiquity, see, e.g., Edelstein, "The Dietetics of Antiquity," esp. 308–16 (for recognition of the special dietetic requirements of the scholar and philosopher); Temkin, *Galenism*, esp. 26, 36–39, 85; Smith, "Development of Classical Dietetic Theory"; and, notably, Foucault, "Dietetics," in *The History of Sexuality*, 2:97–139; also 3:140–41.

30. Caesar wanted "men about me that are fat": *Julius Caesar,* 1.2. And see also Shakespeare's association of thinness, diet, and intelligence: "Fat paunches have lean pates; and dainty bits/Make rich ribs, but bankrupt quite the wits" (*Love's Labour's Lost*, 1.1); and "Methinks sometimes I have no more wit than a Christian or an ordinary man has: but I am a great eater of beef and I believe that does harm to my wit" (*Twelfth Night*, 1.3). The link was proverbial. An early modern English saying pronounced that "The sparing diet is the spirit's feast"; another (attributed to Socrates) judged that "The belly is the head's grave"; an Italian proverb said "Capo grasso, cervello magro"; and Saint Jerome referred to an old Greek adage: "A gross belly does not produce a refined mind": see Tilley, *Dictionary of Proverbs*, e.g., 44, 156, 526.

31. Benedict, *Rule*, 80–81. One rule was to take and to consume what one was given without complaint and even without speech. In another exercise it would be necessary to recover precise his-

the heroically abstinent: was this fasting figure genuine or a fraud? was it quite clear that the faster was not motivated by pride? that such abstinence did not testify to an unreasonable and unwholesome *attention* to the demands of the body? that he or she was not diabolically rather than divinely inspired?[32]

When Saint Francis of Assisi was ill with a fever, his friends urged him to take a little solid nourishment, only to have his eventual backsliding made into a further spectacular public display of self-abasement. Stripping himself naked, and putting a cord round his neck, he commanded a colleague to lead him into the piazza, where he addressed the people: "You believe me to be a holy man, and so do others who, on my example, leave the world and join the Order and way of life of the brothers. But I confess to God and to you that in this sickness of mine I ate meat and broth cooked with meat. . . . Here is the glutton who has grown fat on the meat of chickens."[33] It was just this kind of gesture that might be interpreted as proceeding more from pride than piety. In the fourteenth century Saint Catherine of Siena progressed from a diet of bread, water, and raw vegetables (occasionally supplemented by pus from the suppurating ulcers of a cancer victim's breast) to an announcement that she took nourishment only from the Host. Her friends reminded her that Jesus told his disciples to "eat such things as are set before you" (Luke 10:8), and skeptics suspected that she was in fact sustained by Satan. Carefully watched, Catherine nevertheless satisfied her monitors that she could retain no food in her stomach and that "her body heat consumed no energy."[34] In the seventeenth century those set to watch over Saint Veronica's fasting observed her periodically to gorge, but this was explained as the work of the devil. Pressure was successfully brought to bear to get her to submit to the regular dietetics of her order, of which she ultimately became abbess.[35] So the dietetic moderation to which the civic actor was pervasively enjoined was also, albeit typically on a more ascetic scale, counseled by the Church to its clerics and to the community of believers. The cultures of both civic and sacred institutions possessed ways of understanding, sometimes approving, and sometimes worrying about, the special moral state and the special epistemic claims of the heroically abstinent.

torical distinctions between the practices designated by *abstinence, temperance, fasting,* and related locutions, though, as Bynum points out (*Holy Feast and Holy Fast,* 37–38) the one term *abstinence* came to refer to practices as diverse as refraining from certain types of foods, taking only one meal a day, eating no cooked foods, and eating nothing at all for a period; see also Rousselle, *Porneia,* 167–69.

32. Georgianna, *The Solitary Self,* 25–37; Kleinberg, *Prophets in Their Own Country,* 19, 135–41; Bowman, "Of Food and the Sacred," esp. 111–14. For Thomas Aquinas's debate with himself over whether extreme abstinence counted as virtue or vice, see *Summa Theologica,* 2:1783–92.

33. Quoted in Kleinberg, *Prophets in Their Own Country,* 135 and n. 14.

34. Bell, *Holy Anorexia,* 25–27; see also Camporesi, "The Consecrated Host."

35. Bell, *Holy Anorexia,* chap. 3.

As the examples of Catherine, Veronica, and many other female saints make plain, the gesture of heroic abstinence was at least as available for holy women as it was for holy men. Caroline Bynum has beautifully described differences (as well as similarities) in medieval male and female gestures of holiness, noting the special significance for women of food and its renunciation. Food was pervasively "a powerful symbol" and was therefore central to interpreting the human condition and its eschatological future, especially in endemic conditions of scarcity: "But food was not merely a powerful symbol. It was a particularly obvious and accessible symbol to women, who were more intimately involved than men in the preparation and distribution of food."[36] Women's bodies were, indeed, the source of life and of food, and their acts of giving birth and nursing could be recruited as powerful analogies of Christ's body. Yet male medical writing from Antiquity through the Middle Ages (and beyond) tended to conceive of the female body as colder and wetter than the male body, more liable to corruption, more *organic*. "Although all body," Bynum says, "was feared as teeming, labile, and friable, female body was especially so." Yet, she notes, "women could triumph over organic process." This meant that dominant understandings of women's bodies could count as an obstacle to female gestures of spirituality (and female entitlements to spiritual knowledge), while at the same time they gave grounds for regarding the gestures of the heroically abstinent woman as *specially* powerful.[37]

■ *Temperance and Its Early Modern Meanings* ■

FROM ANTIQUITY TO THE early modern period formal medical texts consistently counseled the prudent person to adopt a dietetics of moderation. Yet strands of sixteenth- and seventeenth-century culture contested the meaning of the temperate life, debating how it was that ancient philosophers had lived and how the modern wise person ought to live. The dominant notes in texts written for a genteel readership remained the prudential commendation of a temperate and moderate course of life and an associated condemnation of fashionable excess. The English humanist Sir Thomas Elyot closely followed Galen in listing the qualities of different foods and their effects on persons of varying temperaments. Gross meat made gross bodily juices, and, while the roast beef of Olde Englande offered suitable victuals for laborers and for others of coarse constitution, "it

36. Bynum, *Holy Feast and Holy Fast,* 24–29 (quoting 29).
37. Bynum, *Resurrection of the Body,* 221, and Smith, "Problem of Female Sanctity," 18–20; see also MacLean, *Renaissance Notion of Woman,* esp. 41–46; Bell, *Holy Anorexia;* Vandereycken and van Deth, *From Fasting Saints to Anorexic Girls;* Grimm, "Fasting Women," 229–30; and Shapin, *A Social History of Truth,* 86–91 (for women's physical and social natures in relation to their effective participation in knowledge making).

maketh grosse bloude, and ingendereth melancoly." (Hare too was proverbially said to be "melancholy meat" but capon was recommended for those whose complexion was that way inclined.) Simply prepared things were best; the simultaneous consumption of a variety of meats was to be avoided; gluttony and drunkenness were worst of all. Abstinence, however, might itself be dangerous, and its practice too must be observed in moderation. After all, both Plato and Galen (Saint Paul was not mentioned here) recommended using "a little wine for thy stomach's sake." Excess in abstinence might be conducive to melancholy.[38]

The mid-sixteenth-century homespun advice manual *La Vita Sobria* by the centenarian Venetian gentleman Luigi Cornaro became the most widely circulated early modern tract celebrating dietary temperance. Like Elyot, Cornaro denounced the routine gluttony of modern patrician society. He ate in quantity "only what is enough to sustain my life": bread, broth (with perhaps an egg), no fruit, all sorts of fowl, veal (but no beef), some fish—all flesh being taken in moderation. While the temperate life was "pleasing to God," its justification here took a largely secular form: this is the way one ought to live if one desired health and a robust old age. The lives of ancient philosophers—Plato, Isocrates, Cicero, and Galen—were recruited as patterns of dietary restraint, but nothing about this version of temperance made it unfit for those "in service of the State" or for the ordinary civic actor: "I am nothing but a man and not a saint."[39]

Montaigne's late sixteenth-century skepticism was targeted at dietary as well as at philosophical *systems:*

> My way of life is the same in sickness as in health; the same bed, the same hours, the same food serve me, and the same drink. I make no adjustments at all, save for moderating the amount according to my strength and appetite. Health for me is maintaining my accustomed state without disturbance. It is for habit to give form to our life, just as it pleases.[40]

"[T]here is no way of life so stupid and feeble as that which is conducted by rules and discipline," and one who attempted to eat and drink by the book was no less liable to go wrong than one who sought to regulate belief and action by the book. The "most unsuitable quality for a gentleman" is "bondage" to system. One should not decline to follow local dietary custom because it conflicted with systemic medical principles: "Let such men stick to their kitchens." In di-

---

38. Elyot, *The Castel of Helth* (1541), 11v, 15v–16r (quoted passage), 20r, 32r–33v, 42r–43r, 53v–54r. For Saint Paul, see 1 Timothy 5:23.

39. Cornaro, *The Temperate Life,* 59–60, 75, 87. Cornaro deplored (112) the fact that so many men then in monastic orders no longer lived the temperate lives originally intended for them and were "for the greater part, unhealthy, melancholy, and dissatisfied."

40. Montaigne, "Of Experience" (comp. 1588), 827.

etary matters, one should conform to rules tested by experience but not be "enslaved" by them. There was indeed a vice of "daintiness," of taking "particular care in what you eat and drink," but that vice might be equally manifest in vigilant temperance or in gormandizing fastidiousness. Over a lifetime, one's sense of pleasure adapted one's stomach to its usual fare, and radical change was always likely to do more harm than good. And even if long life was promised to those who would radically amend their dietary habits, "Is it so great a thing to be alive?"[41]

Francis Bacon's posthumously published *History of Life and Death* (1636) worked subtle but consequential changes both on the dietetic culture handed down from Antiquity and on Cornaro's program of systematic temperance. He agreed that the Pythagoreans and the Church Fathers were unusually long-lived and that their abstemious dietetics was substantially responsible for that longevity. And he claimed that "light contemplations" had similarly beneficial effects in prolonging life: "For they detain the spirits on pleasing subjects, and do not permit them to become tumultuous, unquiet, and morose. And hence all contemplators of nature, who had so many and such great wonders to admire, as Democritus, Plato, Parmenides, and Apollonius, were long-lived." By contrast, what he called "subtle, acute, and eager inquisition shortens life; for it fatigues and preys upon the spirits."[42] Yet Bacon dissented from the ascetic tradition that causally associated dietary abstemiousness with intellectual good: "It is certain also that the brain is as it were under the protection of the stomach, and therefore the things which comfort and fortify the stomach by consent assist the brain, and may be transferred to this place."[43] Most important, Bacon adapted traditional injunctions toward dietary moderation, generally preserving the *form* commending the dietary Golden Mean while altering its content and prescriptive meaning. "Frequent fasting," he announced, "was bad for longevity"; and experience showed that great gluttons "are often found the most long-lived": "[W]here extremes are prejudicial, the mean is the best; but where extremes are beneficial, the mean is mostly worthless." Diets that were too spare were to count as extreme, with all the effects on body and mind that flowed from excess. The gentlemanly actor in society was placed in a position where occasional surfeit was a routine and civically prescriptive fact of life, and accommodating oneself

41. Ibid., 830–32, 843.

42. Bacon, "History of Life and Death," 217, 251, 261, 280. For eighteenth-century medical agreement about the longevity of ancient philosophers and its dietetic cause, see Mackenzie, *History of Health*, 243–44.

43. Bacon, "History of Life and Death," 299. Bacon also disagreed (301–2) with dominant religious and philosophical recommendations of dietary simplicity: a variety of dishes was, he said, *better* for digestion, and daintily sauced foods likewise assisted the making of good bodily juice.

to such circumstances was both politically and dietetically prudent: "With regard to the quantity of meat and drink, it occurs to me that a little excess is sometimes good for the irrigation of the body; whence immoderate feasting and deep potations are not to be entirely forbidden."[44]

So early modern culture worked with, and ingeniously reworked, dietetic traditions ultimately inherited from pagan and early Christian literatures. Sixteenth- and seventeenth-century advice was firmly linked to its ancient sources by the recommendation of prudent moderation and temperance for those wishing to live a healthy, happy, and productive life in society, even as the meaning of what it was to observe a temperate dietetics was modified according to differing conceptions of *how* and *where* the good life was to be lived and according to differing conceptions of *who* the philosopher and the prudent person were. When humanist writers urged a relocation of the ideal life of the mind from cloistered to civic settings, dietary advice was part of that attempted cultural transformation. If study and philosophizing were to be legitimate activities within a civic setting, contributing to civic concerns, then the dietetics of the legitimately learned should be substantially similar to that of the prudent civic actor. At the same time, this attempted respecification continued to offer ways of understanding, and even appreciating, the austere dietetics of the otherworldly intellectual.

By the Renaissance and early modern period, Greek and Roman theories of the humors, temperaments, and complexions had been developed into important reflective understandings of what scholars and philosophers "naturally" were like and, in turn, what effects the life of truth-seeking wrought upon their bodies. In the late fifteenth century the neo-Platonist Marsilio Ficino wrote influentially about the melancholy to which learned people were especially prone, by virtue of their natural constitutions (disposing them toward the philosophical life) and by virtue of the effect their habits had upon humoral balance.[45] In

44. Ibid., 261, 277, 304. This twist in the meaning of dietary moderation helps make sense of the pattern of life that John Aubrey noted (with apparent approval) in Bacon's amanuensis Thomas Hobbes:

> I have heard him say that he did beleeve he had been in excesse in his life, a hundred times; which, considering his great age, did not amount to above once a yeare. When he did drinke, he would drinke to excesse to have the benefitt of Vomiting, which he did easily; . . . but he never was, nor could not endure to be, habitually a good fellow, i.e. to drinke every day wine with company, which, though not to drunkennesse, spoiles the Braine. (Aubrey, "Hobbes," 155)

Eighteenth-century medical dietetics rounded on Bacon's advocacy of occasional excess; see, e.g., Mackenzie, *History of Health*, 125–26 (cf. 207–12): to be "warmed with wine" does indeed assist conversation, and even philosophizing, but "a chearful glass" is not to be confused with surfeit. It was popularly but falsely attributed to Hippocrates that "getting drunk once or twice every month [w]as conducive to health."

45. Ficino, *Three Books on Life* (1489); also Klibansky, Panofsky, and Saxl, *Saturn and Melancholy.*

the early seventeenth century the Jacobean physician Robert Burton's *The Anat-omy of Melancholy* (1628) codified and distributed a picture of scholarly melan-choly: you could identify those who unremittingly pursued truth by their bodily "temper," their countenance, their situation and way of life. The philosophical body was different from the civic citizen's body. Dedication to truth was physi-cally inscribed upon it. Bodily form and mode of life were visible as ways of rec-ognizing a philosopher, and these were also ways by which those meaning to *present* themselves as philosophers might effectively do so. These, then, are the cultural traditions against which stories about early modern philosophers should be understood. The stories that attached to seventeenth- and eighteenth-century natural philosophers emerged from traditions attaching them to spiritual in-tellectuals, with much the same meaning for portraying the status and value of knowledge.

■ *The Dietetics of Early Modern Philosophy* ■

ARISTOTLE WONDERED WHY men of genius tended toward melancholy, and Seneca asked why God afflicted the wisest men with ill health.[46] These questions continued to circulate in the seventeenth century and beyond. The natural phi-losopher Walter Charleton announced that the "finest wits" are rarely committed to "the custody of gross and robust bodies; but for the most part [are lodged] in delicate and tender Constitutions."[47] Dead White Males, that is, were generally Sick White Males. And in seventeenth-century English natural philosophy Rob-ert Boyle was widely recognized as such a one. Poised between the role of the gentleman and that of the Christian scholar, Boyle (and his friends) reflected upon the state and meaning of his special bodily constitution and way of life. Few contemporary commentators on Boyle omitted to mention, and to draw out the cultural significance of, Boyle's disengagement, abstemiousness, and physical delicacy. For some it represented melancholy, the badge of "a hard stu-dent," while others contested his identity as a melancholic on the grounds of its incompatibility with gentlemanly civic obligations. John Evelyn saw Boyle's

46. Aristotle, *Problems*, 953a.10–15; Seneca, *Moral Essays*, 1:29; see also Lepenies, *Melancholy and Society*, 13–16, 31–32; Simon, *Mind and Madness in Ancient Greece*, 228–37.
47. Charleton, *Concerning the Different Wits of Men* (1669), 104–5. I must here set aside the im-portant and related question of the relationship between genius and *mental* illness. The physiological fragility of the learned continued to be described, explained, and dietetically managed through the eighteenth and nineteenth centuries and into recent times; see, among many examples, Ramazzini, "Of the Diseases of Learned Men," in his *Diseases of Tradesmen* (1700), 61–65; Cheyne, *Essay of Health and Long Life* (1724), xiii–xiv, 33–38, 83–87; idem, *The English Malady* (1733), 38; Macken-zie, *History of Health*, 137–40, 155–62, 187–88, 197, 223; Watson, "Sick Scientists," in idem, *Scien-tists Are Human* (1938), 29–32; and see also chapters in this volume by Iliffe (4), Warwick (8), and Winter (6).

fragility as a form of refinement and even as a kind of strength: His body was "so delicate that I have frequently compared him to a chrystal, or Venice glass; which, though wrought never so thin and fine, being carefully set up, would outlast the hardier metals of daily use."[48] The funeral sermon preached over Boyle's corpse by his friend Gilbert Burnet carefully rejected the charge of scholarly melancholy. Boyle was too much the gentleman for that: "To a depth of knowledge, which often makes men morose, . . . Boyle added the softness of humanity and an obliging civility." At the same time, Boyle exercised rigorous stoic bodily control, neglecting all display of "pomp in clothes, lodging and equipage." And, tellingly, over a course of more than thirty years "he neither ate nor drank, to gratify the varieties of appetite, but merely to support nature."[49] Arguably, everyone listening to that sermon in Saint Martin-in-the-Fields recognized the trope. The meaning of this stipulation proceeded from its resonance with a culturally pervasive sensibility causally associating types of bodies and types of minds.

Boyle's contemporary, the Cambridge philosopher Henry More, devoted much attention to the care of his own body and to the dietetic regulation of other philosophers' bodies. More's Platonism here expressed itself in the view that all individuals have within themselves a "Divine Body, or Celestial Matter" the state of which depended upon the management of dietetics and passions.[50] The care of the philosopher's special mind involved the special care of his special body: "[T]here is a *sanctity* even of *Body* and *Complexion,* which the sensually-minded do not so much as dream of."[51] More's early eighteenth-century biographer announced that

> [h]e was of a singular Constitution both for Soul and Body: His very Temperature was such as fitted him for the greatest Apprehensions and Performances; especially when by his Temperance, and most earnest Devotion he had refin'd and purified it. A rich "Æthereal sort of body for what was inward" (to use here his own Pythagorick phrase) he had even in this Life; that is to say, a mighty Purity and Plenty of the Animal Spirits, which he still kept up lucid and defaecate by that Conduct and Piety with which he govern'd himself.[52]

48. Evelyn to William Wotton, 30 March 1696, in Evelyn, *Diary and Correspondence,* 3:351; see also Schaffer's chapter 3 in this volume. For the widely distributed late medieval and early modern delusion that one's body was made of glass, see Speak, "An Odd Kind of Melancholy."
49. Burnet, "Character of a Christian Philosopher," 351, 360–62, 366–67; and see Shapin, *A Social History of Truth,* 151–56, 185–87.
50. Ward, *Life of More* (1710), 83.
51. More, "Preface General," in idem, *Collection of Philosophical Writings* (1662), 1:viii. More was here specifically situating his views in the Pythagorean and early Christian traditions of such writers as Plotinus.
52. Ward, *Life of More,* 82–83. Cf. Boethius's sixth-century claim that "The body of a holy man is formed of pure aether": *Consolation of Philosophy,* 99.

Constitutionally endowed with this purity of animal spirits and a warm complexion, More's way of life distilled his natural inclination to the search for truth. He "had always a great care to preserve his Body as a well-strung instrument to his Soul." He said that his body "seem'd built for a Hundred Years, if he did not over-debilitate it with his Studies." A disciplined body was made to serve the philosophical will, for philosophizing was heroic work. By his abstemiousness he had "reduc'd himself . . . to almost Skin and Bones; and was to the last but of a thin and spare Constitution." At the end of his life More said that there were two things he repented: the first was that, although he had the means to afford it, "he had not lived [at Cambridge] as a Fellow-Commoner," and the second was that he had "drunk Wine." [53]

More's remarkable correspondence with Lady Anne Conway is well known to historians as a rich source for English Cartesianism and theology, but it is also almost uniquely informative about the dietetics of early modern philosophical bodies. Incessantly, More and Conway (as well as her husband and her brother John Finch) exchanged recipes for the diet that would best adjust and manage the bodily heat necessary to high philosophical inquiries, while preventing that heat from flaming over into pathological enthusiasm and "phrensy." This was a task requiring the most painstaking management of the quantity and quality of food and drink and attention to the fine adjustment of consumption in relation to momentary bodily state and the precise nature of intellectual labor. Early modern culture understood thought, emotion, and diet as elements of a reciprocally interacting causal system: just as diet could influence mood and cognition, so the forms and content of intellectual activity could affect humoral balance and dietetic requirements. [54] "Too much small beere and fruit" damped the body's heat; wine and roasted meats stoked its fire. To know Anne Conway was to know that her complexion was warm and to know the risks and capacities that attached to such a complexion. More counseled Conway "to eat such kinde of meat as begetts the finest and coolest blood, and to abstain from all gross food, which many times is the most savoury, but breeds melancholy blood," while her brother warned her against overdoing a cooling diet: "Take heed of overcooling your

53. Ward, *Life of More,* 84–85, 123–24, 230. Apart from bouts of fasting, More's Cambridge diet was not said to be extraordinary: he sometimes did not refrain from meat during Lent (since that abstinence "quite altered the Tone of his Body"), and "His Drink was for the most part the College Small Beer: Which, in his pleasant way of speaking, he would say sometimes, was 'Seraphical, and the Best Liquor in the World'" (ibid., 122; see also More to Anne Conway, 5 April and 5 August 1662, in Conway, *Letters,* 200, 205).

54. For More's extended treatment of the dietetics of "enthusiasm," its relation to melancholy and philosophizing, and its management, see "Enthusiasmus Triumphatus," in his *Collection of Philosophical Writings,* vol. 1, esp. 14, 37, 47. For contemporary views of enthusiasm and its medical management, see Heyd's important *"Be Sober and Reasonable,"* esp. chaps. 2–3, and see also Iliffe, "'That Puzleing Problem,'" 436–39.

selfe for your temper being naturally hott to take perpetuall cool thinges is to cure not your disease but to disturb your temper." [55]

And here again considerations of gender have epistemic pertinence, as pervasive understandings of the female complexion (colder than the male's) could provide a general basis for explaining women's absence in philosophic enterprises while the same humoral scheme allowed a heightened appreciation of Anne Conway's special individual constitution.[56] In warmth of complexion and its bearing on the capacity for and nature of philosophical speculation, More and Conway recognized each other (despite male/female difference) as similarly endowed, facing similar predicaments. The advice More gave to his warm woman friend was advice he took for himself. The dietetic counsel conveyed in their letters and (presumably) in their face-to-face conversations *tuned* each other's philosophic thermostats, while Conway worried that her humor might "prove infectious." [57]

Probably the richest seventeenth- and eighteenth-century sources for portrayals of the disembodied philosopher attach to the person of Isaac Newton, and stories of his disengagement and otherworldliness echo into the twentieth century, painting some of our culture's most vivid pictures of the special body whose mind is wholly given over to truth. The legendary and the portable status of these stories is an index of their topicality. George Cheyne's *Natural Method of Curing Diseases* (1742) noted that, in order to "quicken his faculties and fix his attention," Newton "confined himself to a small quantity of bread." [58] Other contemporaries observed both Newton's abstemiousness and forgetfulness of food—as in the chicken story retold at the outset. He "gave up tobacco" because "he would not be dominated by habits." In London, his niece remarked that Newton "would let his dinners stand two hours": "his gruel, or milk and eggs, that was carried to him warm for supper, he would often eat cold for breakfast." [59] In Cambridge, an amanuensis related that he often went into Newton's

55. Finch to Conway, 27 April 1652; More to Conway, 28 March 1653 and 3 September 1660, in Conway, *Letters*, 63, 75, 164; see also Conway to her husband, 16 September 1664, in ibid., 230.

56. John Finch speculated that Anne's terrible headaches might arise from "the closenesse of the sutures [or pores] in your head which may hinder the perspiring of vapours; but in regard few of your sex have that inconvenience," and instructed her not to cool herself excessively "when you are very hott or sweat in your bed": Finch to Conway, 27 April 1652 and 9 April 1653, in ibid., 63, 79; and for treatment of attempts to cure her headaches see Schaffer, chapter 3 in this volume.

57. More to Conway, 1 May 1654, 5 June, 4 and 27 December 1660, and 31 December 1663; Conway to More, 28 November 1660, in Conway, *Letters*, 96, 164, 181, 184, 220; see also Ward, *Life of More*, 146.

58. Cheyne, *Natural Method of Curing Diseases*, 81. During periods of intense concentration, Cheyne added, Newton took "a little sack and water, without any regulation, . . . as he found a craving or failure of spirits."

59. Quoted in L. T. More, *Isaac Newton*, 129, 132; see also ibid., 206 and especially Iliffe, chapter 4 in this volume.

rooms and found his food untouched: "Of which, when I have reminded him, he would reply, 'Have I?'" "His cat grew very fat on the food he left standing on his tray." Still another (contested) report testified that Newton, like the Pythagoreans, abstained from meat.[60]

In the 1930s L. T. More collected these and other stories—"which the world has so often heard"—and gave them new life. More wrote about "Sir Isaac's forgetfulness of his food when intent upon his studies"; "He took no exercise, indulged in no amusements, kept no regular hours and was indifferent to his food." And More drew out the significance of Newton's abstemiousness, abstractedness, and solitude for his identity and that of the knowledge he produced: "It is little wonder that his contemporaries have passed on to us the impression that he was not a mortal man, but rather an embodiment of thought, unhampered by human frailties, unmoved by human ambition. . . . Passion had been omitted from his nature."[61] More recently, Richard S. Westfall's depiction of "a solitary scholar" identified Newton's disembodiment as that "of a man possessed" by love of truth, wholly other, not responsive to his body's needs, not *there* in his own body.[62]

Later in the eighteenth century, Henry Cavendish became a popular attachment for similar stories, and these stories too were prominently told and retold by his biographers, to similar ends. They were struck by the frugality and disengagement of Cavendish's way of life, and this despite his enormous wealth: "A Fellow of the Royal Society reports, 'that if any one dined with Cavendish he invariably gave them a leg of mutton, and nothing else.'" He was so shy of human contact that in his own house he ordered his spare meals by leaving a note for the housekeeper upon a table. George Wilson's mid-nineteenth-century portrayal of Cavendish's body was as telling for the bodily features omitted as for those to which it drew attention: "he did not love; he did not hate; he did not hope; he did not fear. . . . [A]n intellectual head thinking, a pair of wonderfully acute eyes observing, and a pair of very skilful hands experimenting or recording."[63]

60. More, *Isaac Newton*, 247, 250; Westfall, *Never at Rest*, 103–4 (for the fat cat), 580, 850–51, 866; and see Stukeley, *Memoirs of Newton* (1752), 48, 60–61, 66. Newton's niece contradicted reports of his vegetarianism and, according to More (*Isaac Newton*, 135), "said that he followed the rule of St. Paul to take and eat what comes from the shambles without asking questions for conscience's sake." Andrew Combe's influential mid-nineteenth-century dietetic text also denied Newton's status as an icon of vegetarianism; there was, Combe said, much evidence (including the gout from which Newton suffered) that "he did not usually confine himself to a vegetable diet": Combe, *Physiology of Digestion* (1842), 149.

61. More, *Isaac Newton*, 131–32, 206–7, 247.

62. Westfall, *Never at Rest*, 103. For important treatment of images of Newton's person and mind in the eighteenth and nineteenth centuries, see Yeo, "Genius, Method, and Morality."

63. Wilson, *Life of Cavendish* (1851), 164 (for mutton), 169–70, 185 (for head, eyes, and hands); see also Berry, *Henry Cavendish*, 15, 22. The late twentieth-century circulation of the Cavendish mutton story is indicated by Oldroyd, "Social and Historical Studies of Science," 751, 756 n. 1.

■ *The Dietetics of Modern Philosophy* ■

I STARTED BY JUXTAPOSING Newton and Wittgenstein, so suggesting—intentionally—that the trope portraying truth-lovers' ascetic bodies persisted into the twentieth century. And so it did. In 1925 the fictional medical scientist Max Gottlieb (maximum God-love) explained to Martin Arrowsmith (in Sinclair Lewis's novel) just why the scientist was not as other men:

> To be a scientist [says Dr. Gottlieb]—it is not just a different job, so that a man should choose between being a scientist and being a . . . bond-salesman . . . [I]t makes its victim all different from the good normal man. The normal man, he does not care much what he does except that he should eat and sleep and make love. But the scientist is intensely religious—he is so religious that he will not accept quarter-truths, because they are an insult to his faith.[64]

And in 1949 the iconic scientific intellectual of the twentieth century specified the *constitutional* difference between those who lived for truth and those who lived for the belly:

> When I was a fairly precocious young man [Albert Einstein wrote] I became thoroughly impressed with the futility of the hopes and strivings that chase most men restlessly through life. Moreover, I soon discovered the cruelty of that chase, which in those years was much more carefully covered up by hypocrisy and glittering words than is the case today. By the mere existence of his stomach everyone was condemned to participate in that chase. The stomach might well be satisfied by such participation, but not man insofar as he is a thinking and feeling being.[65]

And the stories of bodily abstraction attached to Wittgenstein as one of this century's most celebrated philosophers focus importantly upon his "otherness" in terms instantly recognizable from the classical and early Christian traditions. So some twentieth-century thinkers—like the Desert Fathers and seventeenth-century natural philosophers—*could be* depicted (to use Franz Kafka's striking phrase) as "hunger artists."[66] Some, but not all, or, I think, even very many.

Some caveats against misunderstanding my interpretation of these stories about disembodied truth-lovers: First, it is obviously *not* the case that these depictions attach uniquely to scientists and philosophers. Insofar as the tropes spec-

64. Lewis, *Arrowsmith,* 267. Gottlieb was here giving voice to a continuing conception of the scientific vocation as a *calling* rather than as a *job,* and it would be valuable to have a study tracking how and when vocation changed its dominant meaning from the former to the latter.

65. Einstein, *Autobiographical Notes,* 3. See also Ezrahi, "Einstein and the Light of Reason," esp. 268–73, for Einstein's representation of the relativistic physicist as lonely saint and magus.

66. Ellmann, *The Hunger Artists,* see esp. 17, 27–32, 63–69.

ify the constitution of truth-lovers, they attach to those who secure whatever body of knowledge is represented in the relevant local culture as a repository of truth and value, whether it be religious, scientific, philosophical, or artistic. It ought therefore to be understood that by focusing upon the bodies of early modern and modern scientific truth-lovers I mean to draw attention to the ways in which pervasive tropes locally attached to specific, highly valued forms of culture.

Second, it is also evidently not the case that the trope of disembodiment is without what might be called a "countertrope." Whenever and wherever the trope of disembodiment works to specify proper knowledge an opportunity is created for its purposeful denial or modification. So in Antiquity some sects of philosophers (for example, the Cynics) played with a carnal presentation, and did so as a way of marking out their philosophical practice from that of the dominant tribes of philosophers.[67] And, as I shall shortly note, late nineteenth- and twentieth-century philosophical voices importantly analyze, interpret, and reject the very idea of disembodiment as the condition for making and recognizing truth. Here I want to say that the presentation of disembodiment just has the character of a cultural *institution*, against which critical voices stake their position, not that disembodiment is the only way of presenting and warranting truth. The sociable, merry, and moderately gormandizing philosopher of the eighteenth century—a perspicuous instance here is "le bon David" Hume—makes a statement about the nature and placement of philosophic knowledge whose meaning is understood against the background of a dominant ascetic ideal.[68]

Third, I want to acknowledge both the possibility and, within limits, the legitimacy of a "realist" psychological and sociological way of talking about the disengagement and ascetic discipline of intellectuals. It might, for example, be plausibly said that abstraction, solitude, and self-denial simply *are the conditions* for innovation or for producing knowledge of a certain character. Truth-lovers are "just like that"—by temperament, or are made so by their way of life. And in this connection I am well aware of recent psychological and psychiatric causal inquiry into creativity, innovation, genius, and mental health. Such realist claims may be legitimate within their own causal idiom, though their legitimacy within

67. The "carnivalesque" inversion of the "proper" relationship between mind and belly was hinted at by Mikhail Bakhtin (*Rabelais and His World,* 171): "Most of the epithets and comparisons applied by Rabelais to spiritual things have what one might call an edible character. The author boldly states that he writes only while eating and drinking, and adds: 'Is that not the proper time to commit to the page such sublime themes and such profound wisdom?'" Indeed, the prologue to *Gargantua* makes explicit reference to the carnal habits of Diogenes the Cynic.

68. A case could also be made for Galileo as a secularizing seventeenth-century natural philosopher associated with convivial connoisseurship (especially in wines), even though stories about him also picked out his "abstemiousness" and tendencies towards melancholy; see Camporesi, *The Magic Harvest,* 51–59.

that idiom cannot count as a sufficient explanation of why these stories circulate and persist, and one has to be careful not to take at face value the historical anecdotes that provide some of their evidence. For example, manuscript evidence indicates that, at the very same time stories about Newton's abstemiousness circulated so widely, deliveries to his London household for a single week showed "one goose, two turkeys, two rabbits, and one chicken." A contemporary observed that Newton had grown so fat in later life that "[w]hen he road in his coach, one arm would be out of the coach on one side and the other on the other." [69] And the officially ascetic monks of the abbey of Saint Riquier in the twelfth century are known to have received yearly from their tenants seventy-five thousand eggs, ten thousand capons, and ten thousand chickens. [70]

However the case may turn out about "real" philosophical bodies and their "real" dietetics, historical engagement with the stories that speak about them, about their meanings and uses, and about the conditions of their circulation, has its own legitimacy and interest. Such stories are culturally significant public *presentations* and *stipulations*. They testify at once to the constitution of knowledgeable bodies and to the status of bodies of knowledge; they represent norms for philosophical knowledge and the philosophical knower. And stories about the normative way of life for the truth-lover could, and did, stably coexist with massive evidence that the ideal might not (always or usually) be realized. That is just the nature of norms in relation to the behavior they both describe and prescribe.

Finally, I want speculatively to explore the possibility that, despite gestures at Einsteinian and Wittgensteinian portrayals, the topic of disembodiment has rapidly been losing its sense and force in late modern culture. While the trope of the absent-minded professor continues with some currency, the very idea that the truth-lover is "not as other people" and, particularly, that he (and now, importantly, she) secures knowledge through denying bodily and material wants seems to many naive or quaintly outmoded. On the one hand, much modern sociology of science was founded on the claim that no special temperament or motives distinguished the scientist from the ordinary run of humanity, while, on the other hand, some of the most popular "realistic" portrayals of the modern scientist (e.g., James Watson's *The Double Helix*) secure their public credibility *as* realistic through free confession of scientists' concern for fame, power, money, and sex. The very idea that Dr. Grant Swinger, Professor Morris Zapp, or, in-

69. Westfall, *Never at Rest*, 580, 866; More, *Isaac Newton*, 127.
70. Durant, *The Age of Faith*, 786. For images of late medieval monks as gluttons, see, e.g., Bakhtin, *Rabelais and His World*, chaps. 4–5.

deed, the author of this chapter would ever pass up a pot of money, or a nice chicken, in the quest for truth is currently risible.[71]

Increasingly, I suggest, heroically self-denying bodies and specially virtuous persons are being replaced as guarantees of truth in our culture, and in their stead we now have notions of "expertise" and of the "rigorous policing" exerted on members by the institutions in which expertise lives. Expertise and vigilance, and the warrants for truth these offer, are, of course, no new things in the twentieth century: the ancients too had the ability to recognize expertise. But they—and, I think, intellectuals through the nineteenth century—had other conceptions of knowledge apart from expertise: conceptions of virtuous and sacred knowledge attached to special persons inhabiting special bodies.[72] So, in an eggshell, the suggestion is that the career of the ascetic ideal in knowledge follows the same career as the notion of sacred knowledge and its warrants. Late modern culture appears to be conducting a great experiment to see whether we can order our affairs without a sacred conception of knowledge, and, thus, without a notion that those who produce and maintain truth are any differently constituted, or live any differently, than anyone else. That is the sense in which it might be thought that all knowledge has the character of expertise: experts don't know *differently;* they just know *more.* W. B. Yeats said that "the passions, when we know that they cannot find fulfilment, become vision."[73] Expertise is not vision.

By the 1880s strands in philosophy itself took a decisive turn against the ascetic ideal, notably in the work of Friedrich Nietzsche and his followers. In *The Gay Science* Nietzsche meant to acquire "a subtler eye for all philosophizing to

71. You'll be hungry by now, so here's my recipe for *Fricassée du poulet épistémologique:* joint one free-range chicken; brown in olive oil; in same pan add chopped garlic and soaked dried ceps (or porcini); add one cup dry vermouth; reduce a little, then slowly braise covered for forty-five minutes; remove chicken to warm plate, then add some soaking water from the mushrooms and a quarter cup of sherry vinegar to the pan; reduce on high heat, pour over chicken, and garnish with chervil or Italian parsley. Serve with "foreign wine." *Bon appetit!*

72. It is in this connection that I want to draw attention to apparently systematic changes in the topical content of intellectual biographies from the period before ca. 1850–ca. 1930 to more recent treatments. Biographical accounts in the earlier period routinely contained sections entitled "Appearance and Manner of Living" or otherwise offered detailed accounts of what intellectuals looked like, how they conducted their personal and social lives, and, indeed, what and how they ate. (For a perspicuous late example, see Stuckenberg's *Life of Kant* [1882], chaps. 4, 6.) And, as I have shown, earlier cultures worked with conceptions of knowledge and the knower in which such details were vitally important. In the space formerly occupied by such conceptions, more modern intellectual biography now confronts a great "problem," that of the narrative and causal relationship between what is "personal" and what is "intellectual." Following Freud, there is a recognized (if controversial) idiom for speaking of the link between the gonads and the mind, but, as the introduction to this volume indicated, the very suggestion that significant stories may be told connecting belly and mind now has the character of a joke.

73. Yeats, "Per Amica," 341.

date." The philosophical tradition against which he revolted inscribed disembodiment at its pathological core:

> [E]very ethic with a negative definition of happiness, every metaphysics and physics that knows some finale, some final state of some sort, every predominantly aesthetic or religious craving for some Apart, Beyond, Outside, Above, permits the question whether it was not sickness that inspired the philosopher. . . . [W]hether, taking a large view, philosophy has not been merely an interpretation of the body, and a *misunderstanding of the body.*

The thrust of Nietzsche's criticism of the ascetic ideal was that the pathologies of philosophy "may always be considered first of all as the symptoms of certain bodies."[74] A healthy philosophy was to proceed from an appreciation of a healthy body. Nietzsche's tactics were well-judged: if one means to subvert existing conceptions of transcendental philosophical knowledge one *should* proceed by way of an attack on the ascetic ideal. What Nietzsche could not know, and what his intellectual heirs still cannot clearly visualize, is the shape of a society that has *wholly* dispensed with those conceptions of knowledge and the knower that lie at the heart of the ascetic ideal.

74. Nietzsche, *The Gay Science,* 34–35; and see Harpham, *The Ascetic Imperative,* chap. 4 (for Nietzschean and Foucauldian topics). For Nietzsche's intense philosophic and personal interest in dietetics, see Chamberlain, "A Spoonful of Dr Liebig's Beef Extract," 15: "No more greasy, stodgy, beer-washed idealistic Christian German food for me! I shall curl up with gut pain, vomit if you don't give me Italian vegetables."

■ ACKNOWLEDGMENTS ■

For critical comments on an earlier version of this chapter I thank Michael Lynch and Charles Rosenberg.

■ REFERENCES ■

Arendt, Hannah. *The Life of the Mind.* 2 vols. in 1. San Diego: Harcourt Brace, 1978.
Aristotle. *Complete Works,* ed. Jonathan Barnes. 2 vols. Princeton: Princeton University Press, 1984.
Aubrey, John. "Thomas Hobbes." In *Aubrey's Brief Lives,* ed. Oliver Lawson Dick, 147–59. Ann Arbor: University of Michigan Press, 1957.
Augustine, Saint. *Confessions.* Trans. Henry Chadwick. Oxford: Oxford University Press, 1991.
Bacon, Francis. "The History of Life and Death . . ." [1636]. In *The Philosophical Works of Francis Bacon,* 5 vols., ed. James Spedding, Robert Leslie Ellis, and Douglas Denon Heath, 5:213–335. London: Longman and Co., 1857–58.
Bakhtin, Mikhail. *Rabelais and His World.* Trans. Hélène Iswolsky. Bloomington: Indiana University Press, 1984.

Bell, Rudolph M. *Holy Anorexia.* Chicago: University of Chicago Press, 1985.

Benedict, Saint. *The Rule of St. Benedict.* Trans. Anthony C. Meisel and M. L. del Mastro. Garden City, N.Y.: Image Books, 1975.

Berry, A. J. *Henry Cavendish: His Life and Scientific Work.* London: Hutchinson, 1960.

Boethius, Ancius Manlius Severinus. *The Consolation of Philosophy.* Trans. I. T., ed. William Anderson. Carbondale: Southern Illinois University Press, 1963 (composed ca. 523).

Bourre, Jean-Marie. *Brainfood: A Provocative Exploration of the Connection between What You Eat and How You Think.* Trans. Charles Ramble. Boston: Little, Brown, 1993.

Bowman, Frank. "Of Food and the Sacred: Cellini, Teresa, Montaigne." *L'Esprit Créateur* 16 (1976): 111–33.

Brewster, David. *The Life of Sir Isaac Newton.* London: John Murray, 1831.

Brown, Peter. *The Body and Society: Men, Women and Sexual Renunciation in Early Christianity.* London: Faber and Faber, 1989.

Burnet, Gilbert. "Character of a Christian Philosopher, in a Sermon Preached January 7. 1691–2, at the Funeral of the Hon. Robert Boyle." In *Lives, Characters, and An Address to Posterity,* ed. John Jebb, 325–76. London: James Duncan, 1833.

Burton, Robert. *The Anatomy of Melancholy.* Ed. Floyd Dell and Paul Jordan-Smith. 1628. Reprint, New York: Tudor Publishing Company.

Bynum, Caroline Walker. *Holy Feast and Holy Fast: The Religious Significance of Food to Medieval Women.* Berkeley and Los Angeles: University of California Press, 1987.

———. *The Resurrection of the Body in Western Christianity, 200–1336.* New York: Columbia University Press, 1995.

Camporesi, Piero. *The Anatomy of the Senses: Natural Symbols in Medieval and Early Modern Italy.* Trans. Allan Cameron. Cambridge: Polity Press, 1994.

———. "The Consecrated Host: A Wondrous Excess." Trans. Anna Cancogne. In *Fragments for a History of the Human Body,* ed. Feher et al., 1:221–37. New York: Zone Books, 1989.

———. *The Magic Harvest: Food, Folklore and Society.* Trans. Joan Krakover Hall. Cambridge: Polity Press, 1993.

Cassian, John. *Conferences.* Trans. Colm Luibeheid. New York: Paulist Press, 1985.

Chamberlain, Lesley. "A Spoonful of Dr Liebig's Beef Extract." *Times Literary Supplement,* no. 4871 (9 August 1996): 14–15.

[Charleton, Walter]. *Two Discourses. I. Concerning the Different Wits of Men: II. Of the Mysterie of the Vintners.* London, 1669.

Cheyne, George. *The English Malady: Or, A Treatise of Nervous Diseases of All Kinds, as Spleen, Vapours, Lowness of Spirits, Hypochondriacal, and Hysterical Distempers, &c.* London, 1733.

———. *An Essay of Health and Long Life.* London, 1724.

———. *Natural Method of Curing Diseases of the Body and Disorders of the Mind.* London, 1742.

Combe, Andrew. *The Physiology of Digestion, considered with Relation to the Principles of Dietetics.* Ed. James Coxe. 10th ed. Edinburgh: Maclachlan and Stewart, 1881.

Conway, Anne. *The Conway Letters: The Correspondence of Anne, Viscountess Conway, Henry More, and Their Friends 1642–1684.* Ed. Marjorie Hope Nicolson. Rev. ed. Sarah Hutton. Oxford: Clarendon Press, 1992.

Cornaro, Luigi. *The Temperate Life* [1558]. In *The Art of Living Long,* ed. William F. Butler, 37–113. New York: Fowler and Wells, 1903.

Desert Fathers. *The Lives of the Desert Fathers (The Historia Monachorum in Aegypto).* Trans. Norman Russell. Oxford: Mowbray; Kalamazoo, Mich.: Cistercian Publications, 1981.

———. *The Sayings of the Desert Fathers (Apophthegmata Patrum).* Trans. Benedicta Ward. London: A. R. Mowbray and Co., 1975.

Diogenes Laërtius. *Lives of Eminent Philosophers.* Trans. R. D. Hicks. 2 vols. Cambridge: Harvard University Press, 1979.

Dodds, E. R. *The Greeks and the Irrational.* Boston: Beacon Press, 1957.

Drury, Maurice O'C. "Conversations with Wittgenstein." In *Recollections of Wittgenstein,* ed. Rush Rhees, 97–171. Oxford: Oxford University Press, 1984.

Durant, Will. *The Age of Faith: A History of Medieval Civilization—Christian, Islamic, and Judaic—From Constantine to Dante: A.D. 325–1300.* New York: Simon and Schuster, 1950.

Edelstein, Ludwig. "The Dietetics of Antiquity [1931]." In *Ancient Medicine: Selected Papers of Ludwig Edelstein,* ed. Owsei Temkin and C. Lilian Temkin, 303–16. Baltimore: Johns Hopkins University Press, 1967.

Einstein, Albert. *Autobiographical Notes.* Ed. and trans. Paul Arthur Schilpp. La Salle, Ill.: Open Court, 1979.

Ellmann, Maud. *The Hunger Artists: Starving, Writing, and Imprisonment.* Cambridge: Harvard University Press, 1993.

Elyot, Thomas. *The Castel of Helth.* London, 1541.

Epictetus. *The Discourses of Epictetus, with the Encheiridion and Fragments.* Trans. George Long. New York: A. L. Burt, [1888?].

Erasmus, Desiderius. *Proverbs or Adages.* Ed. and trans. Richard Taverner. 1569. Reprint, Gainesville, Fla.: Scholars' Facsimiles and Reprints, 1956.

Evelyn, John. *Diary and Correspondence of John Evelyn, F.R.S.* 4 vols. London: Henry Colburn, 1854.

Ezrahi, Yaron. "Einstein and the Light of Reason." In *Albert Einstein: Historical and Cultural Perspectives,* ed. Gerald Holton and Yehuda Elkana, 253–78. Princeton: Princeton University Press, 1982.

Feeley-Harnik, Gillian. *The Lord's Table: Eucharist and Passover in Early Christianity.* Philadelphia: University of Pennsylvania Press, 1981.

Ficino, Marsilio. *Three Books on Life* [1489]. Ed. and Trans. Carol V. Kaske and John R. Clark. Binghamton, N.Y.: Renaissance Society of America, 1989.

Fletcher, Angus. "Iconographies of Thought." *Representations* 28 (1989): 99–112.

Foucault, Michel. *The History of Sexuality.* Trans. Robert Hurley. 3 vols. (vol. 1 = *An Introduction;* vol. 2 = *The Use of Pleasure;* vol. 3 = *The Care of the Self*). New York: Vintage Books, 1988, 1990.

———. "On the Geneaology of Ethics: An Overview of Work in Progress." In *Michel Foucault: Beyond Structuralism and Hermeneutics,* ed. Hubert L. Dreyfus and Paul Rabinow, 2d ed., 229–52. Chicago: University of Chicago Press, 1983.

Galen. *On the Passions and Errors of the Soul.* Trans. Paul W. Harkins. Columbus: Ohio State University Press, 1963.

Georgianna, Linda. *The Solitary Self: Individuality in the "Ancrene Wisse."* Cambridge: Harvard University Press, 1981.

Goody, Jack R. *Cooking, Cuisine and Class: A Study in Comparative Sociology.* Cambridge: Cambridge University Press, 1982.

Gould, Graham. *The Desert Fathers on Monastic Community.* Oxford: Clarendon Press, 1993.

Grimm, Veronika. "Fasting Women in Judaism and Christianity in Late Antiquity." In *Food in Antiquity,* ed. John Wilkins, David Harvey, and Mike Dobson, 225–40. Exeter: University of Exeter Press, 1995.

Grmek, Mirko D. *Diseases in the Ancient Greek World.* Trans. Mireille Muellner and Leonard Muellner. Baltimore: Johns Hopkins University Press, 1989.

Harpham, Geoffrey Galt. *The Ascetic Imperative in Culture and Criticism.* Chicago: University of Chicago Press, 1987.

Heffernan, Thomas J. *Sacred Biography: Saints and Their Biographers in the Middle Ages.* New York: Oxford University Press, 1988.

Heyd, Michael. *"Be Sober and Reasonable": The Critique of Enthusiasm in the Seventeenth and Early Eighteenth Centuries.* Leiden: E. J. Brill, 1995.

Iamblichus. *On the Pythagorean Life.* Trans. Gillian Clark. Liverpool: Liverpool University Press, 1989.

Iliffe, Rob. "'That Puzleing Problem': Isaac Newton and the Political Physiology of Self." *Medical History* 39 (1995): 433–58.

Kleinberg, Aviad M. *Prophets in Their Own Country: Living Saints and the Making of Sainthood in the Later Middle Ages.* Chicago: University of Chicago Press, 1992.

Klibansky, Raymond, Erwin Panofsky, and Fritz Saxl. *Saturn and Melancholy: Studies in the History of Natural Philosophy, Religion, and Art.* London: Thomas Nelson, 1964.

Lepenies, Wolf. *Melancholy and Society.* Trans. Jeremy Gaines and Doris Jones. Cambridge: Harvard University Press, 1992.

Lewis, Sinclair. *Arrowsmith* [1925]. New York: New American Library, 1980.

Mackenzie, James. *The History of Health, and the Art of Preserving It.* 3d ed. Edinburgh, 1760.

MacLean, Ian. *The Renaissance Notion of Woman: A Study in the Fortunes of Scholasticism and Medical Science in European Intellectual Life.* Cambridge: Cambridge University Press, 1980.

Maimonides, Moses. *The Eight Chapters of Maimonides on Ethics (Shemonah Perakim): A Psychological and Ethical Treatise.* Ed. and trans. Joseph I. Gorfinkle. New York: Columbia University Press, 1912 (composed ca. 1160).

————. *The Medical Aphorisms of Moses Maimonides.* Trans. and ed. Fred Rosner and Suessman Muntner. 2 vols. New York: Yeshiva University Press, 1970–71 (composed ca. 1187–90).

Malcolm, Norman. *Ludwig Wittgenstein: A Memoir.* Oxford: Oxford University Press, 1962.

Mendelsohn, Everett. *Heat and Life: The Development of the Theory of Animal Heat.* Cambridge: Harvard University Press, 1964.

Montaigne, Michel Eyquem de. "Of Experience." In *The Complete Essays of Montaigne,* trans. Donald M. Frame, 815–57. Stanford: Stanford University Press, 1965.

More, Henry. *A Collection of Several Philosophical Writings of Dr Henry More.* 2d ed., 2 vols. London, 1662.

More, Louis Trenchard. *Isaac Newton: A Biography.* New York: Scribner's, 1934.

Musurillo, Herbert. "The Problem of Ascetical Fasting in the Greek Patristic Writers." *Traditio* 12 (1956): 1–64.

Nietzsche, Friedrich. *The Gay Science* [1887]. Trans. Walter Kaufmann. New York: Vintage Books, 1974.

————. *The Genealogy of Morals* [1887], trans. Horace B. Samuel. In *The Philosophy of Nietzsche,* 617–807. New York: Modern Library, 1954.

Nussbaum, Martha. *The Therapy of Desire: Theory and Practice in Hellenistic Ethics.* Princeton: Princeton University Press, 1994.

Oates, Whitney J., ed. *The Stoic and Epicurean Philosophers: The Complete Extant Writings of Epicurus, Epictetus, Lucretius, Marcus Aurelius.* New York: Random House, 1940.

Oldroyd, David. "Social and Historical Studies of Science in the Classroom?" *Social Studies of Science* 20 (1990): 747–56.

Onians, Richard Broxton. *The Origins of European Thought about the Body, the Mind, the Soul, the World, Time, and Fate.* 2d ed. Cambridge: Cambridge University Press, 1988.

Osborne, Catherine. "Ancient Vegetarianism." In *Food in Antiquity,* ed. John Wilkins, David Harvey, and Mike Dobson, 214–24. Exeter: University of Exeter Press, 1995.

Plato. *The Collected Dialogues.* Ed. Edith Hamilton and Huntington Cairns. Princeton: Princeton University Press, 1961.

Porphyry. *On Abstinence from Animal Food.* Trans. Thomas Taylor, ed. Esme Wynne-Tyson. London: Centaur Press, 1965.

Ramazzini, Bernardino. *Diseases of Tradesmen* [1700]. Ed. Herman Goodman. New York: Medical Lay Press, 1933.

Rouselle, Aline. *Porneia: On Desire and the Body in Antiquity,* trans. Felicia Pheasant. Oxford: Blackwell, 1988.

Seneca. *Moral Essays.* 3 vols. Trans. John W. Basore. Cambridge: Harvard University Press, 1958.

Shapin, Steven. "'The Mind Is Its Own Place': Science and Solitude in Seventeenth-Century England." *Science in Context* 4 (1991): 191–218.

———. *A Social History of Truth: Civility and Science in Seventeenth-Century England.* Chicago: University of Chicago Press, 1994.

Simon, Bennett. *Mind and Madness in Ancient Greece: The Classical Roots of Modern Psychiatry.* Ithaca: Cornell University Press, 1978.

Smith, Julia M. H. "The Problem of Female Sanctity in Carolingian Europe c. 780–920." *Past and Present* 146 (1995): 3–37.

Smith, Wesley D. "The Development of Classical Dietetic Theory." In *Hippocratica: Actes du Colloque Hippocratique de Paris (4–9 Septembre 1978),* ed. M. D. Grmek, 439–46. Paris: CNRS, 1980.

Smith, William. *A Sure Guide in Sickness and Health . . .* London, 1776.

Solomon, Norman. "Asceticism." In *Historical Dictionary of Judaism,* ed. idem., forthcoming.

Speak, Gill. "An Odd Kind of Melancholy: Reflections on the Glass Delusion in Europe (1440–1680)." *History of Psychiatry* 1 (1990): 191–206.

Spinoza, Benedict de. "The Ethics [1677]." In *On the Improvement of the Understanding, The Ethics, Correspondence,* trans. R. H. M. Elwes, 43–271. New York: Dover Books, 1955.

Stanley, Thomas. *The History of Philosophy.* London, 1687.

Stuckenberg, J. H. W. *The Life of Immanuel Kant.* London: Macmillan, 1882.

Stukeley, William. *Memoirs of Sir Isaac Newton's Life.* Ed. A. Hastings White. 1752. Reprint, London: Taylor and Francis, 1936.

Temkin, Owsei. *Galenism: Rise and Decline of a Medical Philosophy.* Ithaca: Cornell University Press, 1973.

———. "Nutrition from Classical Antiquity to the Baroque." In *Human Nutrition: Historic and Scientific,* ed. Iago Galdston, 78–97. New York: International Universities Press, 1960.

Thomas Aquinas, Saint. *Summa Theologica.* Trans. Fathers of the English Dominican Province. 3 vols. New York: Benziger Brothers, 1947–48.

Tilley, Morris Palmer. *A Dictionary of the Proverbs in England in the Sixteenth and Seventeenth Centuries.* Ann Arbor: University of Michigan Press, 1966.

Vandereycken, Walter, and Ron van Deth. *From Fasting Saints to Anorexic Girls: The History of Self-Starvation.* New York: New York University Press, 1994.

Ward, Richard. *The Life of the Learned and Pious Dr Henry More* [1710]. Abridged and ed. by M. F. Howard. London: Theosophical Publishing Society, 1911.

Wartofsky, Marx W. *Feuerbach.* Cambridge: Cambridge University Press, 1977.

Watson, David Lindsay. *Scientists Are Human.* London: Watts and Co., 1938.

Weber, Max. "Religious Rejections of the World and Their Directions [1915]." In *From Max Weber: Essays in Sociology,* ed. H. H. Gerth and C. Wright Mills, 323–59. London: Routledge, 1991.

Westfall, Richard S. *Never at Rest: A Biography of Isaac Newton.* Cambridge: Cambridge University Press, 1980.

Wilson, George. *The Life of the Honble Henry Cavendish.* London: Cavendish Society, 1851.

Wilson, Grove. *The Human Side of Science.* New York: Cosmopolitan Book Corp., 1929.

Yeats, William Butler. "Per Amica Silentia Lunae." In *Mythologies,* 317–69. New York: Collier Books, 1969.

Yeo, Richard R. "Genius, Method, and Morality: Images of Newton in Britain, 1760–1860." *Science in Context* 2 (1988): 257–84.

Zanker, Paul. *The Mask of Socrates: The Image of the Intellectual in Antiquity.* Trans. Alan Shapiro. Berkeley and Los Angeles: University of California Press, 1995.

A MECHANICAL MICROCOSM

Bodily Passions, Good Manners, and Cartesian Mechanism

PETER DEAR

Of fencing experts he remarked that they were masters of a science or art which when they needed it they did not know how to employ, adding that there was something presumptuous in their seeking to reduce to infallible mathematical formulas the angry thoughts and impulses of their adversaries.
— Cervantes, *El licenciado vidriera.*[1]

■ Introduction ■

THE 1669 EDITION of Sébastien Le Clerc's *Pratique de la geometrie* contains numerous plates designed to augment its instruction to young gentlemen in the mathematical arts. It presents geometry as a practical subject that will assist in such areas as military science and fortification, traditional branches of mixed mathematics in the seventeenth century.[2] It also invokes swordplay as an illustration of the rootedness of geometrical curves in practical operations. The corresponding plate (fig. 2.1) displays at the foot of the page four men engaged in combat; the upper half is occupied by geometrical diagrams that display the lines and arcs traced out by the properly wielded blade.[3]

Half a century before, a young French gentleman, already admitted into the culture to which Le Clerc's treatise was to cater, thought to contribute to its refinement by himself writing a manual on fencing. Fencing, like dancing, was one of the basic social accomplishments expected of a nobleman, and as such it demanded a disciplined treatment. This particular manual has been lost, but there

1. Cervantes, *El licenciado vidriera*, 789. I am informed by Dale Pratt of Brigham Young University that such satire of contemporary fencing manuals can also be found in several works by Cervantes's contemporary Francisco de Quevedo, including *La vida del Buscòn*, which refers specifically to Luis Pacheco de Narváez, *Grandezas de las espadas* (1600).
2. Useful accounts of mathematical education and its topical scope in French Jesuit colleges— which trained gentlemen such as Descartes—appear in Dainville, "L'enseignement des mathématiques," and idem, *La géographie des humanistes*, chap. 1. Motley, *Becoming a French Aristocrat*, chap. 3, examines the academy in this period and its topical focus on practical mathematical arts and gentlemanly accomplishments.
3. The plate may be found reproduced in Harth, *Ideology and Culture*, 254; Le Clerc's treatise is discussed on 251–57. I have used the 1682 edition, Le Clerc, *Pratique de la geometrie*.

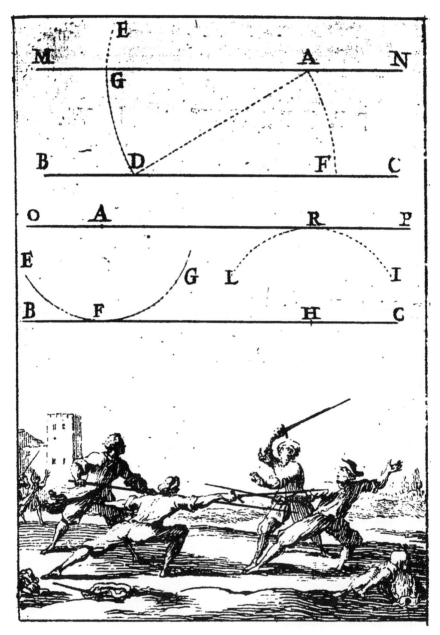

FIGURE 2.1. Swordplay as practical geometry. The action of wielding a sword taken as a proto-typical expression of geometrical construction. From Le Clerc, *Pratique de la geometrie.*

can be little doubt that its author, René Descartes, appreciated the value of "infallible mathematical formulas" in civilizing an art so much associated with "angry thoughts and impulses."[4] Much the same features were found also in the equally gentlemanly, or noble, arts of horsemanship, as contemporary treatises again show. Dressage involved careful, disciplined management of the horse by its rider, but in this case the distance between formalized maneuver and the "angry impulses" of battle was more evident: dressage was dancing on horseback, much as, perhaps, fencing was dancing with swords. Each mimicked violence while in practice eviscerating its affective core, a hypocrisy that Cervantes effectively skewered. Gentlemanly fencers, dancers, and horsemen were automata, going through the motions.[5]

In 1649 Descartes published his *Treatise on the Passions of the Soul.* In it he discusses the influence on the mind of various physiological disturbances, together with observations on the ways in which desirable and undesirable effects may be controlled. No mathematical formulas appear in the course of the exposition, but his mechanistic ontology, the physical instantiation in Descartes's philosophy of mathematical reasoning, underpins the entire discussion. The current chapter is an attempt at making sense of a particular relationship—the Cartesian—between norms of bodily behavior and criteria of intelligibility in seventeenth-century natural philosophy. It does so by investigating the mechanization of natural philosophy and the way in which the resultant structure of intelligible nature constrained (or expressed) the structure of human behavior so intimately connected to it. For Descartes, human behavior was part of human physiology, and that physiology, like the rest of the physical universe, was mechanistic.

■ *Descartes and the Rules of Behavior* ■

RENÉ DESCARTES did not pursue an entirely conventional career for someone of his sort. He was born in 1596, the oldest son of a lawyer, a *conseiller* at the *parlement* of Rennes, who was aiming at the *noblesse de robe,* a very minor bourgeois

4. For other examples of geometrical illustrations of fencing behavior, see the plates, taken from Girard Thibault, *Académie de l'espée* (1628), in Vigarello, "The Upward Training of the Body," 160–65; these pictures integrate the kinds of moves lampooned by Cervantes with familiar Renaissance geometrical overlays on the bodies of the fencers themselves, representing their ideal bodily proportions (for the early seventeenth century, one might point to many examples of the latter in Robert Fludd's works—see Godwin, *Robert Fludd*). See also Motley, *Becoming a French Aristocrat,* 139, on fencing and "the display of the body as a social symbol." On Descartes's fencing treatise see Descartes, *Oeuvres,* 10:533–38. Charles Adam refers to Descartes's probable fencing instruction at La Flèche (as one of the accomplishments of a gentleman) in his "Vie de Descartes," ibid. 12:28.

5. On horsemanship and its pale shadowing of the arts of a former warrior class in the new centralizing French state, see Apostolidès, *Le roi-machine,* 45. See also Motley, *Becoming a French Aristocrat,* 150, on fencing and its increasing stress on "mastering posture and movement of the body."

kind of nobility. Around 1606 he entered the recently founded Jesuit college of La Flèche, already one of the most celebrated schools in France, to study classics, rhetoric, the philosophy of Aristotle, and mathematics. He was there until about 1615, receiving an education that was second to none in Europe at that time, and also receiving reinforcement of his own sense of self—the self of a young French gentleman.[6]

Descartes's treatment at La Flèche took full account of his gentle status. Ordinary scholarship pupils from relatively poor families were consigned to shared dormitories and a strict regime that included effectively constant surveillance; whereas those fewer pupils of a higher status were typically permitted private bedrooms and even, in some cases, a private valet. Neither were they required to attend all the classes prescribed for the others. Descartes had no valet, but he made full use of his private sleeping arrangements. On his arrival, he was put under the wing of one Father Charlet, whose concern for his charge extended to allowing him to remain in bed until quite late in the morning (throughout his life Descartes preferred to rise at around ten o'clock). This was, according to Descartes's seventeenth-century biographer Baillet, in part out of deference to René's (admittedly) delicate health, but also a concession to his intellectual propensities, which supposedly lent themselves naturally to morning meditation.[7] In reality, of course, such indulgence flowed more from Descartes's social position than from a Californian concern for personal growth.

According to Baillet's biography, Descartes's father, Joachim, wanted young René, as well as his less intractable brothers, to follow in his footsteps by entering the legal profession. René actually received a law degree and license at Poitiers in 1616 (although it is unclear whether much more was involved than paying the fee), but he did nothing further in that line. René had been destined, after his departure from La Flèche, to go and serve for a spell in the king's army, but Joachim decided that his son was too young and too weak of constitution for such a life. Baillet says that René was therefore sent to sample life in Paris instead.[8]

Evidently, he soon perked up in the capital: Descartes left France in 1618 for Breda in the United Provinces, and the army of Prince Maurice of Nassau. Military adventure was a known option for young French gentlemen, and Descartes's

6. Gaukroger, *Descartes*, chap. 2, gives a valuable synthetic biographical reconstruction of Descartes's early life based on what little reliable evidence exists. Rodis-Lewis, "Descartes' Life," is a useful short overview of Descartes's career.

7. Baillet, *Vie*, 1:18, 28. On the background to Baillet's book, see Sebba, "Adrien Baillet," esp. 48–57. Snyders, *La pédagogie en France,* discusses the surveillance aspect of Jesuit colleges in this period. For a brief discussion of "meditation" in nondevotional Jesuit pedagogical strategy see Dear, "Mersenne's Suggestion."

8. Baillet, *Vie,* 1:35. See Gaukroger, *Descartes,* 64–65, on the legal excursion.

choice of it, despite his explanations in the *Discourse on Method* of 1637 having to do with the philosophical and moral benefits of learning from experience, looks as much the product of idleness as of vocation.[9] Descartes seems never to have wanted for money. He had his share of the family estates in Poitou, and that seems to have been quite enough to keep him comfortable.[10] He apparently left the rigors of their administration to his brother and other surrogates; throughout his period of residency in the Netherlands (1628–49) he made only occasional visits to France to deal with business matters.

Descartes's father purportedly remarked, following the publication of the *Discourse:* "Only one of my children has given me displeasure. How can I have given birth to a son silly enough to have himself bound in calf!"[11] Joachim Descartes probably never became reconciled to René's choice of letters as his métier. It was very much an occupation associated with clergymen and scoundrels rather than with gentlemen of families aspiring to the *noblesse de robe*.[12] Descartes was making up his own persona, rather than adopting one ready-made.[13] Descartes's apparent contentment with a relatively modest income led Baillet to remark that "it wasn't at all like a needy and grasping gentleman, but like a rich and content philosopher that M. Descartes regarded the goods of the earth."[14] The model of a philosopher here is evidently an antique one, a Stoic or Epicurean, for example, rather than one based on contemporary schoolmen.

Until his permanent move from France to Holland in 1628, Descartes traveled intermittently, visiting Germany and Italy as well as spending extended periods of time in Paris. He spent two years in Paris before going to live in the Netherlands; his claimed reasons for leaving, at a time when he was starting to devote himself seriously to work in philosophy and mathematical sciences, are important in understanding his ongoing process of self-creation. Descartes said that he had gone to the Low Countries in order to be able to control his time; in Paris he had too many visitors and too many social obligations.[15]

Steven Shapin has written of the unusual steps taken by Robert Boyle in the

9. Descartes, *Oeuvres* 6:9.

10. Baillet, *Vie,* 2:459, claims that Descartes, through most of his life, lived on between about six and seven thousand livres of annual rental income.

11. Gaukroger, *Descartes,* 20–23, on Descartes's relationship with his father; quote translated on 23.

12. See, in addition to the previous note, Rodis-Lewis, *L'oeuvre de Descartes,* 1:24–25. The rather déclassé contemporary aura of publishing is noted in Thoren, *The Lord of Uraniborg,* 63. The unorthodoxy of Descartes's career path is noted in Sutton, *Science for a Polite Society,* 59.

13. Compare Goffman, *The Presentation of Self in Everyday Life,* with Greenblatt, *Renaissance Self-Fashioning.*

14. Baillet, *Vie,* 2:459.

15. Descartes, *Oeuvres,* 6:31.

1660s and '70s to establish a refuge of privacy for himself by means of which, at particular times, he would make himself unavailable to visitors so as to be able to pursue his natural-philosophical work undisturbed. Boyle's high social status and pan-European reputation as an experimental philosopher, one who might be expected to have many curiosities to show to his callers, rendered him tempting game for those who had themselves the social standing appropriate for imposing on him unexpectedly. Ordinarily it would have been a serious breach of conduct for a gentleman to refuse to see a qualified caller merely because he was otherwise engaged, but Boyle was able to transcend that norm to a limited extent both because of his particularly high noble status and because he artfully drew on legitimating models of reclusive behavior associated with a religious calling.[16] Descartes seems to have encountered the same problems without having available a comparable solution; instead, he fled.[17]

In effect, Descartes presented the move as his own solution to the same problem as that faced later by Boyle. Vocationally, Descartes saw himself as a philosopher, which, outside a school setting, was an odd thing to be. But the philosopher's role did immediately invoke certain generally understood models of behavior—Baillet explained Descartes's initial decision to leave France by saying that Descartes wanted to be able to enjoy solitude, like all great philosophers, who abandon the courts of princes to enjoy study and meditation away from their own country. Three decades or so earlier, the astronomer Tycho Brahe had provided a similar self-presentation in an account of his own life, describing how he had been obliged to go to a foreign land so as to escape the social shackles that prevent peaceful devotion to one's philosophical calling.[18]

But Descartes's life in the Low Countries, seen in relation to this commonplace picture, presents incongruities. He was not in fact the isolated recluse, the solitary thinker, that intellectual mythography has often suggested and that his successfully promulgated self-image portrayed. He was no more freed of social obligations and the entertaining of visitors in his new home than he would have been if located just a few miles outside Paris.[19] For the most part he lived, during his twenty years' Dutch residence, in major towns (Amsterdam, Utrecht, Leiden), and in those places he hobnobbed with socially elevated people who were

16. Shapin, "The House of Experiment"; idem, " 'The Mind Is Its Own Place.' "

17. It may be valuable to consider by contrast the case of Descartes's popular philosophical friend Mersenne (although he was of humble social origin): he would have been able to avoid callers when he wished because he had religious duties giving him a legitimate claim to solitude.

18. Hannaway, "Laboratory Design and the Aim of Science," 590–91; Thoren, *The Lord of Uraniborg,* 103.

19. Gaukroger, *Descartes,* 187–90, examines various reasons for Descartes's retreat from France and explains it in terms of Descartes's lack of a patron and the constraints this placed on his options.

the local equivalents of those he had left behind: among others, Constantijn Huygens, the father of Christiaan Huygens and a very prominent diplomat and courtier, and the French diplomat Pierre Chanut. Chanut it was who lured Descartes to his death in 1650 (unwittingly, to be sure) by encouraging the Swedish monarch Christina to invite Descartes to Stockholm to tutor her in philosophy. Descartes's keenness to go (he solicited the invitation) indicates his self-perception as the consort of royalty. It should be recalled that one of his greatest works, the *Principles of Philosophy* (1644), had been dedicated to the Princess Elizabeth of Bohemia.[20]

Elizabeth was the eldest daughter of Frederick V, the deposed "Winter King" of the Holy Roman Empire; the family lived in exile in the Netherlands during the period of Descartes's residence there. The philosophical correspondence between Elizabeth and Descartes had begun in 1642, at a time when Descartes had a house a short distance from The Hague, where the princess and other members of her family lived. She began to make a practice of visiting him, sometimes with a party of noble companions who wanted to see the well-known philosopher. Descartes seems to have resented these intruders, and he very soon moved out to a safer distance. From there he could continue to correspond with the princess but to see her only during his own trips to The Hague.[21]

The dedication of the *Principles* was no casual attempt to curry favor. As historians of philosophy well know, throughout the 1640s Descartes and Elizabeth exchanged considerable numbers of letters (thirty-three of Descartes's, and twenty-five of Elizabeth's, survive), besides their personal meetings.[22] Elizabeth treated Descartes not only as her tutor and interlocutor in philosophical discussions, but also as her adviser in medical matters. The latter role she construed quite broadly, often seeking Descartes's advice on ways of relieving her frequent bouts of depression brought on by her family's continuing ill fortunes. Descartes's *Treatise on the Passions of the Soul* (1649) was the direct consequence of his advice to Elizabeth.

Descartes's move to the Netherlands, then, cannot be seen, as he himself asserted, as an attempt to keep his head down. He had not moved to a country where he was freed of social obligations, which, indeed, he seems positively to have courted, even to the point, with Elizabeth, of having to resort to minor evasive actions that he could equally well have taken in France. Evidently, Descartes

20. On Descartes's early acquaintance with Huygens, see ibid., 293; with Elizabeth, ibid., 385–87.
21. Cohen, *Ecrivains français*, 604–7, which uses Samuel Sorbière as its authority about the visits.
22. Most of these letters are translated in Descartes, *The Philosophical Writings*, vol. 3, and in Blom, *Descartes*, with commentary.

liked living in the Low Countries, just as he liked consorting with the people, of his own or higher social status, that he found there.

Descartes's philosophy had to a considerable degree been shaped during his first Dutch sojourn. As a soldier in 1618, he had fallen in with Isaac Beeckman, a schoolmaster and an enthusiast for something he called "physico-mathematics." In Beeckman's usage, this term designated attempts at explaining physical phenomena by reference to submicroscopic particles and their mechanical interactions, a general picture of the world that was to become the centerpiece of Descartes's mature philosophical thought. And when, two decades later, Descartes began to publish his philosophy with the appearance of the *Discourse on Method*, its most enthusiastic sectaries were Dutch.[23]

At first, Descartes clearly had ambitions to enroll in his support the Jesuits, his own teachers. He wanted his philosophy to take the place of Aristotle's in the standard curriculum of Catholic colleges and universities, and he even dedicated the *Meditations* (1641) to the faculty of the Sorbonne. The Jesuits were a particularly tempting target, owing to their control of a hugely influential network of colleges throughout Catholic Europe, and Descartes made sure that he kept on good terms with them. His lack of success on that front, however, was partially offset by the rapid inroads made by Cartesian philosophy into Dutch universities in the years around 1640, during the heyday of Descartes's publication efforts. The Netherlands were particularly fertile ground for his philosophy, just as they were for the establishment of his persona.

■ *Automata and Morals* ■

THE MOST CHARACTERISTIC component of Descartes's philosophy was his mechanistic picture of the physical world. The plausibility to Descartes of Beeckman's version of this ontology may, it has been suggested, owe much to his encounters, during his soldiering, with the clocks, automata, and other mechanical contrivances that were especially favored in that period by some German princes.[24] The construction of artificial people, animals, and ships run by clockwork had become, by the early seventeenth century, a commonplace of expen-

23. Verbeek, *Descartes and the Dutch;* idem, "Regius's *Fundamenta physices*"; Westman, "Huygens and the Problem of Cartesianism."

24. See Mayr, *Authority, Liberty, and Automatic Machinery*, 62–67 on Descartes's mechanistic physiology, chap. 1 on automata and princes; Maurice and Mayr, *The Clockwork Universe*, for many photographs of such devices; Price, "Automata"; Moran, "Princes, Machines." Descartes describes a grotto with automata, similar to one described by Salomon de Caus in a book of 1615, in his posthumously published *Traité de l'homme*, p. 13 of original French ed. (1664) reproduced in Descartes, *Treatise of Man*, i.e., Descartes, *Oeuvres*, 11:130; cf. Descartes, *Discours de la méthode*, 420–22. See also Descartes's remarks in his *Principles of Philosophy*, bk. 4, sect. 203: Descartes, *Oeuvres*, 8:326. Rodis-Lewis, *L'oeuvre de Descartes*, 2:469–72, gives much supplementary material.

sive courtly frivolity. The operation of such devices typically accompanied the marking of the hours in the manner of some elaborate late medieval cathedral clocks (of which the Strasbourg example is the most famous). Descartes's writings often allude to such devices as a means of elucidating his ideas on the animal (including human) organism.

In the fifth part of the *Discourse on Method,* for example, Descartes discusses the action of the heart, explaining how the blood circulates around the body because of the heart's arrangement of valves and its innate heat. The heat rarefies the blood coming into the heart from the veins, and the valves permit the expanded blood to escape only into the arteries, where it cools and condenses again as it travels along. Descartes clarifies the explanatory virtue of his purely qualitative account in the following terms:

> [T]he movement I have just explained follows from the mere arrangement of the parts of the heart (which can be seen with the naked eye), from the heat in the heart (which can be felt with the fingers), and from the nature of the blood (which can be known through observation). This movement follows just as necessarily as the movement of a clock follows from the force, position, and shape of its counterweights and wheels.[25]

The demonstrative form of the account gives it its credibility, according to Descartes, and his paradigm of intelligibility is a clock.

Otto Mayr has noted that although Descartes frequently drew mechanical analogies in his philosophy of nature, these were almost entirely restricted to the realm of living things—plants and animals, including the human body. The workings of the heavens, for example, a prime target for clockwork metaphors since classical antiquity, received no comparisons with automata of any kind.[26] There is thus a sense in which Descartes's mechanistic universe was at its most authentically mechanical when discussing life, a phenomenon to be elucidated in terms of self-contained, self-moving machines.[27] It should always be remembered that a lawlike, mathematically determinate universe need not be specifically mechanical; Descartes's universe was mechanical only insofar as he used machines, and especially automata, as models of intelligibility.

By midcentury, it was no longer an absurd proposition in France that even human behavior should be intelligible in terms of machines (even if not fully

25. Descartes, *Oeuvres,* 6:50; trans. Descartes, *The Philosophical Writings,* 1:136. See, for a particularly lucid account of Descartes's mechanistic physiology, Hatfield, "Descartes' Physiology"; also Gaukroger, *Descartes,* 269–82.

26. Mayr, *Authority, Liberty, and Automatic Machinery,* 63–64.

27. This point is made in Jaynes, "The Problem of Animate Motion." The classic survey of this issue is Rosenfield, *From Beast-Machine to Man-Machine.* See also Vartanian, "Man-Machine."

reducible to them). It is well known, for example, that Blaise Pascal spoke of human "machine" behavior (such as often-repeated religious rituals) as a way of establishing beliefs through habit.[28] Louis XIV's Versailles, slightly later in the century, was sometimes described as a "machine" because of its elaborate courtly ritual and fondness for spectacle that helped constitute the political integrity and power of the king, at the center of the state.[29] But more directly, at the beginning of the century there was already, in the Dutch Republic, what was in some respects a prototypical form of human automation: the organization of its army. The neo-Stoic writings of the Dutch scholar Justus Lipsius had attempted, at the end of the sixteenth century, to renew the virtues of the Roman Empire for the United Provinces, newly emergent from the control of Spain and still fighting for their existence both commercially and militarily. Part of Lipsius's work involved a major treatise, *De militia Romana* (1596), concerning the proper role of armed forces in the constitution of a state, and the manner in which they should be organized; he also wrote at length on Roman military tactics. Lipsius's intention was to promote the Roman model as the right one for the United Provinces. One of the major features of Lipsius's teaching on the military was its stress on the importance of military discipline. This meant the instilling of self-control, restraint, and moderation both in the behavior of individual soldiers and in the collective units made up of them. That was what had made the Roman army so formidable, and should be emulated by the Dutch.[30]

Lipsius was not only a major figure in the intellectual life of this period; his prescriptions, congenial to Prince Maurice of Nassau, were actually put into practice in the remarkably successful armies established by the Dutch. In 1618 Descartes had become a member of one of the two French regiments in Maurice's standing army, and would have seen at first hand the disciplined ethic by which it was governed. Descartes, at that point in his life, was exposed not only to the literal automata beloved of German princes, but also the human automata being drilled and disciplined in the Dutch military encampments.[31]

28. Pascal, *Oeuvres*, 501–2, from "Pensées," on "La Machine"; see on this Keohane, *Philosophy and the State*, 273.
29. This is the central theme of Apostolidès, *Le roi-machine;* see also Revel, "La cour."
30. On neo-Stoicism in this period see Oestreich, *Neostoicism;* on Lipsius, with especial focus on his neo-Stoic natural philosophy, Saunders, *Justus Lipsius.* There is a growing body of literature on Stoic natural philosophy in this period: Barker and Goldstein, "Is Seventeenth-Century Physics Indebted to the Stoics?"; Barker, "Jean Pena"; idem, "Stoic Contributions"; Freudenthal, "Clandestine Stoic Concepts."
31. Oestreich, *Neostoicism,* chap. 5 (on military reform in the Netherlands); see also Gaukroger, *Descartes,* 65–67 on Descartes's military experience. An insightful treatment of Maurice's innovations in regard to drill and its concomitant disciplinary as well as technical efficacy may be found in McNeill, *The Pursuit of Power,* 125–39, which also notes the spread of the new training

It is thus not surprising to find that the highest aim of Cartesian philosophy was proper behavior. In his preface to the French translation of the *Principles of Philosophy* in 1647, Descartes characterizes the philosophy of which his treatise speaks as a tree. "The roots are metaphysics, the trunk is physics, and the branches emerging from the trunk are all the other sciences."[32] Physics, of the kind discussed in the *Principles,* thus gives forth all the special sciences, which have (in the last analysis) to be understood in its terms. Furthermore, these latter "may be reduced to three principal ones, namely medicine, mechanics and morals [*morale*]. By 'morals' I understand the highest and most perfect moral system, which presupposes a complete knowledge of the other sciences and is the ultimate level of wisdom."[33] The study of philosophy, he had observed a little earlier, "is more necessary for the regulation of our morals and our conduct in this life than is the use of our eyes to guide our steps."[34]

"Medicine, mechanics and morals" make a strange triptych nowadays, perhaps, but they made sense to Descartes. The first, medicine, was a lifelong concern of his, one that often provided him with an account of the chief benefit to be derived from a true physics. The prolongation of human life, to a practically preternatural extent, is a recurring theme in his writings, especially in the correspondence, and it is therefore no surprise that Descartes insisted on doctoring himself. It evidently came as a shock to his friends to learn of his demise, in Sweden, at the unduly modest age of fifty-four. Queen Christina sneered that "ses oracles l'ont bien trompé." (One of those oracles was evidently constituted by his teeth, much like a horse; in 1639 he reckoned that their good condition indicated that he had another good thirty years at least, barring accidents.) The red

---

regime to other European armies in emulation of the Dutch; also Parker, *The Military Revolution,* 19–22 (more generally on communal bodily discipline, see also McNeill, *Keeping Together in Time*). Simon Schaffer has drawn my attention to Franz Borkenau's identification, in the 1930s, of Descartes's philosophy with Dutch neo-Stoicism. Borkenau, like Weber, associates the latter with a Calvinist theological outlook (as an expression of bourgeois ideology) with which he then associates Descartes: "coming from the gentry, [Descartes] builds on the presupposition of stoic morality." Borkenau, "The Sociology of the Mechanistic World-Picture," 120. Borkenau links Descartes's insistence on visualizability in his natural philosophy and mathematics (especially his insistence on geometrical representations in mathematics) to "the handicraft basis of production" in this period (ibid., 121); cf. my argument regarding automata in this section. Another Marxist treatment of such issues is Zur Lippe, *Naturbeherrschung am Menschen*; see esp. vol. 2. It is, of course, important not to overemphasize a supposed dominant bourgeois morality in the Netherlands at this time: for a less idealized picture, see Schama, *The Embarrassment of Riches,* chap. 3.
32. Descartes, *Principes,* author's letter, in Descartes, *Oeuvres,* 9:14 (*Principes* has its own, separate pagination within the volume); trans. Descartes, *The Philosophical Writings,* 1:186.
33. Ibid. Note that these three are all comprised under the general heading of "physics" (a contemporary synonym for "natural philosophy"), which would not traditionally have included ethics. Evidently Descartes is intent on establishing "physics" as covering the entire realm of creation.
34. Descartes, *Oeuvres,* 9:3–4; trans. Descartes, *The Philosophical Writings,* 1:180.

wine infused with tobacco that he insisted on treating himself with during his final illness brought up the phlegm but brought down the philosopher.[35]

In the preface to the French *Principles,* Descartes had also proposed that "a nation's civilization and refinement depends on the superiority of the philosophy which is practised there."[36] According to Norbert Elias, this notion of "civilization" arose in France and elsewhere in the early modern period as an aspect of court society and the associated social stratifications that had begun to coalesce in the sixteenth century.[37] Elias finds the French word *civilisation* in use no earlier than the middle of the eighteenth century,[38] but the concept of *civilité,* closely associated with allied terms such as *politesse* and *gentillesse,* or with the ideal of the *honnête homme,* was in common currency to denote a condition both of individuals and of societies. It is not surprising that Descartes, an *honnête homme* born and bred, and an assiduous philosophical courtier, should identify with it.

Good philosophy leads to good behavior, and good behavior is "civilized" behavior. The considerable courtesy literature of the sixteenth and seventeenth centuries, notably of French, Italian, and English provenance, details the specifics of this kind of good manners; its relation to Descartes's philosophy seems, on the face of it, less evident. How could mechanistic philosophy and the "method" lead to civilization and refinement? There can be no doubt, as we have already seen, of Descartes's concern with these latter attributes. But even the *Discourse on Method* displays the attributes expected of an *honnête homme.*

As a part of his carefully controlled self-revelation, Descartes displays a characteristic attitude toward formal learning. His subtle jibes at the various scholarly disciplines encountered at La Flèche make appeal to a fashionable condescension toward pedantry later found exemplified in Molière. A well-known civility manual that first appeared in 1630, called *L'honeste homme: ou, L'art de plaire à la cour,* deals with the matter this way:

> Without it being necessary to get mixed up in all the disputes of philosophy, which would consume perhaps uselessly the entire life of a man who would profit better from studying in the great book of the world than in Aristotle, I judge that it's enough that he have a moderate smattering of the

35. See the presentation of relevant materials in Lindeboom, *Descartes and Medicine,* esp. 94 on Christina and teeth; Gaukroger, *Descartes,* 416, discusses Descartes's death.

36. Descartes, *Oeuvres,* 9:3; trans. Descartes, *The Philosophical Writings,* 1:80. Descartes talks of a nation "plus civilisée & polie."

37. Elias, *The Civilizing Process,* "The History of Manners," part 1, chap. 1; part 2.

38. Ibid., 241 n. 25. *The Oxford English Dictionary* similarly gives no examples of the English word *civilization* prior to the eighteenth century. For additional references on the provenance of the word, see Fox, "Introduction," 29 n. 111.

more agreeable questions that sometimes come up in good society [*les bonnes compagnies*].[39]

In part 1 of the *Discourse,* Descartes recounts the shortcomings of his education and tells us that, as soon as he could, he

> entirely abandoned the study of letters. Resolving to seek no knowledge other than that which could be found in myself or else in the great book of the world, I spent the rest of my youth travelling, visiting courts and armies, mixing with people of diverse temperaments and ranks, gathering various experiences, testing myself in the situations which fortune offered me, and at all times reflecting upon whatever came my way so as to derive some profit by it.[40]

Descartes purports to make a philosophy out of *honnêteté*.

To be *civilisé* in early modern court society was, among other things, to display moderation in one's appetites. Elias's term "civilizing process" is meant to capture the social disciplining of personal behavior, and in particular to emphasize the internalization of that discipline. Elias wanted to show how patterns of behavior became established not as a simple result of active social sanctions directed against deviation, but more profoundly through having people behave in particular ways because, for them, *that's the way you do it.* His argument focuses on the ways in which elaborate codes of courtly behavior extended outward into other social groups during the sixteenth century and beyond so as to generate new kinds of formalized or disciplined behavior. Elias's examples focus to a large extent on material from Germany and the Low Countries, starting with Erasmus's enormously popular work from the early sixteenth century on the instilling of manners in children. Other historians, such as Robert Muchembled, Jacques Revel, and Anna Bryson (for England), have developed Elias's basic insights: such work has involved, inter alia, the use of Dutch genre paintings as especially valuable resources in the tracing of these new bourgeois, and more generally "elite," cultural sensibilities, with their emphasis on restraint and stylization in personal demeanor.[41]

Politesse and self-control formed an integral part of this highly specialized code, and social stratification played a crucial role. What counted as appropriate

---

39. I have used a later edition: Faret, *L'honeste homme,* 31. One of the accomplishments that Faret deems worthwhile is the ability to judge "de la delicatesse des tons de musique" (ibid., 32). Descartes's earliest surviving full treatise (presented to Beeckman as a New Year's gift 1618/19) is the *Compendium musicae,* Descartes, *Oeuvres,* 10:89–141.

40. Descartes, *Oeuvres,* 6:9; trans. Descartes, *The Philosophical Writings,* 1:115.

41. Revel, "The Uses of Civility"; Muchembled, *L'invention de l'homme moderne;* idem, "The Order of Gestures" (genre paintings); Bryson, "The Rhetoric of Status."

behavior in the company of one's social inferiors would not count as appropriate with one's superiors, or, crucially, vice versa. A closeted Louis XIII, for example, showed no qualms about appearing naked before his servants.[42] Appropriate behavior required a delicate assessment of the situation proportional to the delicacy of the new social hierarchy described by Elias. All social life requires situational sensitivity, of course, since the appropriate attitude must always be that "nothing unusual is happening," even though a shift in the details of the social relationship could radically change that assessment. Louis XIII no doubt assumed that his servants would treat his closeted nakedness as routinely proper in much the same way as gynecologists assume that their own characteristic activity will be regarded as routine by women seeking medical examination.[43] When such activity is self-conscious, however, matters are subtly different. The aforementioned courtesy literature of the early modern period shows the self-consciousness of people who wished to pass as proficient in forms of behavior that they evidently regarded as less than routine—Elias speaks of the bourgeois emulation of court society in France.[44] Manners pretend to morality in the more elevated sense, and manners that seem arbitrary or capricious to some appear self-evident to others. The readers of courtesy literature sought counsel on how to conform to an alien morality. Their fear of missteps was a fear of revealing to competent members of court society that they did not possess full membership in that moral community. But that moral community itself had pretensions to an inwardness that was merely expressed in the outward forms of manners. Descartes's various remarks on *morale* (culminating in his *Treatise on the Passions of the Soul* of 1649) show, as do other such contemporary writings, the deep seriousness of literature on manners.

In a world where behavior was so powerfully controlled by one's place in a social system of remarkably precise, and obvious, ordering, what seemed appropriate in social interaction and what seemed appropriate in the dispositions of the natural world—that is, what was regarded as "natural" in each—had certain points of contact that bound them together.[45] Cartesianism in the Netherlands

---

42. Cf. Elias, *The Civilizing Process*, 113–14.
43. Emerson, "Nothing Unusual Is Happening."
44. Elias, *The Civilizing Process*, 29–41 (see esp. 29–33 for basic claim).
45. It ought to be observed at this point that the structure of the present argument is not in the mold of many of the contributions to that hoary classic, Barnes and Shapin, *Natural Order*. Some of those essays set up isomorphisms between ideas about an aspect of the natural world and attitudes toward the social order held by the same people, and infer a causal link from the former to the latter (e.g. Wynne), while others take a practically instrumental view of the function of particular ideas about nature for the furthering of their proponents' social interests (e.g. Barnes and MacKenzie). I shall, by contrast, be attempting to show that appropriate social behavior and appropriate behavior toward nature were fundamentally the same thing for Descartes and many of his contemporaries.

owed its success and its basic point precisely to new forms of personal behavior and their associated sensibilities. The meaning of Descartes's mechanical universe resided in its ability to make sense of those new forms. Thus Dutch society was especially ready to embrace Descartes's work because Cartesianism was a natural philosophy for a bourgeois society.

Descartes's way of formalizing the idea and scope of a machinelike component of human behavior involved restructuring the established concept of "souls." The part-theological, part-psychological genre into which Descartes's writings on the soul and the passions fall has been examined by Nannerl Keohane.[46] She identifies a number of important themes that reappear in seventeenth-century French discussions of love, a principal concern in treatments of the passions. An Augustinian conceptualization predominated: it involved a distinction between two kinds of love, the one pure, selfless, and directed toward God, the other self-interested—*amour-propre*. The *littérateur* Guez de Balzac, a correspondent of Descartes, emphasized the Augustinian distrust of self-love and advocated participation in the wider community as the proper condition of humanity: "Each individual is not enough even to be one unless he tries to multiply himself in certain ways with the help of many; and to consider us in general, it seems that we are not so much whole bodies as disconnected parts that are reunited in society."[47] Descartes expressed similar sentiments in a letter of 1645 to Elizabeth, as also in *Passions of the Soul*.[48] In the wider context of an ongoing discussion of Seneca, Descartes observes that "one could not subsist alone and is, indeed, one of the parts of the earth, and more particularly, of this state, of this society, of this family, to which one is joined by one's residence, one's fealty, and one's birth."[49]

Descartes's varied strategies in his writings depended to a large extent on the cultural audience to which he wished to make appeal. In France, which remained the reference point of his intellectual world, the competing principles of the elites of the sword and the robe, of the erudite, humanist court speaking French and the academic, disputatious university speaking Latin, required different modes of presentation. Both were adopted by the Jesuits, but they were clearly distinct in style.[50] Descartes's switching between French and Latin for his publications, and the increasing formalism of the *Principles of Philosophy* and *Meditations*, with their original Latin versions in the 1640s, as contrasted with the

46. Keohane, *Philosophy and the State,* chap. 6; see also Levi, *French Moralists,* chap. 10.
47. Quoted in Keohane, *Philosophy and the State,* 200 (from Balzac's *Aristippe*).
48. See the discussion in ibid., 204–8.
49. Descartes to Elizabeth, 15 September 1645, in Descartes, *Oeuvres,* 4:290–96; cf. Blom, *Descartes,* 151 (my translation deviates from Blom's).
50. Fumaroli, *L'âge de l'éloquence,* esp. 247–56.

easy style of the French *Discourse on Method* of a few years earlier, seems to mirror that cultural tension. The humanist stress on the importance of rhetoric and the centrality of imitation in rhetorical pedagogy dominated Jesuit education, however, as did the basic assumption that, as Juan Luis Vives had observed in the sixteenth century, "a true imitation of what is admirable is a proof of the goodness of the natural disposition."[51] Descartes's philosophical project furthered the humanist assumptions of the court much as did Jesuit pedagogy, even while maintaining a link between the two cultures.

The portions of Descartes's projected natural philosophy that were to deal with the nature and relationship of the human body and soul had been outlined in the *Discourse*. Descartes seems never to have finished his intended account (at least, it does not survive), although the *Passions of the Soul* goes some way toward filling the gap. But in objecting to the views of his erstwhile disciple, the Dutchman Regius, in 1641, Descartes insisted that there is only one kind of soul, the human rational soul. All other vegetative and animal properties are due to the arrangement of bodily parts.[52] Furthermore, the synopsis in the *Discourse* presents Descartes's position quite clearly. Perhaps the most telling passage concerns the aforementioned question of the differences between automata and humans. He wants to stress not only that the living human body, like other animal bodies, is a kind of elaborate automaton of the sort that God would be capable of fabricating, but also that human beings are not *just* automata. He says:

> if any such machines had the organs and outward shape of a monkey, or of some other animal that lacks reason, we should have no means of knowing that they did not possess entirely the same nature as these animals; whereas if any such machines bore a resemblance to our bodies, and imitated our actions as closely as possible for all practical purposes, we should still have two very certain means of recognizing that they were not real men.[53]

The first way lies in seeing that they were unable to use language. Descartes notes that an automaton could be made to pronounce words,[54] and even to respond to certain stimuli, as, for example, "if you touch it in one spot it asks you what you

51. Vives, *Vives: On Education,* 194.
52. Consecutive letters to Regius, May 1641, in Descartes, *Oeuvres,* 3:369–70, 370–72 (cf. Descartes, *Treatise of Man,* 114). On souls in Renaissance philosophy, see Park, "The Organic Soul." For an argument stressing the role of more general "philosophical" reasons in Descartes's rejection of a vitalist conception of life, rather than reasons derived from within the anatomical tradition and based on empirical grounds of practical intelligibility, see Sloan, "Descartes, the Sceptics, and the Rejection of Vitalism."
53. Descartes, *Oeuvres,* 6:56; trans. Descartes, *The Philosophical Writings,* 1:139–40.
54. Bedini, "The Role of Automata," discusses the long tradition of talking statues that this example invokes.

want of it"; nonetheless, he is convinced that "it is not conceivable that such a machine should produce different arrangements of words so as to give an appropriately meaningful answer to whatever is said in its presence, as the dullest of men can do."[55] Presumably the difficulty would lie in the sheer number of possible conversations in which the automaton might be required to participate.[56] One might, of course, object that if we imagine God as the automaton's artificer, His omnipotence would allow any number of diverse stimulated linguistic responses to be "programmed" into the machine. Descartes's second means of identifying the nonhuman impostor, however, may be seen as addressing that difficulty.

> Secondly, even though such machines might do some things as well as we do them, or perhaps even better, they would inevitably fail in others, which would reveal that they were acting not through understanding but only from the disposition of their organs. For whereas reason is a universal instrument which can be used in all kinds of situations, these organs need some particular disposition for each particular action; hence it is for all practical purposes impossible for a machine to have enough different organs to make it act in all the contingencies of life in the way in which our reason makes us act.[57]

Descartes's notion of the equality or even superiority of machines to humans in particular tasks resembles the sociologist Harry Collins's idea of "behavior-specific action," or "machine-like action." This is a kind of action the description of which is entirely exhausted by a specification of its characteristic behavioral coordinates—the physical motions in space and time. Machines can in principle emulate this kind of action perfectly, but they cannot adapt to new circumstances in which the "same" action might now correspond to a different set of behavioral coordinates. So too with Descartes's automata.

Descartes's own explanation of the distinction is expressed in terms of the capacity of human beings to use reason, the expression of their immaterial *res cogitans*. Reason allows a person to adapt to changing circumstances—"all the occurrences of life"—in appropriate ways. By contrast, Collins makes his demarcation on an importantly different level. Rather than speaking of "reason," a faculty possessed by individuals acting so as to produce behavior appropriate to the

---

55. Descartes, *Oeuvres*, 6:56–57; trans. Descartes, *The Philosophical Writings*, 1:140.

56. This argument is very similar to the (much more elaborate) one given in Collins, *Artificial Experts*, chap. 14: see below.

57. Descartes, *Oeuvres*, 6:57; trans. Descartes, *The Philosophical Writings*, 1:140. See also Descartes, *Discours de la méthode*, 423–25, for further material from Descartes's correspondence reiterating the same point at greater length, and further references.

situation, Collins highlights social life, by definition shared among individuals. It is that social life which creates the ever-shifting context within which appropriate action is created and validated. A machine cannot take part in human social life, according to Collins, and hence is excluded from the possibility of genuinely human action—"acting through understanding," in Descartes's phrase. Descartes's talk of "understanding," or "reason," is an expression of Collins's Wittgensteinian notion of "form of life," and it is fundamentally social.

■ Passions ■

IN HIS *Treatise on the Passions of the Soul* in 1649, Descartes relates his humanizing category of "reason" to internal emotional states by explaining these states mechanistically. The little book had originally been written for the benefit of Princess Elizabeth, who apparently suffered from bouts of depression brought on by her family's ill-fortune. As many commentators have observed, Descartes's treatise fits squarely within a genre of writing on the passions that had both classical antecedents (notably Epicurean and Stoic) and modern exemplars, often directly integrated with theological issues.[58] The closest precedents to Descartes's own approach were Stoic, again paralleling the neo-Stoicism of Justus Lipsius, and like Lipsius, Descartes stressed the goal of self-mastery.

The "passions" are disturbances that affect the mind; they are so called because the mind is affected by them as a patient, not produced by it as an agent. Specifically, Descartes is concerned with the kind of passions that, while often triggered by physiological events, are not experienced as external sentiments, as of pain in a limb or the heat of a fire, but as purely internal, emotional states, such as sadness or joy.[59] The passions affect the mind by means of a naturally established relation between their particular physical manifestations and the character of the mind's apprehension of them through their effect on the flow of (entirely material) "animal spirits" through the pineal gland in the brain—so that the dryness of the throat will serve to produce an active desire to drink, for example. But they can also themselves be affected by the mind, again by the redirection of spirits through the pineal gland. The individuality of Descartes's book stems in large part from its stress on the mind's capacity for controlling the passions to its own advantage. The mind, in this view, is not to be trained merely to

58. See Descartes, *Les passions de l'âme,* "Introduction," esp. 21–32; Rodis-Lewis, introduction to Descartes, *The Passions of the Soul;* and refs. in n. 46, above. Gaukroger, *Descartes,* chap. 10, is an exemplary discussion of the *Passions,* both the arguments therein and the context for the text's production.

59. Descartes identifies three different kinds of passions strictly speaking; only the kind in which the mind is described as acting on itself is of relevance in his discussions of passions qua inner emotional states or urges: Descartes, *Passions,* arts. 17–27.

restrain the tendency of the passions to cloud judgment; it is also to be trained to use the passions actively for positive ends. The otherwise rather similar Stoic doctrine sought only the first goal, that of subduing the passions.[60] Thus pleasure can be gained from the proper handling of the passions (Descartes instances theatrical performances to show that even apparently unpleasant passions such as fear or sadness can sometimes engender pleasure), or courage may be summoned up when needed.[61] And while Juan Luis Vives, whose own discussion of the passions Descartes cites, had couched his account of the physiological disturbances that correspond to each passion in terms of orthodox Galenic humoral theory, Descartes describes them in mechanistic terms.[62] The mind controls the passions, without disdaining them, through the use of reason and the will to control the machine of the body.

Descartes's understanding of the essence of human beings was inseparable from his perception of machines. Since Descartes regarded the human body as a kind of machine, so that the action of the heart, as well as all other organs and parts, was subject to the same mechanical necessity as that of a clock, it might appear that a mechanical physics of the kind outlined in the *Principles* could have no connection with manners and morality—which inhabited a realm categorically different from that concerning the physical springs of bodily behavior. But Descartes's account of the passions of the soul shows how the mechanistic human body, described in greatest detail in the *Traité de l'homme* (ca. 1633) and the brief *Description du corps humain* (late 1640s),[63] could interact with the mind so as to produce behavior that was an amalgam of the two substances. That behavior was therefore neither wholly mechanical nor wholly rational; morality consisted of an appropriate accommodation of the two. In the *Description*, Descartes begins by saying: "There is no more fruitful exercise than attempting to know ourselves. The benefits we may expect from such knowledge not only relate to ethics, as many would initially suppose, but also have a special importance for medicine."[64]

The self to be understood through such means was the self that Descartes described in the *Passions.* The body acts on the mind through the passions, and the mind reacts on the passions by its control of the body through the pineal

60. On the general, and considerable, neo-Stoic elements of Descartes's moral views, see esp. Levi, *French Moralists,* esp. 241–48 (up to and including the *Discourse*), and chap. 10 on the *Passions* and attendant correspondence with Elizabeth, which provides further references. On neo-Stoicism in this period, see above, n. 30.

61. Descartes, *Les passions de l'âme,* part 3, "Des Passions particulieres."

62. Vives, *The Passions of the Soul;* see also Noreña, *Juan Luis Vives.*

63. Both unpublished in his lifetime.

64. Descartes, *Oeuvres,* 11:223; trans. Descartes, *The Philosophical Writings,* 1:314.

gland. Unlike animals, which lack incorporeal, rational souls, human beings can therefore endeavor to control even those parts of their behavior that are directly caused by the actions of the body.[65] As Vance G. Morgan usefully puts it, the passions "are a unique manifestation of the mind-body union in the human being, and their primary purpose is to aid in the preservation of that union"[66]—by naturally inclining us toward actions that will usually be beneficial.[67] Many of the passions, however, represent involuntary physical expressions of mental states (such as joy, desire, hate, and so forth). Mastering the passions therefore requires self-discipline.

> For anyone who has lived in such a way that his conscience cannot reproach him for ever having failed to do anything he judged to be best (which is what I call following virtue here) derives a satisfaction with such power to make him happy that the most vigorous assaults of the Passions never have enough power to disturb the tranquillity of his soul.[68]

Gestures and facial behavior were classic expressions of the passions and accordingly receive much treatment in Descartes's treatise. Thus people flush, tremble, cry, turn pale, laugh, weep, sigh, and in many other ways seem to betray their inner feelings.[69] In particular, "there is no Passion which is not manifested by some particular action of the eyes."[70] Descartes goes on to discuss the subtleties of expression discernible in the "movement and shape of the eye" as well as in facial behavior generally and notes the frequent difficulty found in making sharp distinctions between them. "It is true that there are some that are quite recognizable, like a wrinkled forehead in anger and certain movements of the nose and lips in indignation and mockery, but they do not seem to be natural so much as voluntary."[71] Descartes now makes the bridge between physical expressions of the passions and the place of such expressions in social life: "And in general all the actions of both the face and the eyes can be changed by the soul, when, willing to conceal its passion, it forcefully imagines one in opposition to it; thus one can use them to dissimulate one's passions as well as to manifest

65. Cf. Descartes, *Les passions de l'âme*, art. 138.
66. Morgan, *Foundations of Cartesian Ethics*, 165. Ibid., chap. 5 is a particularly clear account of Descartes on the passions; see also Rorty, "Cartesian Passions."
67. See the Sixth Meditation (Descartes, *Oeuvres*, 7:84–88) for Descartes's explanation of how this relation can sometimes go awry.
68. Descartes, *Les passions de l'âme*, art. 148, trans. in Descartes, *The Passions of the Soul*.
69. See in particular ibid., arts. 114–36. For the role that music was taken to play in inciting the passions, and the operational control that music theorists such as Marin Mersenne thought might thereby be attained over the passions, see Duncan, "Persuading the Affections"; Montagu, *The Expression of the Passions*, 55.
70. Descartes, *Les passions de l'âme*, art. 113 (trans. Voss).
71. Ibid.

them."[72] Not only can the inward manifestation of the passions be controlled (their "assault" on the soul), but their outward manifestation as well. Such issues of gesture and facial behavior were important matters in seventeenth-century social interaction; Elias and others have documented the self-conscious concern with which people handled them.[73] The courtesy and civility literature of the period discusses at great length these issues and the importance of their mastery for social interaction.

Descartes's account of the passions exploits a rigid distinction between them and reason which is rooted in a mechanistic ontology effectively *defined* by the mind-body distinction. It is this distinction, and this ontology, which allows the individual to make his or her own persona. Elizabeth, whose woes and unhappiness largely prompted Descartes to expatiate on these matters at such length, owed her difficulties to a social standing the ordinary expectations of which were thwarted—she was a princess in exile. A milkmaid would not have grieved overmuch for want of a kingdom, but Elizabeth did. Descartes comforted her by advising a moral stance that portrayed her troubles as manageable and her determined behavior as consonant with her status.

It is noteworthy that Descartes does not discuss sexual differences when giving his physiological accounts of the passions. At a less explicit level, one might expect gender differences to appear in the presentation of examples, but these are determinedly rooted at the level of generic human behavior regarding the passions themselves, gender playing only an incidental modulating role.[74] It is this relative gender neutrality at the physiological and mental, if not social, levels in Descartes's philosophy that led some philosophical women in the seventeenth century to adopt Cartesianism as a badge of emancipation.[75]

72. Ibid. The edition by Stephen Voss (Descartes, *The Passions of the Soul*) includes plates from Charles Le Brun, *Conférence sur l'expression générale et particulière* (Paris, 1696), which depict (for artists) typical facial expressions corresponding to a conventional set of passions. See Ross, "Painting the Passions"; Montagu, *The Expression of the Passions*, esp. chap. 1 on Le Brun's theory and its avowed indebtedness to Descartes.

73. Elias, *The Civilizing Process*, and refs. in n. 41, above; Bremmer and Roodenburg, esp. articles by Burke, "The Language of Gesture," Roodenburg, "The 'Hand of Friendship,'" and Muchembled, "The Order of Gestures."

74. As, for example, Descartes, *Les passions de l'âme*, arts. 82, 168, referring to the passions of a brutish man for the woman he wants to rape and to the "honorable" passion of a woman to protect herself from such treatment. These examples tend to escape gendered specificity in regard to the accounting of particular passions; such cases are illustrated by a variety of examples besides those just given, and none is identified with women rather than men. Thus *Passions*, art. 168, on "honorable passions," is also illustrated by examples in which men are the subjects.

75. See Harth, *Cartesian Women*. Harth (67–78) discusses the dedicatee and prime inspirator of the *Passions*, Elizabeth, noting (75–76) Elizabeth's ironic greater readiness than that essayed by Descartes to invoke her physiological characteristics as a woman in discussing the management of her melancholy passions. Gaukroger, *Descartes*, 468 n. 93, comments on art. 147 of *Passions*, regarding

Thus, for a mechanist like Descartes, and for the legion of his followers that sprang up in the 1630s and '40s (including the young Christiaan Huygens), nature was made intelligible through the idea of machinelike action. And at the heart of Descartes's philosophical project lay the assumption that this kind of intelligibility also serves to make people intelligible: the behavior of people and the behavior of machines are in large measure semantically identical. The residual differences between them that the Cartesian philosophy maps out are to be accounted for in terms of the distinction between the behavior of machines and the behavior of reasoning agents, a distinction that helps to define what machines and reason really are. In a strong sense, as I have argued, Cartesian "reason" is a form of socialization.

■ *Cutting a Figure* ■

DESCARTES'S OWN bodily self-presentation is a little difficult to recover, insofar as, whatever his actual behavior in such matters, he did not speak of them at any length. The reports of others, however, provide appropriately responsive accounts of what he achieved, regardless of what he may have intended. One small exception to Descartes's silence suggests an intention that casts light on his practice, however: in a letter to his friend Chanut in 1649, commenting on a visit to Paris, Descartes complains that his friends wanted him only so as to show off his face, implying that they did not want him for his mind.[76] Evidently he thought that his value and virtue did not lie in visible bodily attributes.[77] And yet, as readers of the *Discourse on Method* know, Descartes cared about the figure that he cut in the world. Baillet (who relied for much of his material on Chanut) elaborates on similar themes. Descartes had noted in the *Discourse* that it was a good rule of life to fit in with the customs of the society in which one lived, and to avoid excess in one's judgments as measured by local community standards.[78] Baillet characterizes Descartes's behavior in similar fashion, noting that Descartes liked to wait until new fashions had become common before adopting them himself, "so as to avoid eccentricity." More positively, Baillet says that Descartes "was never negligent [in his appearance], and he avoided above all appearing dressed like a philosopher. When he withdrew to Holland, he abandoned the

the case of a man outwardly grief stricken but secretly glad about the death of his wife, that it shows a "low view of women" by Descartes insofar as it does not seem plausible that Descartes might have presented the story the other way around, with the woman secretly glad of her husband's death. However, this case could just as easily be read as critical of the grieving husband for hypocrisy.

76. See Gaukroger, *Descartes,* 411.

77. Although, as mentioned in ibid., 333, Descartes did indicate to Constantijn Huygens in 1637 his worry about the graying of his hair; see also Lindeboom, *Descartes and Medicine,* 94–95.

78. Descartes, *Oeuvres,* 6:22–23.

sword for the cloak, and silk for wool."[79] Again, this portrait of Descartes as an artful recluse conforms with his self-image as it appears in the biographical format of the *Discourse*.

Descartes had wanted the *Discourse* and the essays on optics, geometry, and meteorology that it prefaced to be published anonymously. His idea was to be able to observe the responses to it from a safe distance—"hiding behind the curtains," as he put it, just as, a few years earlier, he had spoken in a letter about appearing in public "wearing a mask," like an actor in a play (that is, literally adopting a *persona*).[80] So no name appears on the title page of the *Discourse* and its essays. But such anonymity did not prevent the reader from learning who, in a generic sense, had written it. The famous autobiographical format of the first three parts of the *Discourse* establishes that the author is a gentleman, well and properly educated at the famous Jesuit college of La Flèche. The author, that is, told his readers everything they needed to know about him without revealing his individual identity. By this means, Descartes attempted to continue in the state that he celebrated in the *Discourse* itself, that of being "able to lead a life as solitary and withdrawn as if I were in the most remote desert."[81]

An apparent result of this affectation to solitude was that when Descartes had returned to Paris from a stay in Germany in 1624, he came equipped with a rumor that associated him, undesirably, with the Rosicrucians. This was around the time of the "Rosicrucian scare" that swept Paris, briefly, in 1624.[82] The Rosicrucians (who probably did not actually exist as an organized group) were reputed to be devoted to secret knowledge of a Paracelsian and alchemical kind, concerned particularly with medicine, and Descartes had himself, somehow, by now become known as an independent-minded savant—and, indeed, a rather reclusive one. However, the evident unorthodoxy of Rosicrucianism, confirmed by official condemnation, was not something with which a gentleman would wish to associate himself. One of the most notorious attributes of a Rosicrucian

79. Baillet, *Vie*, 2:447.
80. See his remarks on this subject in a letter to Mersenne, 27 February 1637, Descartes, *Oeuvres*, 1:347–48; on the "philosopher behind the mask" see the discussion of Descartes's remark (from the early *Cogitationes privatae*) in Gouhier, *Les premières pensées de Descartes*, 67–68 n. The role of personal modesty as an appropriate moral-rhetorical stance is discussed in Fumaroli, "Rhétorique et philosophie," 42–43, quoting relevant remarks from Castiglione's *The Courtier* concerning the use of the first person.
81. Descartes, *Oeuvres*, 6:31; trans. Descartes, *The Philosophical Writings*, 1:126. The allusion to deserts invokes the persona of a late-antique Christian hermit like Saint Anthony (cf. Shapin, "The Philosopher and the Chicken," chapter 1 in this volume); in another letter to Chanut, in 1648, while visiting Paris, Descartes speaks of returning soon "to the innocence of the desert that I left, where I was much happier" (Descartes, *Oeuvres*, 5:183; cf. Gaukroger, *Descartes*, 411).
82. See discussion in Shea, *The Magic of Numbers and Motion*, chap. 5, "Descartes and the Rosicrucian Enlightenment."

was invisibility: the Parisian "Rosicrucian scare" had involved the appearance of posters around the city proclaiming its covert infiltration by these mysterious magicians. Descartes, in order to scotch the idea that he was himself a Rosicrucian, therefore made a point of walking around the streets of the city every day— so as to *be seen*.[83]

The frugal, eremitic life that Descartes nonetheless imagined for himself seems to have left its mark on his management of the brute bodily proprieties of a philosopher. Baillet comments on his abstemiousness (while noting that his lack of enthusiasm for wine did not harm his good fellowship at a convivial table), and on his care in choosing food on the basis of medical and physiological considerations, which led him to favor root vegetables and fruit.[84] It is clear that the cultural image of the "faceless philosopher" that Descartes had favored remained after him to color accounts of his life. Baillet served him up as the obverse of a glutton—albeit someone who also slept a lot, perhaps itself an appropriately disembodied state.[85] Baillet excuses Descartes's unduly carnal begetting of a child by a serving maid with the observation that it was difficult for "a man who was almost all his life involved in the most recondite anatomical investigations" to lead the life of a celibate.[86] The daughter, Francine, who was to die at the age of five, thus appears as the accidental by-product of properly intellectual pursuits.

■ *Freedom and Autonomy* ■

IN THE *Principles of Philosophy*, there is a section headed: "The supreme perfection of man is that he acts freely or voluntarily, and it is this which makes him deserve praise or blame."[87] This freedom was precisely what machines lacked.

> The extremely broad scope of the will is part of its very nature. And it is a supreme perfection in man that he acts voluntarily, that is, freely; this makes him in a special way the author of his actions and deserving of praise for what he does. We do not praise automatons for accurately producing all the movements they were designed to perform, because the production of these movements occurs necessarily. It is the designer who is praised for

83. Baillet, *Vie*, 1:107–8, and Gouhier, *Les premières pensées de Descartes*, chap. 7.

84. Baillet, *Vie*, 2:448. Baillet continues: "This anchorite's regimen was not invariable in M. Descartes's conduct"; sometimes, he would eat eggs (ibid.).

85. Ibid., 2:449.

86. Ibid., 2:89. On Francine and the maid, Hélène, see Gaukroger, *Descartes*, 294–95. For an interesting discussion of the affective connotations of anatomical research and its associated perceptions of the human body in this period, see Camporesi, *The Anatomy of the Senses*, chap. 5.

87. *Principia*, bk. 1, sect. 37: Descartes, *Oeuvres*, 8A:18; trans. Descartes, *The Philosophical Writings*, 1:205.

constructing such carefully-made devices; for in constructing them he acted not out of necessity but freely. By the same principle, when we embrace the truth, our doing so voluntarily is much more to our credit than would be the case if we could not do otherwise.[88]

Praise and blame attach only to the actions of a free agent, and machines are not free. Such a point would not need to be made if matters were not already being represented in ways that made the opposite look increasingly plausible. Automatic machines, one might say, represent a model of predestination where the omniscience of the clock maker stands in for that of God. But the metaphor (if that is all it is) also abandons explicitly any semblance of free will, the chief function of Descartes's separation of the mind from the body.

As we saw above, Descartes counted judgment and will as the two primary properties of the mind, the *res cogitans.* "Judgment" is a faculty for using reason, typically on material provided from outside; the will, however, is the faculty that renders the human soul truly free.[89] There are traces here of Averroism, wherein the "active" soul, which instantiates reason, is unitary and unchanging; everyone participates in it since reason is the same for all. Human psychic individuation, according to Averroës, results from the "passive" soul, which is shaped by the individual's unique experiences.[90] Since Descartes often speaks of the will, or volition, as the central autonomous characteristic of the soul, rather than of judgment or reason, one can see the will for Descartes as serving the same role as did the passive intellect for Averroës; the freedom of the individual will is the only identifying mark of nonmaterial human individuality, and the extent of its scope is therefore all-important.

It is therefore unsurprising that Thomas Hobbes, a more uncompromising materialist than Descartes or any of his other critics, took issue with Descartes in the Third Objections to the *Meditations* on just this point. Hobbes, regarding Descartes's discussion of the possibility of error in human knowledge, complains that it should be noted "that the freedom of the will is assumed without proof, and in opposition to the view of the Calvinists."[91] Hobbes was all in favor of causal determinism (of a specifiably mechanical sort) as the primary criterion of intelligibility in philosophy, and he attempts to buttress his position by equating

88. Ibid.
89. See esp. the Fourth Meditation, Descartes, *Oeuvres,* 7:58–62, and discussion in Morgan, *Foundations of Cartesian Ethics,* 144–45.
90. Kessler, "The Intellective Soul," for the fortunes of this doctrine in the Renaissance. Gaukroger, *Descartes,* 391–92, makes a similar observation regarding Descartes's remarks on "intellectual memory" and its restriction to universals.
91. Descartes, *Oeuvres,* 7:190; Descartes, *The Philosophical Writings,* 2:133.

it with Calvinist theology. Hobbes wants Calvinist predestination to be fully in-stantiated by automata, whereas Descartes wants to leave an escape route;[92] the lack of freedom exhibited by machines exempts them from moral judgments that are unquestionably appropriate for human beings.

However, while final causes in the Aristotelian sense are noticeably absent from Descartes's construal of his machine paradigm, they cannot be said to be absent from the paradigm itself: the machine metaphor does not determine the nature of machines. Clocks have a function, but Descartes does not regard that function as descriptive of what they are in themselves *as machines*: like Boyle, he would have thought that he "had fairly accounted for it, if, by the shape, size, motion &c. of the spring-wheels, balance, and other parts of the watch [he] had shown, that an engine of such a structure would necessarily mark the hours."[93] Descartes's vision of natural philosophy differs signally from that of scholastic orthodoxy. In the *Principles of Philosophy* Descartes argues that "[i]t is not the final but the efficient causes of created things that we must inquire into."[94] He explains that "[w]hen dealing with natural things we will, then, never derive any explanations from the purposes which God or nature may have had in view when creating them and we shall entirely banish from our philosophy the search for final causes."[95] This is because

> we should not be so arrogant as to suppose that we can share in God's plans. We should, instead, consider him as the efficient cause of all things; and starting from the divine attributes which by God's will we have some knowledge of, we shall see, with the aid of our God-given natural light, what conclusions should be drawn concerning those effects which are apparent to our senses.[96]

We can only understand what God has made, not why He has made it. And the criteria of intelligibility to be applied to this task are rooted in the perceived

92. For more on Hobbes's materialism and its implications, see Sarasohn, "Motion and Morality."

93. Boyle, *Hydrostatical Discourses,* quoted in Shapin and Schaffer, *Leviathan and the Air-Pump,* 216. The present point is discussed at greater length in Dear, *Discipline and Experience,* chap. 6, sect. 1.

94. *Principia,* bk. 1, sect. 28: Descartes, *Oeuvres,* 8A:15; Descartes, *The Philosophical Writings,* 1:202. On final causes and Descartes's mechanical explanations, Rodis-Lewis, "Limitations of the Mechanical Model"; see also Osler, *Divine Will and the Mechanical Philosophy,* 212–13, noting some of the subtleties of Descartes's attitudes toward final causes; also Garber, *Descartes' Metaphysical Physics,* 273–74.

95. *Principia,* bk. 1, sect. 28: Descartes, *Oeuvres,* 8A:15; Descartes, *The Philosophical Writings,* 1:202; the final clause is an addition found in the French *Principes* (Descartes, *Oeuvres,* vol. 9).

96. *Principia,* bk. 1, sect. 28: Descartes, *Oeuvres,* 8A:15–16; Descartes, *The Philosophical Writings,* 1:202.

intelligibility of automata. That is the fundamental point of Descartes's mechanical philosophy. Automata, most familiarly clocks, came to exemplify the transparency of nature to the understanding. And yet they were themselves artificial contrivances.[97]

Descartes's kind of natural knowledge had thus become, rather than an identification of purposes in nature, an attempt at characterizing the "rules" of nature—or what in the seventeenth century are increasingly called *laws* of nature.[98] The rules governing nature's behavior take the place of the purposes for the sake of which that behavior occurs: we want to know how to handle a sword, not why we are fighting. Indeed, fencing is swordplay without instrumental purpose, just as dancing is perambulation without a destination or dressage horsemanship without a battle to win; it is just what a gentleman does.

The amorphousness of our view of the seventeenth century's philosophical mechanism is due to a failure to examine the contemporary meaning of contrivance itself. The distinction between art and nature was dissolved by *fiat:* the possibility of saying that art is a matter of manipulating rather than overriding nature relied on shifting the focus from ends to means. Things in the world became made things, and specifying the agency that had made them revealed no essential difference. The import of mechanism was therefore at root methodological rather than ontological. Descartes presented this idea through ontological talk, but he made it work through contrivance. In the end, God could not tell human beings what His purposes were, and human beings could not tell each other what their own purposes were.

Charles Le Brun, the midcentury artist and writer on facial expressions as manifestations of the passions, claimed to derive the physiological warrant for his teachings from Descartes's *Treatise.* He also produced many of the paintings that decorated the palace at Versailles, as well as producing the design for the (imperfectly realized) gardens. Sébastien Le Clerc, the geometrician who had used the gentlemanly art of fencing to illustrate his lessons, was himself responsible for the plates that illustrated Charles Perrault's 1677 *Le labyrinthe de Versailles.*[99] Le Brun, like Le Clerc, surely knew the significance of courtliness and the meaning of the outward signs of kingly authority, just as he claimed to know

97. Rossi, "Hermeticism," esp. 252–53, considers the mechanistic view of the world as a criterion of intelligibility.

98. Milton, "The Origin and Development"; idem, "Laws of Nature"; Oakley, *Omnipotence, Covenant, and Order,* 77–92; Zilsel, "The Genesis of the Concept of Physical Law"; also Funkenstein, *Theology and the Scientific Imagination,* 192–93; Ruby, "The Origins of Scientific 'Law,'" on medieval precedents.

99. Montagu, *The Expression of the Passions,* 43–45; Perrault, *Le labyrinthe de Versailles.*

the meaning of the outward signs of inner emotion and their propriety.[100] Social
life, formalized through manners, meant letting people see each other as automata under the control of reason; so automata is what they became.[101] Alongside
the discipline of the Dutch, we are left with the image of Louis XIV amusing
himself, toward the close of the century, by spending hours playing with a small
automaton.[102]

100. Apostolidès, *Le roi-machine*, esp. 86–92.
101. Porter, *Trust in Numbers*, esp. chap. 4, discusses the analogous sense in which the cultural
dominance of statistics in public affairs tends to remake a society in its own image.
102. Apostolidès, *Le roi-machine*, 138.

■ ACKNOWLEDGMENTS ■

I thank Jacques Revel, Simon Schaffer, Harry Collins, and Hal Cook for their
valuable comments on earlier versions of this chapter.

■ REFERENCES ■

Apostolidès, Jean-Marie. *Le roi-machine: Spectacle et politique au temps de Louis XIV.* Paris:
   Éditions de Minuit, 1981.
Baillet, Adrien. *Vie de Monsieur Descartes.* 2 vols. Paris, 1691.
Barker, Peter. "Jean Pena and Stoic Physics in the 16th Century." In Ronald H. Epp, ed., *Spin-
   del Conference 1984: Recovering the Stoics (Southern Journal of Philosophy,* supplement).
   18 (1985): 93–108. (Supplement has separate pagination.)
———. "Stoic Contributions to Early Modern Science." In *Atoms, Pneuma, and Tranquillity:
   Epicurean and Stoic Themes in European Thought,* ed. Margaret J. Osler, 135–54. Cam-
   bridge: Cambridge University Press, 1991.
Barker, Peter, and Bernard R. Goldstein. "Is Seventeenth-Century Physics Indebted to the
   Stoics?" *Centaurus* 27 (1984): 148–64.
Barnes, Barry, and Steven Shapin, eds. *Natural Order: Historical Studies of Scientific Culture.*
   Beverly Hills, Calif.: Sage, 1979.
Bedini, Silvio A. "The Role of Automata in the History of Technology." *Technology and
   Culture* 5 (1964): 24–42.
Blom, John J. *Descartes: His Moral Philosophy and Psychology.* New York: New York University
   Press, 1978.
Borkenau, Franz. "The Sociology of the Mechanistic World-Picture." *Science in Context* 1
   (1988): 109–27.
Bremmer, Jan, and Herman Roodenburg, eds. *A Cultural History of Gesture.* Ithaca: Cornell
   University Press, 1991.
Bryson, Anna. "The Rhetoric of Status: Gesture, Demeanour and the Image of the Gentleman
   in Sixteenth- and Seventeenth-Century England." In *Renaissance Bodies: The Human
   Figure in Renaissance Culture c. 1540–1660,* ed. Lucy Gent and Nigel Llewellyn, 136–53.
   London: Reaktion Books, 1990.
Burke, Peter. "The Language of Gesture in Early Modern Italy." In Bremmer and Rooden-
   burg, 71–83.

Camporesi, Piero. *The Anatomy of the Senses: Natural Symbols in Medieval and Early Modern Italy.* Trans. Allan Cameron. Cambridge: Polity Press, 1994.

Cervantes, Miguel. *El licenciado vidriera.* In *The Portable Cervantes,* trans. Samuel Putnam, 760–96. New York: Viking, 1951.

Cohen, Gustave. *Ecrivains français en Hollande dans la première moitié du XVIIe siècle.* Paris: Éduard Champion, 1920.

Collins, H. M. *Artificial Experts: Social Knowledge and Intelligent Machines.* Cambridge: MIT Press, 1990.

Cottingham, John, ed. *The Cambridge Companion to Descartes.* Cambridge: Cambridge University Press, 1992.

Dainville, François de. "L'enseignement des mathématiques dans les Collèges Jésuites de France du XVIe au XVIIe siècle." *Revue d'histoire des sciences* 7 (1954): 6–21, 109–23.

———. *La géographie des humanistes.* Paris: Beauchesne, 1940.

Dear, Peter. *Discipline and Experience: The Mathematical Way in the Scientific Revolution.* Chicago: University of Chicago Press, 1995.

———. "Mersenne's Suggestion: Cartesian Meditation and the Mathematical Model of Knowledge in the Seventeenth Century." In *Descartes and His Contemporaries,* ed. Roger Ariew and Marjorie Grene, 44–62. Chicago: University of Chicago Press, 1995.

Descartes, René. *Discours de la méthode,* ed. Étienne Gilson. Paris: J. Vrin, 1930.

———. *Oeuvres complètes.* 13 vols. Ed. Charles Adam and Paul Tannery. Paris, 1897–1913.

———. *Les passions de l'âme.* Ed. Geneviève Rodis-Lewis. Paris: J. Vrin, 1966.

———. *The Passions of the Soul.* Ed. and trans. Stephen Voss. Indianapolis: Hackett, 1989.

———. *The Philosophical Writings of Descartes.* Ed. and trans. John Cottingham, Robert Stoothoff, and Dugald Murdoch (with Anthony Kenny). 3 vols. Cambridge: Cambridge University Press, 1985–91.

———. *Treatise of Man.* Trans. Thomas Steele Hall. Cambridge: Harvard University Press, 1972.

Duncan, David Allen. "Persuading the Affections: Rhetorical Theory and Mersenne's Advice to Harmonic Orators." In *French Musical Thought, 1600–1800,* ed. Georgia Cowart, 149–75. Ann Arbor: U.M.I. Research Press, 1989.

Elias, Norbert. *The Civilizing Process.* Trans. Edmund Jephcott. Oxford: Blackwell, 1994.

Emerson, J. P. "Nothing Unusual Is Happening." In *Human Nature and Collective Behavior,* ed. T. Shibutani, 208–22. Englewood Cliffs: Prentice-Hall, 1970.

Faret, Nicolas. *L'Honeste homme: ou, l'Art de plaire à la cour.* Lyons: A. Cellier, 1661.

Fox, Christopher. "Introduction: How to Prepare a Noble Savage: The Spectacle of Human Science." In *Inventing Human Science: Eighteenth-Century Domains,* ed. Christopher Fox, Roy Porter, and Robert Wokler, 1–30. Berkeley and Los Angeles: University of California Press, 1995.

Freudenthal, Gad. "Clandestine Stoic Concepts in Mechanical Philosophy: The Problem of Electrical Attraction." In *Renaissance and Revolution: Humanists, Scholars, Craftsmen and Natural Philosophers in Early Modern Europe,* ed. J. V. Field and Frank A. J. L. James, 161–72. Cambridge: Cambridge University Press, 1993.

Fumaroli, Marc. *L'âge de l'éloquence: Rhétorique et "res literaria" de la Renaissance au seuil de l'époque classique.* Geneva: Librairie Droz, 1980.

———. "Rhétorique et philosophie dans le Discours." In *Problématique et réception du Discours de la Méthode et des essais,* ed. Henry Méchoulan, 31–46. Paris: J. Vrin, 1988.

Funkenstein, Amos. *Theology and the Scientific Imagination from the Middle Ages to the Seventeenth Century.* Princeton: Princeton University Press, 1986.

Garber, Daniel. *Descartes' Metaphysical Physics.* Chicago: University of Chicago Press, 1992.

Gaukroger, Stephen. *Descartes: An Intellectual Biography.* Oxford: Clarendon Press, 1995.

Godwin, Joscelyn. *Robert Fludd: Hermetic Philosopher and Surveyor of Two Worlds.* London: Thames and Hudson, 1979.

Goffman, Erving. *The Presentation of Self in Everyday Life.* Garden City, N.Y.: Doubleday, 1959.

Gouhier, Henri. *Les premières pensées de Descartes: Contribution à l'histoire de l'anti-renaissance.* Paris: J. Vrin, 1958.

Greenblatt, Stephen. *Renaissance Self-Fashioning: From More to Shakespeare.* Chicago: University of Chicago Press, 1980.

Hannaway, Owen. "Laboratory Design and the Aim of Science: Andreas Libavius versus Tycho Brahe." *Isis* 77 (1986): 585–610.

Harth, Erica. *Cartesian Women: Versions and Subversions of Rational Discourse in the Old Regime.* Ithaca: Cornell University Press, 1992.

———. *Ideology and Culture in Seventeenth-Century France.* Ithaca: Cornell University Press, 1983.

Hatfield, Gary. "Descartes' Physiology and Its Relation to His Psychology." In Cottingham, 335–70.

Jaynes, Julian. "The Problem of Animate Motion in the Seventeenth Century." *Journal of the History of Ideas* 31 (1970): 219–34.

Keohane, Nannerl O. *Philosophy and the State in France: The Renaissance to the Enlightenment.* Princeton: Princeton University Press, 1980.

Kessler, Eckhard. "The Intellective Soul." In Schmitt et al., 485–534.

Le Clerc, Sébastien. *Pratique de la geometrie, sur le papier et sur le terrain.* Paris: Jombert, 1682.

Levi, Anthony. *French Moralists: The Theory of the Passions 1585 to 1649.* Oxford: Clarendon Press, 1964.

Lindeboom, G. A. *Descartes and Medicine.* Amsterdam: Editions Rodopi N.V., 1978.

Maurice, Klaus, and Otto Mayr, eds. *The Clockwork Universe: German Clocks and Automata, 1550–1650.* Washington: Smithsonian Institution; New York: N. Watson Academic Publications, 1980.

Mayr, Otto. *Authority, Liberty, and Automatic Machinery in Early Modern Europe.* Baltimore: Johns Hopkins University Press, 1986.

McNeill, William H. *Keeping Together in Time: Dance and Drill in Human History.* Cambridge: Harvard University Press, 1995.

———. *The Pursuit of Power: Technology, Armed Force, and Society since A.D. 1000.* Chicago: University of Chicago Press, 1982.

Milton, John R. "Laws of Nature in the Seventeenth Century." In *The Cambridge History of Seventeenth-Century Philosophy,* ed. Michael Ayers and Daniel Garber. Cambridge: Cambridge University Press, forthcoming.

———. "The Origin and Development of the Concept of the 'Laws of Nature.'" *Archives européennes de sociologie* 22 (1981): 173–95.

Montagu, Jennifer. *The Expression of the Passions: The Origin and Influence of Charles Le Brun's* Conférence sur l'expression générale et particulière. New Haven: Yale University Press, 1994.

Moran, Bruce T. "Princes, Machines and the Valuation of Precision in the 16th Century." *Sudhoffs Archiv* 61 (1977): 209–28.

Morgan, Vance G. *Foundations of Cartesian Ethics.* Atlantic Highlands, N.J.: Humanities Press, 1994.

Motley, Mark. *Becoming a French Aristocrat.* Princeton: Princeton University Press, 1990.

Muchembled, Robert. *L'invention de l'homme moderne.* Paris: Fayard, 1988.

———. 1991. "The Order of Gestures: A Social History of Sensibilities under the Ancien Régime in France." In Bremmer and Roodenburg, 129–51.

Noreña, Carlos G. *Juan Luis Vives and the Emotions.* Carbondale: Southern Illinois University Press, 1989.

Oakley, Francis. *Omnipotence, Covenant, and Order: An Excursion in the History of Ideas from Abelard to Leibniz.* Ithaca: Cornell University Press, 1984.

Oestreich, Gerhard. *Neostoicism and the Early Modern State.* Cambridge: Cambridge University Press, 1982.

Osler, Margaret. *Divine Will and the Mechanical Philosophy: Gassendi and Descartes on Contingency and Necessity in the Created World.* Cambridge: Cambridge University Press, 1994.

Park, Katharine. "The Organic Soul." In Schmitt et al., 464–84.

Parker, Geoffrey. *The Military Revolution: Military Innovation and the Rise of the West, 1500–1800.* Cambridge: Cambridge University Press, 1988.

Pascal, Blaise. *Oeuvres complètes.* Ed. Louis Lafuma. Paris: Éditions du Seuil, 1963.

Perrault, Charles. *Le labyrinthe de Versailles.* 1677. Reprint, with an afterword by Michel Conan, Paris: Éditions du Moniteur.

Porter, Theodore M. *Trust in Numbers: The Pursuit of Objectivity in Science and Public Life.* Princeton: Princeton University Press, 1995.

Price, Derek J. de Solla. "Automata and the Origins of Mechanism and Mechanistic Philosophy." *Technology and Culture* 5 (1964): 9–23.

Revel, Jacques. "La cour." Paper presented at a Center for 17th and 18th Century Studies and William Andrews Clark Memorial Library, UCLA, workshop on "Civility, Court Society and Scientific Discourse: Reframing the Scientific Revolution," 22–23 November 1991.

———. "The Uses of Civility," trans. Arthur Goldhammer. In *A History of Private Life,* ed. Roger Chartier, 3:167–205. Cambridge: Harvard University Press, 1989.

Rodis-Lewis, Geneviève. "Descartes' Life and the Development of His Philosophy." In Cottingham, 21–57.

———. Introduction to Descartes, *Passions.*

———. "Limitations of the Mechanical Model in the Cartesian Conception of the Organism." In *Descartes: Critical and Interpretive Essays,* ed. Michael Hooker, 152–70. Baltimore: Johns Hopkins University Press, 1978.

———. *L'oeuvre de Descartes.* 2 vols. Paris: J. Vrin, 1971.

Roodenburg, Herman. "The 'Hand of Friendship': Shaking Hands and Other Gestures in the Dutch Republic." In Bremmer and Roodenburg, 152–89.

Rorty, Amélie Oksenberg. "Cartesian Passions and the Union of Mind and Body." In *Essays on Descartes' Meditations,* ed. idem, 513–34. Berkeley and Los Angeles: University of California Press, 1986.

Rosenfield, Leonora Cohen. *From Beast-Machine to Man-Machine: Animal Soul in French Letters from Descartes to La Mettrie.* 1941. Reprint, New York: Octagon Books, 1968.

Ross, Stephanie. "Painting the Passions: Charles Le Brun's *Conférence sur l'expression.*" *Journal of the History of Ideas* 45 (1984): 25–47.

Rossi, Paolo. "Hermeticism, Rationality and the Scientific Revolution." In *Reason, Experiment, and Mysticism in the Scientific Revolution,* ed. M. L. Righini Bonelli and William R. Shea, 247–73. New York: Science History Publications, 1975.

Ruby, Jane E. "The Origins of Scientific 'Law.'" *Journal of the History of Ideas* 47 (1986): 341–59.

Sarasohn, Lisa T. "Motion and Morality: Pierre Gassendi, Thomas Hobbes and the Mechanical World-View." *Journal of the History of Ideas* 46 (1985): 363–79.

Saunders, Jason Lewis. *Justus Lipsius: The Philosophy of Renaissance Stoicism.* New York: Liberal Arts Press, 1955.

Schama, Simon. *The Embarrassment of Riches: An Interpretation of Dutch Culture in the Golden Age.* Berkeley and Los Angeles: University of California Press, 1988.

Schmitt, Charles B., Quentin Skinner, Eckhard Kessler, and Jill Kraye, eds. *The Cambridge History of Renaissance Philosophy.* Cambridge: Cambridge University Press, 1988.

Sebba, Gregor. "Adrien Baillet and the Genesis of His *Vie de M. Des-Cartes.*" In *Problems of Cartesianism,* ed. Thomas M. Lennon, John M. Nicholas, and John W. Davis, 9–60. Kingston: McGill-Queen's University Press, 1982.

Shapin, Steven. "The House of Experiment in Seventeenth-Century England." *Isis* 79 (1988): 373–404.

———. "'The Mind Is Its Own Place': Science and Solitude in Seventeenth-Century England." *Science in Context* 4 (1991): 191–218.

Shapin, Steven, and Simon Schaffer. *Leviathan and the Air-Pump: Hobbes, Boyle, and the Experimental Life.* Princeton: Princeton University Press, 1985.
Shea, William R. *The Magic of Numbers and Motion: The Scientific Career of René Descartes.* Canton, Mass.: Science History Publications, 1991.
Sloan, Phillip R. "Descartes, the Sceptics, and the Rejection of Vitalism in Seventeenth-Century Physiology." *Studies in History and Philosophy of Science* 8 (1977): 1–28.
Snyders, Georges. *La pédagogie en France aux XVIIe et XVIIIe siècles.* Paris: Presses Universitaires de France, 1965.
Sutton, Geoffrey V. *Science for a Polite Society: Gender, Culture, and the Demonstration of Enlightenment.* Boulder, Colo.: Westview Press, 1995.
Thoren, Victor E. *The Lord of Uraniborg: A Biography of Tycho Brahe.* Cambridge: Cambridge University Press, 1990.
Vartanian, Aram. "Man-Machine from the Greeks to the Computer." In *Dictionary of the History of Ideas,* ed. P. P. Wiener, 3:131–46. New York: Scribner's, 1973.
Verbeek, Theo. *Descartes and the Dutch: Early Reactions to Cartesian Philosophy 1637–1650.* Carbondale: Southern Illinois University Press, 1992.
———. "Regius's *Fundamenta physices.*" *Journal of the History of Ideas* 55 (1994): 533–51.
Vigarello, Georges. "The Upward Training of the Body." In *Fragments for a History of the Human Body,* ed. Michel Feher et al., 2:148–96. New York: Zone Books, 1989.
Vives, Juan Luis. *The Passions of the Soul: The Third Book of De anima et viva.* Trans. with an introduction by Carlos G. Noreña. Studies in Renaissance Literature, vol. 4. Lewiston: Edwin Mellen Press, 1990.
———. *Vives: On Education, a Translation of the De tradendis disciplinis.* Trans. Foster Watson. Cambridge: Cambridge University Press, 1913.
Westman, Robert S. "Huygens and the Problem of Cartesianism." In *Studies on Christiaan Huygens: Invited Papers from the Symposium on the Life and Work of Christiaan Huygens, Amsterdam, 22–25 August 1979,* ed. H. J. M. Bos, M. J. S. Rudwick, H. A. M. Snelders, and R. P. W. Visser, 83–103. Lisse: Swets and Zeitlinger, 1980.
Zilsel, Edgar. "The Genesis of the Concept of Physical Law." *Philosophical Review* 51 (1942): 245–79.
Zur Lippe, Rudolph. *Naturbeherrschung am Menschen.* 2d ed. 2 vols. Frankfurt am Main: Suhrkamp, 1981.

THREE

REGENERATION

The Body of Natural Philosophers in Restoration England

SIMON SCHAFFER

■ Angels' Knowledge ■

IN HIS *Essay Concerning Human Understanding* John Locke put forward "an extravagant conjecture" about angels. Though "angels of all sorts are naturally beyond our discovery," Locke guessed these "spirits of a higher rank" would have perfect memories, incorruptible bodies, and, above all, remarkable eyesight. "What wonders would he discover, who could so fit his eyes to all sorts of objects as to see, when he pleased, the figure and motion of the minute particles in the blood, and other juices of animals, as distinctly as he does, at other times, the shape and motion of the animals themselves?" In summer 1686 he had seen these sights for himself through the glasses of the Dutch microscopist Antoni van Leeuwenhoek. As a physician and natural philosopher, Locke wrote that "it is not to be doubted that spirits of a higher rank," equipped with these "microscopical eyes," would "have as clear ideas of the radical constitution of substances as we have of a triangle."[1] But despite the advantages of such angelic capacities, Locke consoled his readers that God adapted actual human senses "for the neighbourhood of the bodies that surround us, and we have to do with." Virtuosi armed with prosthetic microscopes would be like angelic strangers, claiming superior insight into essences yet "in a quite different world from other people." No one would "make any great advantage by the change," Locke judged, "if such an acute sight would not serve to conduct him to the market and exchange." Their capacities would be profitless.[2]

From the late 1660s Locke's arguments helped him answer patriarchalism, rationalism, and dogmatism. By showing the limits of what could be certainly known, he could show tyranny unjustified. His allies, such as Lord Shaftesbury, critic of divine right monarchy, and Thomas Sydenham, critic of medical dogmatism, agreed that those who arbitrarily imposed their own body of knowledge

---

1. Locke, *Essay Concerning Human Understanding,* 201–2, 381, 428; for Locke and Leeuwenhoek see Dewhurst, *John Locke,* 272–73.

2. Locke, *Essay Concerning Human Understanding,* 201–2; for Locke on the microscope's limits see Wilson, *The Invisible World,* 240–41. Locke reckoned teaching about spirits should always precede teaching about bodies; see Locke, *Some Thoughts about Education,* 245.

would destroy the commonwealth.[3] Locke's concerns with the bodily limits of knowledge were not unprecedented. Many held that human senses were defective because when Paradise was lost, so was the perfectly knowing body. Humanity could not now attain the prelapsarian certainties of Adam, first of natural philosophers. To claim otherwise was to propose regeneration, a new birth in the spirit and the restoration of a divinely validated constitution. In works republished from the 1640s, Francis Bacon had notably made much of the effects of the Fall on natural philosophy and civil society. In *Paradise Lost* (1667) John Milton sang of Adam's knowledge and the Fall as an account of idolatry's origins and the Commonwealth's defeat.[4] For some, Adam was not merely the original natural philosopher; he was also, more portentously, the original of monarchy: "for by the divine appointment, as soon as he was created, he was monarch of the world." Locke's principal opponents, patriarchal royalists, held that in the person of the restored king the heritage of Adam had also been reestablished.[5]

The Restoration order was to be secured through the correct attribution of power to bodies. Idolatry attributed spiritual power to merely material objects; enthusiasm exaggerated the significance of these spirits; atheism banished spirit from the world and exaggerated bodies' powers. In a long treatise summarizing the most authoritarian version of royalism developed in the wake of the Restoration, the Tory physician Nathaniel Johnston invoked "the great Leviathan of Government" with which Thomas Hobbes had adorned his notorious frontispiece, "the figure of a Giant's Head, on all parts of which are swarms of all Orders . . . delineated as the Picts are supposed to have painted their Bodies, or Trajan's Column is engraven." Johnston's elaborate analogy indicated just how political iconography, in imperial Rome as in Restoration London, dramatized the right ordering of the body politic. The physician's treatise gave a typical, if peculiarly detailed, specification of this order:

> the common people I may not unfitly call the common Digestor, like the Stomach that affords nourishment to this great Behemoth . . . the Mechanical Traders are the Bowels and lacteal vessels which transmit nourishment in the inferiour Belly; the Merchants are the Vena Porta . . . the wealthy are the Fat and Plumpy Parts.

3. Ashcraft, *Revolutionary Politics*, 110–11, 124–25; Romanell, *John Locke and Medicine*, 106–14; Cunningham, "Thomas Sydenham," 183–84.

4. For Restoration metaphysics see Jose, *Ideas of the Restoration*, 54, 70, and McKeon, *Politics and Poetry in Restoration England*, 247–48, 253–54; for Bacon and the Fall see Rossi, *Francis Bacon*, 129, and Hill, *Intellectual Origins*, 89; for Milton and natural philosophies, see Hill, *Milton and the English Revolution*, 378–80.

5. Ashcraft, *Revolutionary Politics*, 187–88 (for the citation on Adam and patriarchy).

At the head, inevitably, stood the Sovereign, "the portion of divine spirit and ubiquitary power that presides and governs all."[6] Shows of royal power were revived at the Restoration in coronation and court ritual. They were designed to make the sovereign's still shaky spiritual power seem self-evident, to reassert the legitimacy of the restored political body after a regicide that had dramatized its fragility and a profusion of alternative bodily accounts of power and right that flourished under the Commonwealth. Since the political order and that of the body were so closely identified, grotesque or ribald bodies were easily identified with subversion, danger, and criticism. Adamites and Ranters, Diggers and Familists had all notoriously used the body images of regenerate Eden to propose radical accounts of the social order.[7] The histrionics of Charles II's coronation parade, orchestrated by John Ogilby in April 1661, dramatized these lessons (fig. 3.1). Scriptural and classical typologies were there designed to present the monarch as Adam or as Christ, the spiritual healer of the nation's wounds and the embodiment of its restored health. On the very first arch under which the king passed on his way to Whitehall Charles would have seen a figure of "Rebellion," a grotesque body, declare herself "Hell's daughter," "making the Vulgar in Fanatique Swarms Court Civil War and dote on Horid Arms," then dispatched by "Monarchy, in a large Purple Robe, on her head London." The insecurity of the new regime was almost as patent as its triumph: in crowning the monarch it seemed proper to remind him of the immense threat so recently defeated, to use the contrast between angelic and monstrous bodies to make this lesson clear. Thus, in such shows body language and gesture had crucial, if fluctuating, political meanings.[8]

After the spectacle of the royal entry, the most salient example of bodily show in the name of political authority was the ceremony of the royal touch, dramatically revived at the Restoration and typically staged in the grandiose Banqueting House at Whitehall, where the monarch touched sufferers of the King's Evil amid Anglican pomp and court splendor (fig. 3.2). As its greatest historian, Marc Bloch, observes, "the idea of the royal miracle was related to an entire conception of the universe." The management of the miraculous and the supernatural capacity of human bodies was at issue in these regal ceremonies. In the 1630s royal surgeons had staged public trials of those "delinquents" who claimed they could reproduce the royal gift. The suppression of the practice by

6. Hobbes, *Leviathan*, 1; Johnston, *The Excellency of Monarchical Government*, 4–5; Schaffer, "Godly Men and Mechanical Philosophers."

7. Hill, *World Turned Upside Down*; Williams, "Magnetic Figures."

8. Ogilby, *The Entertainment of His Most Excellent Majestie Charles II*, 17, 47; Reedy, "Mystical Politics"; Hammond, "The King's Two Bodies," 19–20.

FIGURE 3.1. "The King's Procession from the Tower to Whitehall, 22 April 1661: The First Arch Erected At Leadenhall Street." From Ogilby, *The Entertainment of His Most Excellent Majestie Charles II.*

London Printet for Dorman Newman at the kings Armes in the Poultry &.   F. H. van. Howe Sculp:

FIGURE 3.2. "The Manner Of His Majesties Curing The Disease CALLED THE KINGS–EVIL." From a 1679 broadside reprinted in Crawfurd, *The King's Evil*.

the Commonwealth and its resuscitation by the restored regime, accompanied by unprecedented bureaucratic regulation of applicants for the touch, all drew attention to the significance of this spiritual act. Sufferers must be certified by local priests and churchwardens, then eventually present themselves to the king's surgeon in Covent Garden on Wednesday and Thursday afternoons, before being admitted in batches of up to two hundred to the Whitehall ceremony. In the four years from summer 1660, almost twenty-three thousand subjects were thus processed. Though charisma was routinized, authorities such as the physician John Browne argued that faith was necessary for a successful cure, and that the king's "Sacred hands are sweetn'd with that sacred salutiferous gift of Healing which both supports the Body Politick and keeps up the Denizens and Subjects thereof in vigor and courage." The king's surgeon Richard Wiseman spelled out the rather broad criteria that applicants must satisfy before being admitted to the ceremony, acknowledged that "the infidelity of many in this fantasticall Age and the want of opportunity of others, doth deprive them of this easy and short remedy," and thus offered naturalistic accounts of how the Evil might be cured without recourse to the admittedly wondrous action of the King himself. "All that I pretend for is, first, the attestation of the Miracles, and secondly a direction for such as have not opportunity of receiving the benefit of that stupendous Power." In the discourse of the royal touch subjects had immediate experience of the wide range of bodies' capacities, the role of spiritual agencies in the security of the royal regime, and the rights of others to judge these agencies' workings. "'Tis not lawful for ordinary persons to assume so great a power," wrote one royalist physician in 1665, "their skill in healing is not the Gift of God as appeareth by the quality of the persons who are generally ignorant and prophane, for which very cause God will not reveal his counsels to them."[9]

As Locke's angelic conjecture indicated, these powers also mattered in natural philosophy and to its regenerate practitioners. Locke accounted experimenters' idealized bodies as occupants of a higher and apparently incommensurable realm. Locke's critic, the Cambridge Calvinist John Edwards, deemed that in a future state "a Vertuoso shall be no rarity." In Restoration England, some natural philosophers did claim that their capacities were above those of their fellows and that they thus might have special access to nature's properties; others reckoned that such failings could rather be corrected by the use of such artificial instruments as microscopes and telescopes. Robert Hooke said instruments would recoup the loss of perception through a "universal cure of the Mind." He urged the Baconian theme that "as at first, mankind fell by tasting of the forbidden Tree of Knowledge, so we, their Posterity, may be in part restor'd by the same way . . . by tasting too those fruits of Natural knowledge, that were never yet forbidden." In his *History of the Royal Society,* Thomas Sprat judged that Christ's miracles were simply "Divine Experiments of his Godhead which may be call'd Philosophical Works perform'd by an Almighty Hand." According to the West Country divine Joseph Glanvill, the standing of Robert Boyle's experiments was scarcely lower: "had this great Person lived in those days when men Godded their Benefactors, he could not have miss'd one of the first places among their deified Mortals."[10]

This chapter recounts some episodes in the 1660s when natural philosophers worked with regenerate bodies, invested as it seemed with some rare or novel spiritual power. A blind Dutchman, Jan Vermaasen, could discriminate colors by

9. Bloch, *Rois Thaumaturges,* 373, 385; Crawfurd, *The King's Evil,* 95–97, 105–7, 121; French, "Surgery and Scrophula"; citations from Browne, *Adenochoiradelogia,* sig. Bb3v; Wiseman, *Severall Chirurgicall Treatises,* 250, 246; [T. A.], Χειρεξοκη, 4–5. For the routinization of charisma and the traditions that sustain hereditary monarchy, see Weber, *Social and Economic Organization,* 363–73.

10. Edwards, *A Compleat History,* 744–45; Hooke, *Micrographia,* sig. bv; Sprat, *History of the Royal Society,* 352; Glanvill, *Plus Ultra,* 93; for patriarchs and naturalists, see Fisch, "The Scientist as Priest," and Hutton, "Edward Stillingfleet, More and the Decline of *Moses Atticus.*"

touch: did this show that all colors were due solely to bodies' shapes? A Cambridge graduate Arthur Coga had his spirits restored with an infusion of sheep's blood: did this show how to regenerate human bodies? An Irish healer Valentine Greatrakes cured the King's Evil by stroking: were all cures, even those effected by Jesus Christ and Charles Stuart, explicable in purely bodily terms? Answers depended on status and location. The natural philosophers tried to make these remarkable and regenerate bodies into matters of fact. In so doing, they relied on their own status as exemplary bodies capable of judging all others. In 1672 Robert Boyle claimed that stories of the vulgar about their own bodies were less reliable than his own observations of inanimate ones. The collective body of natural philosophers was therefore supposed to be indicative of how all would naturally behave, yet it was also peculiarly trustworthy and thus quite unlike many of these bodies. Coga was induced to provide telling testimony of his own state at the Royal Society; Boyle carefully compared Vermaasen's senses with his own and tried painstakingly to replicate Greatrakes's striking cures. Over against this world of refined bodies and reliable facts about their wondrous capacities was the grotesque and ribald body of the plebeians, from whom the virtuosos energetically sought to distance themselves. Their acutest enemies could counter that wondrous bodies were really grotesque, sites of carnivalesque charlatanry.[11]

Vermaasen's marked cards and Coga's drinking bouts were topics of philosophic experiment, then of widespread satire. One royalist historian argued that Greatrakes "began and set up among the Ignorant and Rude part of mankind . . . How much more agreeable had it been for him . . . to have repaired to some Reverend Divines and Physicians than to chat with his Wife and some two or three old Women." Striving to escape old wives' tales and street cant, the virtuosos were all too often depicted as vulgar, not least by court pens whom they otherwise wished to convince. Though they stipulated that their collective capacities were uniquely elevated, it was not obvious how they could ever escape the legacy of the Fall. In coffeehouses and print shops it was easy to exploit comparisons between regenerate human organs and sensitive philosophical instruments. "Every tapster" allegedly heard one venomous critic of Sprat and Glanvill, the physician Henry Stubbe, "his tongue at work in every coffee-house," point out in 1670 that since "our Eyes were Telescopes of God Almighty's making" and all knew their own eyes were unreliable, how deceptive must be the natural philosophers' more mundane instruments? Stubbe "was more prone to suspect their dioptrick Tubes than their integrity." In what follows, the problem of natural

---

11. For Boyle see Shapin, *Social History of Truth,* 265; for the grotesque body, see Stallybrass and White, *Politics and Poetics of Transgression,* 93.

philosophers' integrity is explored through their allegedly regenerate, often grotesque, body techniques.[12]

■ *Adam Needed No Spectacles* ■

BY THE EARLY 1660S many English natural philosophers agreed that different colors could be ascribed to the different shapes of bodies' surfaces. Boyle argued, for example, that the high reflective power of red bodies must be due to
the structure of their surfaces. If these "asperities" could be felt, then the blind
might be able to tell colors apart. Boyle backed up his stories about color with
tales of remarkable human capacities. His friend John Finch, a Cambridge graduate appointed Pisan anatomy professor by the Grand Duke of Tuscany in 1659,
told Boyle of the Dutchman struck blind by smallpox when a child. Finch was
a typically curious philosophical tourist equipped with his own theory of vision
debated at first hand with his Cambridge tutor Henry More. Finch went to Maas-
tricht in autumn 1661 to interview the blind Jan Vermaasen, who could tell playing cards by touch and whose deeds would obviously provide suitable matter
with which to regale his Medicean patron. Finch spent several days testing his
subject, wondered whether Vermaasen had marked the cards and collected the
packs to bring back to Tuscany. Then he began trying colored ribbons. When
Vermaasen fasted he could not only distinguish colors but managed to teach
Finch the trick. The blind seer sometimes erred. He confused white with black
and red with blue, explaining to the Englishman that the former pair of colors
were both very rough, the latter both very smooth. So Finch also saved the ribbons for the Grand Duke and then went to London. Boyle was fascinated but
skeptical. Surely humans' fingers were so much less sensitive than the retina that
asperities would be indistinguishable by touch alone. Had Finch blindfolded the
Dutchman? Finch checked his letters to the Grand Duke for answers to Boyle's
inquiries, and handed over three samples of the colored ribbons, "which I keep
by me as Rarities," Boyle reported. Finch confirmed that he had blindfolded Vermaasen and tested him at night. Boyle remained cautious. Perhaps, by analogy
with dyestuffs, Vermaasen used the ribbons' smell. His loss of the skill after any
big meal might suggest so. Perhaps, on the other hand, the rarity of the gift was
to be compared with those "Critical and Experienc'd Palats" who could tell wines
apart but not teach others to do so. The Royal Society's secretary Henry Oldenburg helpfully reported that those who had met Vermaasen confirmed that he
could "play at cards, wth better succes, yn any man he played wth, as also, to dis-

---

12. Lloyd, *Wonders no Miracles*, 18; Stubbe, *Plus Ultra Reduced to a Non Plus*, 40–41; for Stubbe's
auditors see Jacob, *Henry Stubbe*, 84; for an exemplary critique of instrumental artifice and body capacities, see Cavendish, *Observations upon Experimental Philosophy.*

tinguish men and women from one another by feeling their hands, or necks, and to discriminate ye severall colors of haire, and lastly to discerne the beauty of women by their voyce." Boyle concluded that "I would gladly have had the Opportunity of Examining this Man my self," for then he would have decisive, well-attested histories in support of his corpuscular stories about the production of colors.[13]

Not all were convinced. When he met Vermaasen on a visit to The Hague, for example, Christiaan Huygens noticed he marked his cards and could not tell colors apart unless accompanied by "his brother, whom he brought with him." So Huygens wanted Boyle to check whether Finch had been alone with the blind man. Robert Moray, Huygens's correspondent and Boyle's colleague, agreed that there must be "some deception or charlatanry" involved. But Boyle stayed firm: Finch "took extraordinary care not to be imposed upon." Trustworthy witnesses warranted otherwise implausible human capacities.[14] In 1668 Boyle was told of "a blind man at Copenhagen" who "can as well distinguish colours by his touch" as could Vermaasen. Vermaasen's tactile achievements fitted a major philosophical tradition. Descartes had debated with his empiricist critics about the ideas the blind had of colors: "it is pointless to cite the testimony of a blind philosopher." Newton later urged that "as a blind man has no idea of colours, so we have no idea of the manner by which the all-wise God perceives and understands all things."[15] The Dutchman's deeds also had a famous literary afterlife. In satires of the 1720s devoted to varying conditions of the human body his achievements were attributed by John Arbuthnot to Martinus Scriblerus and by Jonathan Swift to a Lagado academician: "he it was, that first found out the Palpability of Colours, and by the delicacy of his Touch, could distinguish the different Vibrations of the heterogeneous Rays of Light."[16] The Scriblerians' source was Boyle's treatise of color, published in English in 1664, packed with the testimony given by trustworthy subjects about the condition of their own persons. It was targeted against scholastic doctrines of the origin of color. The schools distinguished between real colors which inhered in bodies and imaginary ones, produced by some change in the light they reflected. Boyle agreed with Descartes in his denial of this distinction: no color inhered in external bodies. He also agreed that color could

13. Boyle, *Experiments and Considerations Touching Colours*, 42–49; Malloch, *Finch and Baines*, 34–38; Oldenburg to Boyle, 10 October 1665, in Oldenburg, *Correspondence*, 2:558.

14. Huygens to Moray, 19 August 1664, Moray to Huygens, 9 September 1664, and Boyle to Oldenburg 1664, in Huygens, *Oeuvres complètes*, 5:107–8, 113, 558.

15. Colepresse to Boyle, 30 August 1668, in Boyle, *Works*, 6:555; Descartes, August 1641, in Descartes, *Oeuvres*, 3:432; Newton, *Mathematical Principles*, 549.

16. Kerby-Miller, *Memoirs of Martinus Scriblerus*, 167; Swift, "Travels into Several Remote Nations," 176–77.

sometimes be produced without any modification of light, but solely through the power of imagination or some physical disturbance of the human body. "This the most ingenious *Des Cartes* hath very well observ'd but . . . he seems not to have exemplified it by any unobvious or peculiar observation." Boyle made good this lack.[17]

In a series of well-attested stories from reliable, genteel informants, more credible than blind Dutch organists, Boyle set out evidence that different human bodily states prompted the impression of color. Just before succumbing to a hysterical fit, his learned sister Katherine Lady Ranelagh had seen her friends' clothes "dyed with unusual Colours." "A Lady of unquestionable Veracity" had the same experience after suffering a blow to her eye. "An ingenious man" who visited plague victims reported that when patients had visions of "glorious Colours" this was a reliable symptom of the disease's onset. Boyle vouchsafed his own experiences. During coughing fits he was plagued by visions of bright flames. A philosophical friend had once consulted him after gazing at the sun through a telescope and thence suffering from the fantasy of a bright image whenever he looked upon any white object. So Boyle tried the same, with the precaution of shading the telescope's eyepiece with a colored glass. With his "Discompos'd Eye" he would long after see "adventitious colours," though his other eye remained unaffected. Thus, "without any change in the Object, a change in the Instruments in Vision may for a great while make some Colours appear Charming, and make others Provoking." The bodily sufferings of the trustworthy proved that their own sensations could not be trusted.[18]

During the early 1660s Boyle and his colleagues worked to devise techniques for correcting such errant human testimony by experimental trials with instruments. Boyle never claimed he possessed superior sensory capacity. On the contrary, he presented himself as a wonderfully weak vessel. In 1660 Boyle characteristically blamed "the unhappy Distempers of my eyes" for his failure to master geometry and for the fact that he had neither penned his pneumatic experiments nor checked his amanuenses' transcriptions of them. Glass was an important theme in these discussions of bodily failure, for it represented both human frailty and its plausible correction. As Gill Speak has suggested, the illusion that one's body was made of glass and thus fragile could be developed from a care for chastity and purity. Glassy bodies would all too easily reveal their secrets to others; alternatively, they could easily deceive by bending vision. Thus, in an early moral essay Boyle wrote that sin made men mistake "by the fals optic glass of the Sensuall Appetite," while in a contemporary piece on natural theol-

17. Boyle, *Experiments and Considerations Touching Colours*, 13.
18. Ibid., 13–20.

ogy he enthused about the capacity of microscopes to "discern otherwise invisible objects."[19]

Boyle seems to have endorsed the claims of his Dorset neighbor the physician Nathaniel Highmore that glasses could reveal the very smallest particles. In 1664 Boyle prefaced his presentation of the "Conjectural" Vermaasen story with the tentative claim that improved microscopes, "as I fear are more to be wish'd than hop'd for," would reveal the asperities he guessed produced colors, just as telescopes revealed unimagined landscapes on the moon.[20] Here Boyle drew on the important resource of Restoration experimental philosophy, most famously stated in the preface to Hooke's *Micrographia* (1665), that instruments provided close analogues of prelapsarian capacities. The Yorkshire physician Henry Power wrote in 1664 that with the relevant "Engines" the "faculties of the soul of our Primitive Father Adam" might be reproduced and surpassed. Hooke agreed: microscopes were "artificial organs added to the natural" and extended the "dominion of the senses." Instrument makers could restore men to Eden. At this conjuncture of political and moral Restoration, the incapacities of the human body were therefore simultaneously the reason why instruments were needed, the source of understanding of the way these instruments worked, and the subject at which experimental investigation should be directed.[21]

At the opening of his notorious assault on scholastic dogmatism (1661), Glanvill made the connection between moral regeneracy and experiment explicit. Because of his elevated spiritual state, Adam was an exemplary experimental philosopher who could see the essences of things: "Adam needed no Spectacles. The acuteness of his natural Opticks . . . shew'd him much of the Coelestial magnificence and bravery without a Galilaeo's tube." Glanvill ingeniously used the most recent matters of fact established by natural philosophers to represent the first man as possessed of perfect senses, and he used Adam's status to validate these experimental facts against the pagan schoolmen. Thus Adam could directly sense the earth's motion, the circulation of the blood, and the effect of the moon on the tides. Recent observations were retrospectively confirmed by the sights of Eden. Adam saw magnetic effluvia, "which may gain the

19. Speak, "An Odd Kind of Melancholy"; Boyle, *New Experiments Physico-Mechanicall,* sig. [A5r] (for his lack of geometry), and sig. [A6r] on his amanuenses; Boyle, "Of Sin," in Harwood, *The Early Essays and Ethics,* 147.

20. Boyle, *Experiments and Considerations Touching Colours,* 40–41; for microscopy and Highmore see Frank, *Harvey and the Oxford Physiologists,* 94–95, 101; Kaplan, "*Divulging of Useful Truths in Physick,*" 34–37.

21. Hooke, *Micrographia,* sig. [b2r–v]; Power, *Experimental Philosophy,* sig. [a4r]; for Boyle on the displacement of humans' bodily experiences by instruments, see Shapin and Schaffer, *Leviathan and the Air-Pump,* 218; for Hooke on instruments and the Fall, see Dennis, "Graphic Understanding," 321.

more credit . . . that by the help of Microscopes [observers] have beheld the sub-
tile streams issuing from the beloved Minerall." Lest his readers miss the implication, Glanvill stressed that Adam's remarkable visual acuity was not the result
of a specially miraculous dispensation. Using a Cartesian account of vision, Glanvill argued that all action worked by propagation of moving matter and all sense
data were due to motions in the sensorium. So all objects were observable in
principle. Adam differed from the rest of humanity only through his peculiarly
elevated "faculty of Animadversion," the sensitivity of spirits in his retina. For
this reason, his powerful soul could see stars that otherwise were detectable only
through a telescope. Glanvill prefaced this argument with an acknowledgment
of its conjectural status. Not everyone agreed with him. Power differed from
Glanvill and Highmore: "our Modern Engine the Microscope" could not reveal
the smallest corpuscles. "I am sure [Highmore] had either better eyes, or else
better glasses than ever I saw (though I have look'd through as good as England
affords)." But Power did assert that Adam's capacities could now be *surpassed*
because "the Constitution of [his] Organs was not divers from ours, nor different from those of his Fallen Self." Experimental philosophers, he claimed, "may
well be placed in a rank specifically different from the rest of grovelling Humanity." They were confessedly of the fallen, so they needed instruments. But armed
with these tools they became regenerate, and, according to some, would see what
Adam saw. Tests on experimenters' own senses became a moral duty.[22]

■ *The Blood of the Lamb* ■

ADAM HAD BEEN ABLE to sense the circulation of his own blood. Restoration
natural philosophers were less fortunate. The working of the blood was as important in symbolic and practical accounts of the human body as was vision.
William Harvey had argued in 1651 that the blood was "celestial" because of its
"analogy with the heavens as being the instrument and deputy of heaven." In
an essay of 1665 Boyle described the human body as "an hydraulical or rather
hydropneumatical engine, that consists not only of solid and stable parts, but of
fluids, and those in organical motion," and these "fluids, the liquors and spirits,"
had within them the power to "conduce the recovery or welfare of the body."[23]

22. Glanvill, *Vanity of Dogmatizing*, sigs. B2v–B4v and 5–9; for Glanvill's propaganda, see Cope,
Joseph Glanvill; for Power's criticism see his *Experimental Philosophy*, sigs. [a4r], b2r; ibid., 155, for
the attack on Highmore and 191–92 on the rank of experimental philosophers; for the contrast between Power and Glanvill see Wilson, *Invisible World*, 63–66.

23. Harvey, *Exercitationes Generatione Animalium*, is discussed in Henry, "Boyle and Cosmical
Qualities," 129; Boyle, "Free Inquiry into the Vulgarly Received Notion of Nature," is discussed in
Kaplan, "*Divulging of Useful Truths in Physick*," 71. For the iatrochemistry of blood see Davis, *Circu-
lation Physiology*, 173–76.

Experiments on the action of blood were central to the projects of self-experimentation and moral and physical regeneration. Boyle already canvassed these kinds of experiments in an essay on poisons he wrote in Dorset in the early 1650s.[24] Boyle argued that through the wisdom of God's plan even His most apparently poisonous creatures, such as vipers, "may not onely be made harmles but healthfull." He cited the iatrochemist J. B. van Helmont for authority that "there lurkes in divers noxious simples a very searching and powerful Activity, which . . . can be reconciled to nature or directed to some use where its efficacy may be safe and helpfull." In 1660, when Boyle published a series of trials in which animals were killed within the receiver of his air-pump, he stated that such extinctions demonstrated "that even living Creatures (Man always excepted) are a kinde of curious Engines, fram'd and contriv'd by nature (or rather the Author of it) much more skilfully than our gross Tools and unperfect Wits can reach for." The theological parentheses were crucial. Souls made humans different from animal machines, so direct trials on men might be immoral, facts about animals might not be true of humans, and work on human subjects might imitate an original divine act. In a short extract from his poisons essay published in 1663, he denied the problem of inference and accepted the trouble of human experiment: "the greater number of Poysons being such both to Man and Bruits, the liberty of exhibiting them, when, and in what manner we please, to these (which we dare not do to him) allows us great opportunities of observing their manner of operation and investigating their Nature, as our selves have tryed."[25]

If the distinction with animals was linked to the possession of a soul, then this raised the possibility of trying spirits' effects. Boyle was well used to dosing his own servants, and then making these paternalist gestures into experiments. In his poisons essay he recorded the successful effects of giving laudanum to a servant who was suffering from nosebleeds. In his work on vipers he combined trials on dogs with tests on humans. Boyle agreed with Helmont that snakebite was lethal because of spiritual causes, a conjecture later to be tested by Tuscan natural philosophers and in Paris by Locke with Huguenot apothecaries there. "The venom of Vipers consists chiefly in the rage and fury wherewith they bite, and not in any part of the Body," Boyle stated. A dog fed the head, tail, and gall of a viper survived: the animal "liked his entertainment so well, that he would afterwards, when he met me in the Street, leave those that kept him to fawn on and

24. Oldenburg took a copy of Boyle's poisons essay, "Observations out of Mr B Essay of turning Poisons into Medicins," Royal Society MSS, MM.1 fols. 74r–88v, and mentioned it in Oldenburg to Boyle, 29 August 1657, in Oldenburg, *Correspondence*, 1:133; Boyle printed an extract from the essay in *Some Considerations*, part 2, 57–60.

25. Royal Society MSS, MM.1 fols. 74r, 87v; Boyle, *New Experiments Physico-Mechanicall*, 380; idem, *Some Considerations*, part 2, 57. Compare Oster, "The 'Beame of Divinity.'"

follow me." Boyle did not try this on men. But in summer 1650 he hired a trusting man in order to convince a skeptical physician that a cure for viper poison was to hold a hot iron near the bite. The cure worked, and the victim "hath divers time got money by repeating the Experiment; though otherwise, by the casual bitings of Vipers, he hath been much distrest, and his Wife almost kill'd." [26]

These dramatic projects were repeated at regular intervals and with increasingly reliable techniques from the mid–1650s. In 1656 at Oxford Boyle's colleagues placed bets on the life of a dog injected with poison by Christopher Wren. Boyle "caus'd [the animal] to be whipp'd up and down the Neighboring garden." The dog thrived and publicity spread. Boyle complained that "I could not long observe how it far'd with him. For this Experiment, and some other tryals I made upon him, having made him famous, he was soon after stoln away from me." [27] The next year in London Wren and the physician Timothy Clarke injected an emetic antimony salt into a dog. Boyle soon proposed repeating these experiments with humans; infusion trials thus increasingly resembled judicial acts: "If it could be done, without either too much danger or cruelty, tryal might be made upon some humane Bodies especially those of Malefactors." Such an attempt had already been made in autumn 1657 at the French embassy, where the ambassador sponsored an infusion into "an inferior Domestick of his that deserv'd to have been hangd." The servant soon collapsed under the ministrations of Wren and Clarke "either really or craftily," and the London experimenters abandoned their work. In early 1659 Wren went back to Oxford, leaving Clarke as the sole Londoner who could perform this trick. [28]

During 1665–66, when the Royal Society's London meetings were suspended because of the Plague, fellows learned from the Oxford physician Richard Lower how to manage these transfusion techniques. [29] The possibility of transmutation of species was canvassed. Lower urged his trials should be developed to

26. Royal Society MSS, MM.1 fol. 86r; Boyle, *Some Considerations*, part 2, 58–59; for subsequent trials on vipers see Knoefel, *Francesco Redi on Vipers;* Tribby, "Cooking (with) Clio and Cleo"; Dewhurst, *John Locke*, 55.

27. For work in Oxford, Boyle, *Some Considerations*, part 2, 62–63; for 1656 see Frank, *Harvey and the Oxford Physiologists*, 332 n. 38, and Oster, "The 'Beame of Divinity,'" 157.

28. Boyle, *Some Considerations*, part 2, 67, for London work in 1657. Identification of the French ambassador is in Clarke to Oldenburg, 1668, Oldenburg, *Correspondence*, 4:366. It is not clear that Clarke and Wren administered the embassy trial nor that it preceded the antimony salt experiment. Boyle confirms that Clarke and Wren were the operators, but he dates this before the news from the French embassy. Compare Frank, *Harvey and the Oxford Physiologists*, 171–72.

29. Boyle to Lower, 26 June 1666, and Lower to Boyle, 6 July 1666, in Lower, *Tractatus de Corde*, 177–84 and [Lower], "The Method Observed in Transfusing the Bloud"; Oldenburg asserts priority in "An Account of the Rise and Attempts of a Way to Conveigh Liquors," 130; for the transmission from Oxford to London see Frank, *Harvey and the Oxford Physiologists*, 173–78; Hall, *Promoting Experimental Learning*, 34–37.

exchange blood between "old and young, sick and healthy, hot and cold, fierce and fearful" animals. In September 1666, Boyle sent Lower a set of queries about the change of habits and nature displayed by animals whose blood had been transfused, even involving real changes in species, "at least in Animals near of Kin." [30] The Royal Society appointed a number of curators, including the expert anatomist Edmund King, to try transfusion in animals. King, and Boyle's personal physician Thomas Coxe, successfully demonstrated a transfusion between two dogs at Gresham College on 14 November 1666. That same evening Samuel Pepys joked in a tavern about this "pretty experiment" and the possibility "of the blood of a Quaker to be let into an Archbishop, and such like." Alehouse ribaldry and philosophical debate highlighted transfusion's moral significance. King and Coxe started with two sheep to get Lower's technique perfect. Victims were sometimes eaten to test their meat's consistency. In spring 1667 Lower moved to London to join in, while King and Coxe extended their program to try the possible changes of habits between different species and the role of corrupt blood in transmitting disease. At his own house King bled a sheep until it seemed "past hopes of Recovery," then revived it with much calf's blood. The recipient animal completely recovered and "ran with great violence" at a nearby dog. He also reported during May on the dangers of bleeding from a dog one of whose jugulars had already been cut, since then blood would swiftly coagulate. The sheep he used seemed to avoid its fellows and soon died. Boyle visited King to witness a transfusion from a lamb to a fox, which also later died. King told Boyle that this disaster might be due to the thinness of its "spirits, more apt to fly away" after infusion from the sheep. Animals such as a bitch from which thirty ounces of blood were drawn, and which subsequently whelped, were displayed as "a piece of remarkable Curiosity" in London coffeehouses.[31]

These programs received fresh impetus as a result of a fierce priority dispute involving the claims of a Paris transfuser, Jean Denis, which erupted onto the

30. [Lower], "The Method Observed in Transfusing the Bloud," 357; Boyle, "Tryals proposed to Dr Lower."

31. The London committee was appointed on 26 September 1666, and King reported on 14 November; see Birch, *History of the Royal Society,* 2:115, 123; Pepys comments in *Diary,* 7:370–71. King's reports are "An Account of an Experiment of Transfusing of Bloud out of one Animal into Another," Royal Society, RBC 2. 297–98 (21 November 1666); "The Account of an Experiment made by Dr King at his own House," RBC 2. 320–21 (4 April 1667); "A Further Account of some Experiments of Transfusion made by Dr King at his own House," RBC 2. 322–25 (30 May and 9 June 1667); Coxe's trials are reported in "Account of an Experiment made by Cox of Bleeding a Mangy into a Sound Dog," RBC 2. 321–22 (4 April 1667); see also "Considerations about Transfusion of Bloud from the Veine of an Animal to the Veine of Another," Royal Society Classified Papers 14 (1) no. 6; Oldenburg published a summary of the work of King and Coxe as "An Account of an Easier and Safer Way of Transfusing Blood"; he referred to "remarquable Curiosity" in "An Account of More Tryals of Transfusion," 521.

pages of the *Philosophical Transactions* during summer 1667.[32] In June 1667 Denis staged two transfusions from lambs into humans. In a pamphlet summarized both in the *Journal des Savants* and in London, he contested arguments against human transfusion. Diseases, notably "Madness, Dotage and other Maladies," which all conceded were based in the blood, could well be managed this way. So armed, Denis set out to recruit suitable subjects. The advantage of animals as donors, the Parisian physician argued, was that their regimen could be carefully invigilated prior to transfusion, and that they led more moral lives:

> the blood of Animals is less full of impurities than that of men, because debauchedness and iregularity of eating and drinking are not so ordinary to them as to us. Sadness, Envy, Melancholy, Disquiet and generally all the passions, are as so many causes which trouble the life of man, and corupt the whole substance of the blood: Whereas the life of Brutes is much more regular, and less subject to all those miseries which we ought to consider as sad consequences of the prevarication of our first Parents.

Transfusion, like optical and philosophical instruments, might reverse some of the Fall's effects.[33]

Denis's remarkable pamphlet was distributed in a spurious number of the *Philosophical Transactions,* and Lower swiftly complained about the Parisian seizure of priority for animal transfusion (fig. 3.3). When Oldenburg resumed management of the journal in late September, he published an advertisement reminding his readers of the work of Lower and King in 1665–67. On 17 October 1667 its secretary moved that the society try human transfusion, "it having been already practised at Paris," Lower was made a fellow of the society and the president, Lord Brouncker, ordered the society's physicians to comply with Oldenburg's proposal.[34] At the month's end the fellows contacted Dr. Allen, the manager of Bedlam, to see if it were possible to procure an inmate for their work. King was ordered to prepare a written version of his recipe for animal transfusion, adapted for humans, since the Londoners noticed waspishly that Denis had not "thought fit to describe the manner they used in France for Men; nor any body else come to our knowledge." King's method was read at the society on

32. Brown, "Jean Denis and the Transfusion of Blood"; Maluf, "History of Blood Transfusion"; Peumery, *Les Origines de la Transfusion Sanguine;* Nicolson, *Pepys' Diary,* 84–89; Oldenburg's summary is "An Extract of a Letter of M Denis."

33. Denis's pamphlet was translated as "A Letter Concerning a New Way of Curing Sundry Diseases by Transfusion of Blood," published spuriously in London on 22 July 1667, and reproduced in Farr, "The First Human Blood Transfusion."

34. For Oldenburg's troubles see Hall and Hall, "The First Human Blood Transfusion"; Oldenburg's riposte to Denis is described in Oldenburg to Boyle, 24 September 1667, in Oldenburg, *Correspondence,* 3:480, and published as "An Advertisement Concerning the Invention"; his motion at the society is recorded in Birch, *History of the Royal Society,* 2:201.

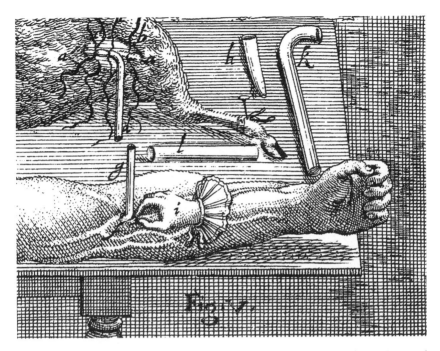

FIGURE 3.3. Equipment for transfusion from a sheep to a man. From Johann Sigismund Elsholtz, *Clysmatica Nova* (Berlin, 1665). Boyle received a copy of this book in early October 1665. Its appearance prompted Oldenburg to defend the priority of English transfusers at the end of 1665.

24 October and published the same week. Oldenburg prefaced it with a letter from a Montpellier physician which urged caution because each subject's blood was spiritually specific, so their mixture must produce fermentation. This fermentation might be therapeutic in small doses, as in the cases treated in Paris by Denis in June but he had subsequently tried transfusion with disastrous consequences on a Swedish nobleman suffering from a "very violent Feaver." After two transfusions from a calf the Swede soon died. The experimenters blamed his "hot and dry Flesh." Oldenburg added that it was for this very reason that "the Philosophers in England," despite having good instruments and an excellent method, were scrupulous about human life and fearful of the law. So Oldenburg amended King's protocol. Where the anatomist insisted that "we have been ready for this Experiment for six months, and wait for nothing but good opportunities," Oldenburg added the phrase "and the removal of some considerations of a Moral nature."[35]

35. For the committee contacting Bedlam see Birch, *History of the Royal Society*, 2:202. For King's method and the Montpellier account see [Oldenburg], "An Account of More Tryals of Trans-

By the end of October, despite Oldenburg's otherwise unbridled enthusiasm for the establishment of English priority, these moral considerations were made manifest in London. Allen refused to let any Bedlamite pass into the transfusers' hands. He was invited to a meeting with Hooke, Clarke, Lower, and King to discuss the problem, to no avail. While Lower, King, and Clarke continued to demonstrate their virtuosity in managing venous flow in dogs, thus producing the symptoms of dropsy and making the dogs' heads swell, and rapidly developed a successful account of the aeration of blood in the lungs, they were still frustrated by the lack of human subjects.[36] The transfusion program stalled for three weeks. Lower helped resuscitate it. During November the society's cosecretary, John Wilkins, tried to persuade him to become the official anatomy curator. It may have been during discussions between Wilkins and Lower that they decided on a possible human subject for transfusion, Arthur Coga. In contrast to Denis's victims, whose ailments were indubitably somatic, the Londoners assumed they should find a subject whose spirits were disturbed, to be regenerated by lamb's blood. Coga, now in his thirties and a Cornishman like Lower, was just such a one. Between 1656 and 1660 he studied at Pembroke College, Cambridge, where his brother later became master. After 1663 Coga himself had apparently read lessons at Wilkins's Church, Saint Lawrence Jewry, the London headquarters of Broad Church theology. These credentials made him a fit witness to his own condition. Coga was also demonstrably disturbed in spirit. King told Boyle that Coga "speaks Latin well when he is in that company he likes his brain in sometimes a little too warm." Oldenburg reported that Coga was "lookt upon as a very freakish and extravagant man . . . an indigent person." Wilkins confessed to Pepys in a London tavern on 21 November 1667 that Coga was "a little frantic, . . . poor and a debauched man." Earlier the same day Lower announced at the society that Coga had agreed to take a guinea for the trial.[37]

So on the morning of Saturday 23 November 1667, a crowd of over forty witnesses assembled at Arundel House, probably in the rooms that the society planned to assign Lower for his anatomical curatorship. The audience included members of the Howard family, who owned Arundel House, as well as many divines, physicians, and members of Parliament. The anatomists let blood run for

---

fusion." The original version of King's report is Royal Society Classified Papers 12 (1) no. 12. More details of the Swedish case are given in a letter from [Gadrois], "A Relation of some Trials."

36. Birch, *History of the Royal Society,* 2:203–4.

37. Details of Coga from Venn and Venn, *Alumni Cantabrigienses,* 1:365; King to Boyle, 25 November 1667, in Boyle, *Works,* 6:646–67; Oldenburg to Boyle, 25 November 1667, in Oldenburg, *Correspondence,* 3:611; and from the remarks of Wilkins to Pepys in Pepys, *Diary,* 8:543 (21 November 1667). See also Nicolson, *Pepys' Diary,* 76–79. Lower's appointment as curator is discussed in Birch, *History of the Royal Society,* 2:206 (5 November 1667), 212 (16 November 1667).

about a minute into a bowl through a silver pipe fixed with a quill to a lamb's carotid artery to calibrate the quantity of blood coming from the animal. Coga, who had already drunk two cups of sack, immediately took some of his own arterial blood on a knife "and tasted it, and finding it of a good relish, he went the more couragiously to its transmission into his veins." Throughout the trial, Coga was one of its directors. King tied Coga's arm, cut the skin over an adequately sized vein, opened it with a lance and let out seven ounces of blood. He inserted another silver pipe into the incision and joined the two pipes with three or four quills. It took almost a minute before the lamb's blood passed along the pipes "and then it ran freely into the Man's Vein for the space of 2 minutes at least." Coga told the operators that the blood did not feel hot, unlike the testimony of the French subjects. King reckoned this was because the arterial blood had time to cool in its long passage from the lamb, "as to come in a temper very agreeable to Venal Blood." The anatomists estimated that in this time they had transfused about nine or ten ounces into Coga. When, at length, Coga announced that "he was not willing to have any more blood" they stopped, pulled the pipe from his vein, tied up his arm, and let more blood run from the sheep. This was to "assure the Spectators of an un-interrupted course of bloud." Coga was "well and merry, and drank a glass or two of Canary, and took a pipe of tobacco in the presence of forty or more persons." During the exercise, the operators repeatedly assured their public, Coga was no more "disorder'd . . . both by his own confession, and by appearance to all bystanders, than any . . . of those yt were in the room wth him." One of the witnesses asked Coga why he had not taken blood from some other animal. Coga's answer soon became a topic of London wit: "Sanguis ovis symbolicam quandam facultatem habet cum sanguine Christi; quia Christus est agnus Dei": "sheep's blood has some symbolic power, like the blood of Christ, for Christ is the lamb of God." So Coga went home, to be subjected to much further scrutiny and interview by the keen transfusers.[38]

Coga's own testimony, and his capacity to provide it, were crucial in the representation of this trial. Pepys and Wilkins judged that since "the man is a healthy man," he would "be able to give an account of what alteration, if any, he doth find in himself, and so may be usefull." On Monday morning, two days after the experiment, Oldenburg and Brouncker visited Coga's lodgings. He was still in bed but confirmed his good health. His landlord reckoned that Coga had become "more composed than he had been before." Encouraged to provide a

38. Descriptions of the transfusion of 23 November 1667 are taken from King, "An Account of the Experiment of Transfusion," and the original version, which differs in some respects, in Royal Society Classified Papers 12 (1) no. 16; King to Boyle, 25 November 1667, and Oldenburg to Boyle, 25 November 1667, in Oldenburg, Correspondence, 3:611.

written account of his own state, Coga "did urge us to have the Experiment repeated upon him again in three or four days," though King guessed this might be
to earn another guinea. The anatomists planned "a further time when we hope
to take a more exact Account, and if it be thought by the Physicians of the Society to order more blood to be taken from this Man, we suppose we may let
in proportionable a larger Quantity which perchance may very much improve
the Experiment."[39] Oldenburg swiftly spread the news of the success round
Europe. He stressed English priority over the French, mentioned the cure of
Coga's "hypochondriacal disorder," and promised major cures for "frenzy" and
a printed account of the protocol. Coga himself was presented at the society's
next meeting on 28 November, bringing with him a Latin account of his own regenerate state: "he finds himself much better since, and as a new man." On Saint
Andrew's Day, two days later, Coga was with Wilkins in a nearby tavern after the
society's annual meeting. Pepys met him there, found him "cracked a little in his
head, though he speaks very reasonably and very well."[40]

In early December Oldenburg discussed with Boyle "those Experiments, that
met wth so much difficulty and contradiction at first [which] may at last prove
very beneficiall to the Health of Men." King read a report on them and Oldenburg published it. The managers decided that a new transfusion should now be
staged at a proper meeting of the society, on 12 December. The transfusers aimed
to get a much more exact trial, by designing new pipes, by weighing the sheep
before and after the operation, and by assigning an assistant to time the operation. Because of the spectacle, this proved hard. "The presence of a strange crowd
both of forrainers and domesticks," according to King, produced "more trouble
in the Operation." He took what he guessed was about ten ounces of blood from
Coga's arm, spilling some in the process. Because of the sheep's struggles, the
quills kept slipping out of the receiver. After seven minutes, King reckoned four-
teen ounces had been transfused. This was a much larger quantity than in November.[41] Coga reported himself "somewhat feverish," though Oldenburg, for

39. Pepys, *Diary,* 8:541; Oldenburg to Boyle, 25 November 1667, in Oldenburg, *Correspondence,* 3:611; King's remarks in Royal Society Classified Papers 12 (1) no. 16.

40. Oldenburg to Sluse, 25 November 1667, in Oldenburg, *Correspondence,* 3:617; Birch, *History of the Royal Society,* 2:216 (28 November 1667); Pepys, *Diary,* 8:554 (30 November 1667).
King's doubts of Coga's report are mentioned in [King], "An Account of the Experiment of Transfusion," 559. News of Coga's trials was transmitted to the Netherlands in March 1668; see Huygens,
Oeuvres complètes, 6:194.

41. Oldenburg to Boyle, 3 December 1667, in Oldenburg, *Correspondence,* 4:6; King's report is
his "An Account of the Experiment of Transfusion." The trial of 12 December is described by King
in Royal Society Classified Papers 12 (1) no. 18 and by Oldenburg to Boyle, 17 December 1667, in
Oldenburg, *Correspondence,* 4:59. See also Birch, *History of the Royal Society,* 2:225 (12 December
1667).

one, judged that "that may justly [be] imputed to his disordering himselfe by intemperate drinking of wine." These disorders foreshadowed critical problems with transfusion and its celebrated subject. A week later Coga was summoned once again to the society to give an account of himself. The diagnosis of drunkenness was publicly confirmed. Oldenburg told Denis that though Coga remained "very well in body, . . . the wildness of his mind remains unchanged—or perhaps we ought rather to consider that the improvement of his brain is prevented by the incurable intemperance of which he is guilty." While Coga spent his time, and the money King paid him, in the taverns, the society's physicians were ordered to find other subjects from the London hospitals, and Oldenburg contemplated transfusions into artificially infected beasts "without scruple and without danger." Lower complained that though he "had decided to repeat the treatment several times in an effort to improve [Coga's] condition of mind," Coga "consulted his instinct rather than the interests of his health, and completely eluded our expectations."[42]

Coga's ravings in winter 1667–68 presaged trouble. The Cambridge naturalist John Ray heard in January 1668 that "the effects of the transfusion are not seen, the coffee houses having endeavoured to debauch the fellow, and so consequently discredit the Royal Society and make the experiment ridiculous." On 9 January King presented the society with his account of Coga's second transfusion, but this was never published. Oldenburg tried to encourage Clarke to publish his own transfusion methods, but confessed to Boyle that "they come short of such effects." In February, Clarke reported to the society that he had found a suitable madwoman for the experiment, but was told to consult her parish officers lest the cost of maintaining her fell on the society itself. Despite several ingenious proposals for a "circular transfusion" between two dogs, no further human trials were staged. At the end of the year, when he finished his major book *De Corde*, dedicated to Boyle, Lower summarized the project. Poisoned and feverish subjects should never be given blood, by analogy with the recalcitrance of moldy wine; but spiritually disordered humans, "the composition of whose brains is not yet spoilt," would surely benefit.[43]

These benefits were widely canvassed by the society's publicists. The trials

42. Birch, *History of the Royal Society,* 2:227 (19 December 1667); Oldenburg to Boyle, 17 December 1667, and Oldenburg to Denis, 23 December 1667, in Oldenburg, *Correspondence,* 4:59, 77; Lower, *De Corde,* 189–90. The transfusion is mentioned by Skippon to Ray, 13 December 1667, in Ray, *Correspondence,* 22.

43. Skippon to Ray, [? January 1668], in Ray, *Correspondence,* 23; Birch, *History of the Royal Society,* 2:236 (9 January 1668), 250 (20 February 1668). Oldenburg complains about Clarke in Oldenburg to Boyle, 3 December 1667, in Oldenburg, *Correspondence,* 4:6. Lower's comments are in *De Corde,* 191. For the appearance of the book see Frank, *Harvey and the Oxford Physiologists,* 208–9.

with Coga followed on Sprat's energetic propaganda in his *History of the Royal Society* at the end of summer 1667. Glanvill's *Plus Ultra,* an even bolder manifesto for the new philosophy, was printed in June 1668 soon after the abandonment of the transfusion work, a program which Glanvill noisily lauded. Then in the midst of a fierce fight for authority between the College of Physicians and the apothecaries, Henry Stubbe, former collaborator of Boyle and Wren in Oxford in the late 1650s and veteran of the "Good Old Cause" of radical republicanism, launched an unprecedentedly violent attack on Sprat and Glanvill and their claims for the society's worth. Human transfusion played a major role in his polemic.[44] From summer 1668 Stubbe and his audiences among physicians and wits noisily debated such trials as those on Coga, trials which seemed risible when they were not lethal. Stubbe denied the matter of fact: "Sure I am that the Transactions report an Untruth, in saying that Coga was ever the better for it." Stubbe also sponsored, and printed, a letter addressed to the society from "the meanest of your Flock, Agnus Coga," in which the Lamb represented himself as the experimenters' "Creature, for he was his own man till your Experiment transform'd him into another Species" and then boldly complained about his loss of finance from the society. Stubbe also circulated a long set of verses "On Agnus Coga his Povertie who by Philosophical Transfusion of Sheeps bloud became Heartless and Shiftless." What the versifier called "Articial Larrifice" was favorably compared with alchemy; much was made of Coga's newfound poverty; and the unfortunate victim was recommended to use as clothing a tapestry which hung in the society's room, principally because of its subject: "Ba-lam and his Asse." Lambs joined learned asses as the wits' favorite topic. In his brilliant satire against the experimenters, *The Virtuoso* (1676), Thomas Shadwell made one of his more choleric characters exclaim that "if the blood of an ass were transfus'd into a virtuoso, you would not know the emittent ass from the recipient philosopher."[45]

Stubbe made sure to insert Coga's letter in his answer to Glanvill. He argued that Cesalpino, not Harvey, discovered circulation of the blood. He pulled rank on Glanvill by recalling his own participation in the Oxford transfusion projects of 1657–58, when he translated Boyle's essays into Latin and saw Wren inject a

44. For the publication of Sprat's *History,* and Glanvill's *Plus Ultra,* see Hunter, *Establishing the New Science,* 45–68; Steneck, "The Background to *Plus Ultra*"; Stubbe's polemics are discussed in Jones, "Mid–Seventeenth Century Science," and Jones, *Ancients and Moderns,* 244–62. His politics are analyzed in Jacob, *Henry Stubbe,* 78–108, and Hill, *The Experience of Defeat,* 264–69. The medical connection is outlined in Cook, "Physicians and the New Philosophy."

45. Stubbe's denial is in *Plus Ultra Reduced to a Non Plus,* separately paginated with *Legends no Histories,* 133, and Coga's letter at 179; the letter and the verses are in British Library MSS, Egerton 29482 fols. 162–63. See Nicolson, *Pepys' Diary,* 167–69. For Shadwell see *The Virtuoso,* 51. For other satires on human transfusion, see Maehle, "Literary Responses to Animal Experimentation."

dog with opium at Boyle's lodgings. He praised the skill of his schoolfellow Lower, whose work at Oxford in spring 1665 was here given a detailed summary. But Stubbe was keen to argue the danger of human transfusion, citing deaths in Paris and Danzig. Above all, he touched on the basic issue of the character of the experimental body. Coga was crazy. This was why Lower and Glanvill both reckoned he was such a good subject. But Stubbe made Coga's case ludicrous. Regeneration was too public a term to escape the vagaries of Restoration wit. In the taverns and the stationers' shops Stubbe dramatized the flaw in the society's program: "To argue from the cures of Madmen, or from what they suffer without hurt, is not for a Physician, but for one that deserves to be sent to Bedlam." By seeking to make Coga's bodily condition a witnessable matter of fact, the virtuosos ran the risk of turning him into a scandalous case of vulgar raving.[46]

■ *"He Is No Philosopher": The Saint as Subject* ■

NATURAL PHILOSOPHERS worked hard to establish that Vermaasen and Coga both possessed bodies now capable of wondrous deeds. Some who tried Coga were veterans of earlier episodes of spiritual and visionary regeneration. Two decades before, for example, the transfuser Thomas Coxe had interrogated a celebrated seer and abstinent Sarah Wight during her eight weeks' fast for hard evidence of her spiritual state. In much-publicized episodes of saintly transports and spectacularly embodied faith, virtuosos, priests, physicians, and the regenerate themselves all debated the roles of mundane bodies and divine spirits. Dreams and visions were elements of holy carnivals in which spiritual dramas could be enacted by the plebs: after the Restoration, such histrionics were treated by their rulers with extreme suspicion. Most important in explicating such events was the notion of mediating aromatic fluids which leaked through regenerate bodies and were seen as signs of specially constituted morality and corporeality.[47] These spirituous fluids were granted the same capacities, powerful and witnessable, as those of light or blood. Their status as experimental facts became salient in the case of Valentine Greatrakes, an Anglo-Irish veteran of the Parliamentary army. From spring 1662 Greatrakes began curing neighbors and visitors around his estates in county Waterford of a range of afflictions, especially the King's Evil, by the laying on of hands. Prolonged stroking and invocation of the name of Christ allowed the "stroker" to move and then expel the seat of disease

---

46. Glanvill's defense of transfusion is in *Plus Ultra*, 9–19. The complex bibliography of Stubbe's polemic is worked out in Spiller, *"Concerning Natural Experimental Philosophie,"* 33–34. Stubbe's discussion of transfusion is in *Plus Ultra Reduced to a Non Plus*, 116–35.

47. For Coxe and Wight see Nuttall, *James Nayler*, 10; Dailey, "Visitation of Sarah Wight," 445. For holy carnival see Cohen, "Prophecy and Madness"; Dailey, "Visitation of Sarah Wight," 449; Mack, *Visionary Women*.

The true and liuely Pourtraicture of *Valentine Greatrakes* of *Affane* in ẙ County of *Waterford*, in ẙ Kingdome of *Ireland*.

FIGURE 3.4. Valentine Greatrakes, the Irish stroker. From the frontispiece to Greatrakes, *A Brief Account of Mr Valentine Greatraks.*

from his patients' bodies.[48] Greatrakes recalled a series of visions that prompted him to take up spiritual cures. Local church courts asked him for his warrant, to which "he answerd he had none but his own strong imagination." Greatrakes ignored ecclesiastic bans and payment for his cures. "Take notice, That as God gave me the several gifts from time to time, he alwayes sent Patients, that applyed themselves to me, for I never sought after any" (fig. 3.4).[49]

In spring 1665 news of his work reached Lord Conway's house at Ragley Hall in Warwickshire, a rallying point for the Platonists Henry More and Ralph Cudworth and their friend the Dean of Connor, George Rust. Lady Conway, John Finch's sister, was a celebrated sufferer of chronic, intense headaches. Eminent physicians, including the Oxford professor Thomas Willis, had failed to cure her. Her husband reckoned Greatrakes might succeed. In July 1665 Conway tried mobilizing his Irish contacts such as Michael Boyle, former chaplain to the English army in Ireland recently elevated to the Dublin see. The archbishop interviewed Greatrakes in Dublin but was told by the former commonwealthman that "he would not goe [to Ragley] for any Lord whomsoever." The divine was skeptical, especially of reports from "those who were on the same side with [Greatrakes] in the last wars." An hour's talk convinced him that Greatrakes's cures might be real, but that the healer was a "man under a strong delusion, and infinitely misguided by his owne imagination (for he now dreames of nothing but converting the Jew and Turk)." This mix of enthusiast credulity and promise of regeneration provided key topics for experimental philosophy in Restoration Britain. The healer's body was made available for test when, through the offices of Dean Rust, Greatrakes eventually reached Ragley on 26 January 1666. "I hope in God," prayed Greatrakes, "I may be an instrument in his hands to free the lady from those distempers."[50]

Spectacular faith healers, especially from Ireland, were not new. In summer 1663 a pub audience in Oxford was entertained by an Irish Catholic named Blake with exorcisms and cures performed in full rite. He worked at Chester and, more dangerously, in the confines of chapel of his coreligionist, Queen Catherine, at Saint James's. Earlier, under the Commonwealth, wondrous cures and claims to inspiration had been common. The Ragley group recalled that in 1654 one

48. McKeon, *Politics and Poetry in Restoration England,* 208–15; Jacob, *Robert Boyle,* 164–76; Duffy, "Valentine Greatrakes"; Kaplan, "Greatrakes the Stroker."

49. Greatrakes, *A Brief Account of Mr Valentine Greatraks,* 22, 36, 26. His magic and the Lismore ban are mentioned by Beale to Oldenburg, 4 September 1665, in Oldenburg, *Correspondence,* 2:496.

50. The Dublin interview is described by Boyle to Lord Conway, 29 July 1665, in Conway, *Correspondence,* 262–63; the role of Orrery and Rust by Lord Conway, August 1665, and Greatrakes to Rawdon, 9 December 1665, in ibid., 265–67.

Matthew Coker had promoted radical religion and millenarian reconciliation in company with his successful cure of women's paralyzed limbs, an earl's bladder, and "a man raging mad." In his discussions with Lady Conway about Coker, More invented a vocabulary for the safe description of such work and warned against "fanaticks or Melancholists" who claimed divine commission. Miracles must be sudden and universal and they must teach some new doctrine. Significantly, just the same criteria were used in the 1650s by the moderate ecclesiastic Thomas Fuller to show that the royal touch itself was only partly miraculous: "I say partly because a complete Miracle is done presently and perfectly, whereas this [royal] Cure is generally advanced by Degree." Coker's cures also took time and worked only with some diseases, and his doctrine that all religions would soon be unified under Christ was no news and so needed no wonders to prove it. So More set out a different account of Coker's successes. By "long temperance and devotion," Coker's "blood and spiritts . . . is become sanative and healing." More reckoned that Coker's "healing infection" was worth trying during one of Lady Conway's fits. Even if this "power partly naturall and partly devotionall" were "due to God," it must not be confused with the actions of Christ and his apostles. Two years after Coker's work, More printed an attack on the ambitions of magical healers, *Enthusiasmus Triumphatus,* in which he argued that "the Spirits of Melancholy Men" might become "a very powerful Elixir." This "sanative and healing contagion" showed that cures by stroking "might be no true Miracle." Only by this careful account of healers' bodies and spirits could moral order and religious truth be secured.[51]

More's philosophical allies developed the model of "sanative virtue" to describe well-attested spiritual cures—Conway used this term in early 1666 for Greatrakes's work. But in time of intense millenarian activity they could not exercise a monopoly on the valuable commodity which Greatrakes's cures afforded other rival groups. In plague-stricken London, from which the Royal Society had fled, Oldenburg correlated information on this new wonder. The West Country naturalist John Beale reported that Greatrakes had tried and failed to raise the dead and heal the blind: "it cannot be denyed yt innumerable great cures are done, and tis not possible yt all exspectations should be satisfyed, wthout a general restauration equivalent to a Resurrection." In London Thomas Sydenham, now an eminent physician but, like Greatrakes, a former soldier in the Parliamentary army, was equally confident: he told Oldenburg that "now he hath no

51. Blake is mentioned in Thomas, *Religion and the Decline of Magic,* 240. On Coker see More to Lady Conway, 7 and 18 June 1654, in Conway, *Correspondence,* 100–104. Coker's career and confinement are in *A Short and Plain Narrative of Mat. Coker,* and [Coker], *A Whip of Small Cords.* More extends his account in *Enthusiasmus Triumphatus,* 40–41. For the partly miraculous royal touch, see Fuller, *Appeal of Injured Innocence,* cited in Bloch, *Rois Thaumaturges,* 425 n. 2.

more reason to doubt it yn to doubt whether he is a man or some other Animal." In the autumn of 1665 Boyle still remained skeptical, despite some good reports from "amongst his country people" in Ireland.[52]

The reports changed when Greatrakes came to Warwickshire at the start of 1666. He failed to cure Lady Conway, but crowds flocked to Ragley, where more than one thousand were touched, and at Worcester he met "great crowds, where I was like to be bruised to death." Henry Stubbe, the Conways' local physician, talked with Greatrakes at Ragley and wrote Boyle a long letter about him. Stubbe's tract, *The Miraculous Conformist,* dedicated to Willis, was published in Oxford just as the stroker reached London.[53] In mid-March Oldenburg reported to Boyle on Greatrakes's work in the capital, and at the same time the metropolitan high churchman David Lloyd completed his *Wonders no Miracles,* the fiercest attack on Greatrakes's probity and politics ever published. In the first week of April Boyle himself went to London to witness Greatrakes's cures and try his patients. In the bookshop of the stationer John Flesher, More showed Boyle his *Enthusiasmus Triumphatus,* a text written soon after his comments on Coker and devoted to the proper interpretation of spiritual working. The two men conversed in some detail about their accounts of Greatrakes's cures.[54] "However it looks at London, it was laughed at in the University," Cambridge dons reported of the Irishman's cures. Conway's brother, Heneage Finch, claimed they "made the greatest faction and disturbance between clergy and laymen than anyone these 1000 years." According to Rust, the stroker's deeds became "the great discourse now at the Coffee Houses."[55]

It was hard to distinguish between the matters of fact about Greatrakes's cures and the reasons for their efficacy. Some commentators reckoned they were miraculous and so stressed the divine power that his success manifested. It was reported in February 1666 that Greatrakes was "the man of miracles, the seventh brother, who opens the eyes of those that have been blind for many years and cures cancers in the breast, which he seldom fails in." Greatrakes himself claimed that he had been given his gift "to convince this Age of Atheism" and "to abate

52. Conway to Rawdon, 9 February 1666, in Conway, *Correspondence,* 268 (for Greatrakes's "sanative virtue and a natural efficiency"). For Sydenham see Oldenburg to Boyle, 18 September 1665; for Beale see Oldenburg to Boyle, 10 October 1665; for Boyle see Moray to Oldenburg, 10 October 1665; in Oldenburg, *Correspondence,* 3:512–13, 556, 561.

53. For Greatrakes at Ragley and Worcester see Greatrakes, *Brief Account,* 39; for Stubbe at Ragley, see Stubbe, *Miraculous Conformist,* 1; Jacob, *Henry Stubbe,* 51.

54. Oldenburg to Boyle, 13 March 1666, in Oldenburg, *Correspondence,* 3:59; for More's meeting with Boyle see More, *Enthusiasmus Triumphatus,* 51, and for More in London see More to Lady Conway 28 April 1666, in Conway, *Correspondence,* 273.

55. Cambridge's view is reported in Thomas, *Religion and the Decline of Magic,* 241; Heneage Finch in Greatrakes to Conway, 24 April 1666, in Conway, *Correspondence* 272; Rust in Glanvill, *Blow at Modern Sadducism,* 84–85.

the pride of the Papists, that make Miracles the undeniable Manifesto of the truth of their Church." He made much of the public evidence of his own power. Protestants could "do such strange things in the face of the Sun which [Catholics] pretend to do in Cells." More picked up the hint. "What Greatrakes did was done in the face of the World, seen and attested by physicians, Philosophers and Divines of the most penetrating and accurate judgment. . . . What ridiculous shams and cheats the Miracles of the Roman Church are, is too well known." One enthusiast commented that "we cannot but conclude that God is about to do some great work in the World, and this is but the forerunner of some great things which we may shortly see." Beale said Greatrakes's cures were "convincing evidence of the powerful name of our Lord Jesus, in a season that needed some evidence that all revelations were not fanatical." [56]

These comments pointed toward the role of divine spirit in Greatrakes's body. Thus there were those, according to Rust, who "adore him as an Apostle." But on the other hand, many observers deemed it as important to provide corporeal accounts of the cures. Such an account might be used to demolish their credibility. David Lloyd, for example, complained about the immodesty of Greatrakes's actions when he tried his touch on women. He also argued that Greatrakes's strokes worked on patients' imaginations and deluded them into cures. Rust denied this: "to say that this impulse . . . is but like Dreams that are usually according to mens constitutions doth not seem a probable account of the Phaenomenon." When he first encountered Greatrakes, Beale also appealed to bodily work in order to cast doubt on his efficacy. He judged that in the case of a dropsical Irishwoman "ye jogging of ye coach 28 m[iles] in ye country was as great a means of ye cure as his violent stroking her naked belly, but perhaps the continuance of his warm hand on a ladies belly for 14 days together . . . might do much." Greatrakes's failure to cure the king's favorite, Sir John Denham, on 3 March 1666, was explained the same way: "Denham is now stark mad, which is occasioned (as is said by some) by the rough striking of Greatrakes upon his limbs." [57]

The dominant concern in spring 1666 was whether divine power could be experienced through Greatrakes's strokes. Any account of Greatrakes that made too much of his body would run the risk of atheist materialism. Lloyd reckoned that Greatrakes and his backers were stressing the material capacities of stroking

56. McKeon, Poetry and Politics, 212; Greatrakes, Brief Account, 30–31; More, Enthusiasmus Triumphatus, 53; Beacher, Wonders if not Miracles, 4–5; Beale to Boyle, 7 September 1665, cited in Duffy, "Valentine Greatrakes," 261.

57. Lloyd, Wonders no Miracles, 17; Rust to Glanvill (March–April 1666), in Glanvill, Blow at Modern Sadducism, 85, 89; Beale to Oldenburg, 4 September 1665, in Oldenburg, Correspondence, 2:496; Walsh to Slingsby, 17 April 1666, in Conway, Correspondence, 252 n. 26 on Denham.

to do down royal power: "When it appeared that [the king] could do no more than other men, he should be no more than other men." Certainly Greatrakes's works were easily absorbed into the sensitive routines of the royal touch. A Berkshire man afflicted with the King's Evil first treated at Oxford by Thomas Willis was, as stipulated, scrutinized by the king's surgeons there, then on Willis's recommendation sent to London to be stroked, successfully as it proved, by the Irish wonder-worker.[58] This was why the divines and virtuosos had to provide a compelling story that defended talk of body without a lapse into subversion or atheism. The dangers were dramatized by Stubbe's tract. More noted "the resentment of serious men" against the book: "they look upon his management of the matter [as] not so advantageous for Relligion." Stubbe cleverly used Willis's new doctrine of morbific and sanative fermentations to argue that Greatrakes succeeded because of his body's "particular Ferments, the Effluvia whereof, being introduced sometimes by a violent Friction, should restore the Temperament of the Debilitated parts, reinvigorate the Bloud, and dissipate all heterogeneous ferments out of the Bodies of the Diseased." So the physician used medical doctrine to speak of Greatrakes's body and thus to explicate his power, then claimed that Christ and the apostles had performed their miracles the same way: "If he doth the things that never man did, except Christ and the Apostles &c. judge what we are to think." Stubbe's readers were to think all miracles were mere passions of body explicable without divine intervention. Boyle was warned that Stubbe's materialist story was designed to show that Christ's miracles, like Greatrakes's, "might be merely the result of his constitution which same may be affirmed of others that have performed real miracles." Against the claim that miracles would always teach some new truth, Stubbe argued that "men may learn to try Miracles by the truth, not the Truth by miracles." Most experimental philosophers and divines resisted this trial. They sought different accounts of the interaction of body and spirit. Greatrakes's cures were real and therefore affected bodies, not miraculous but surely spiritual.[59]

The Ragley solution was to stress Greatrakes's regenerate body. More's model of "sanative contagion" provided some criteria for telling truly miraculous action: it must be sudden and universal and teach new doctrine. Greatrakes failed on all three counts. Rust summed up: "I am convinced it is not miraculous. . . . He pretends not to give Testimony to any Doctrine, the cure seldom succeeds without reiterated touches, his Patients often relapse, he fails frequently, he can

---

58. Lloyd, *Wonders no Miracles*, 14; Greatrakes, *Brief Account*, 87–88.
59. More to Lady Conway, 17 March 1666, in Conway, *Correspondence*, 269; Stubbe, *Miraculous Conformist*, 10, 27; Coxe to Boyle, 5 March 1666, in Jacob, *Henry Stubbe*, 52. For Stubbe and Willis, see Kaplan, "Greatrakes the Stroker," 179; Jacob, *Henry Stubbe*, 51.

do nothing where there is any Decay in Nature, and many Distempers are not at all obedient to his touch." More agreed: "Although he cur'd all those diseases, yet he did not succeed in all his applications. Nor were his cures always lasting."[60] Though no miracle worker, he was not a common sinner. Rust judged that the stroker was "free from all design, of a very agreeable conversation, not addicted to any Vice, nor to any Sect or Party." More defined Greatrakes as a moral regenerate. He discharged his public offices with justice, equity, and goodwill. He persecuted no sect, not even the Catholics, despite their barbarous treatment of the Irish Protestants: "Truly he seems to me such an exemplar of candid and sincere Christianity, without any pride, deceit, sourness, or superstition." This exemplary morality was accompanied by bodily effects quite characteristic of the regenerate whose virtuous odors were well-understood signs of their spiritual state. Like one of the anointed, monarch or saint, Greatrakes's body smelled good. Conway noticed that the stroker's hand and bosom filled the room with the scent of flowers. Rust discovered that Greatrakes's urine smelt of violets. More confirmed that Greatrakes's "Body as well as his Hand and Urine had a sort of herbous Aromatick Scent." The combination of virtue and perfume seemed enough to evince the way Greatrakes's sanative contagion worked on "that subtil morbifick matter which, by the application of his hand, would become Volatil."[61]

More and Boyle used the same resource to establish their own interpretations of Greatrakes: the condition of their own bodies and their own status as subjects. Divines and experimental philosophers could use what they knew of themselves to calibrate Greatrakes's powers. More himself was a seventh son, and therefore a possible candidate for the status of divine healer. As such, it was important for More to investigate his own odors. He announced that his own urine also smelt of violets. When a Cambridge student, he had once set his shoes by the fire, thus prompting his roommate to exclaim at the "mighty smell of Musk or Civet." His colleague George Rust argued that the role of the philosopher was crucial in explaining Greatrakes's powers. Rust told Glanvill that Greatrakes wrongly believed that his power was "an immediate gift from heaven; and 'tis no wonder, for he is no Philosopher." The claim was that only philosophers could properly adjudicate the genuine capacities of themselves or other men. More had warned that no one should tell Matthew Coker why his cures worked,

60. Rust to Glanvill, in Glanvill, *Blow at Modern Sadducism,* 85; More, *Enthusiasmus Triumphatus,* 51.

61. Rust to Glanvill, in Glanvill, *Blow at Modern Sadducism,* 85; compare Rust's comments in Greatrakes, *Brief Account,* 60–61. For investigations of Greatrakes's smell see Stubbe, *Miraculous Conformist,* 11; More, *Enthusiasmus Triumphatus,* 51–52. For the odors of sanctity, see Albert, *Odeurs de Sainteté,* 183; Classen, Howes, and Synnott, *Aroma,* 52–54.

"lest it come to his knowledge, to the disturbance of him in his way, and the weakening of him in his present facility." By contrast, Rust reckoned that a philosopher might well have excited Greatrakes to act as stroker, otherwise "the Gift of God had been to no purpose." He guessed that Greatrakes might have been told of the powers he could exert: "Perhaps some Genius who understood the sanative vertue of his Complexion, and the readinesse of his minde might give him notice of that which otherwise might have been for ever unknown to him." The ambiguity of the term "Genius" was deliberate. Rust might mean a divine spirit, but he might well mean a spiritual divine. Such men could describe, interpret, and then activate or interrupt the work of spirit on matter.[62]

In his notes of March 1666 Boyle also carefully explained the difference between miraculous and natural spiritual action and insisted that Greatrakes's body evinced the latter just because its behavior could be judged by analogy with others. Boyle agreed with Stubbe that the age of miracles had not passed: "I rejoice in the appearing of a protestant that is enabled and forward to do good in such a way, especially in an age where so many do take upon them to deride all that is supernatural." Boyle wove together issues of credibility and self-experiment. Supernatural events were by no means ruled out in his world, but he held that Greatrakes was not a miracle worker because he did not meet the criteria that More, among others, had laid down. Greatrakes refused some patients and failed with others, sometimes his cures relapsed, and in any case his strokes were so violent that there was "a great resemblance betwixt the operations of his hand and the actions of physical agents." Boyle had a good candidate for these agents. He reckoned that the "sanative and perhaps anodyne steams of his body" could cure by reinvigoration and "occasioning a great and therefore probably sometimes a lucky commotion in the blood and spirits" by heightening the passions of patients impressed by the healers' "extraordinary and public" performances.[63]

This was a story that mixed accounts of friction and corpuscular effluvia with the effects of "exalted imagination." The combination was designed to scotch Stubbe's attempts to give true miracles materialist explanations, to preserve Greatrakes's spiritual status, and to render his work comparable with that of other bodies. Boyle needed now to give a role to the specific individual capacity of Valentine Greatrakes. He had already collected stories of "wonderful and extraordinary" cures effected by sympathy, by invisible effluvia, and by transplantation. He printed accounts of successes with sympathetic powders, of a

62. More describes his own body in *Enthusiasmus Triumphatus,* 52; Rust to Glanvill, in Glanvill, *Blow at Modern Sadducism,* 87–89. More warns against telling Coker in a letter to Lady Conway, in Conway, *Correspondence,* 102.

63. Boyle to Stubbe, 9 March 1666, in Boyle, *Works,* 1:lxxvi–lxxxv; see Kaplan, "Greatrakes the Stroker," 180–81; Jacob, *Robert Boyle,* 168–69, 175; Macintosh, "Locke and Boyle on Miracles," 206.

cure of the King's Evil effected when a dog licked the sufferer's ulcers and thus had them transplanted to its own body, and of Harvey's successes in curing tumors by the touch of a dead man's hand. Boyle reckoned that these cures effected "by subtle effluvia" could be compared with Greatrakes's work. Oldenburg agreed and printed a summary of these stories about "the effects of touch and friction" in the *Philosophical Transactions.* Before he met Greatrakes Boyle asked about the role of prayer in the cures, whether Greatrakes could cure "Men of different religions," and about the details of the stroke. Was it true that "ye Effluvia of Mr Gr[eatrakes] his Body are well sented," and did Greatrakes's clothes preserve his virtues? This raised the sensitive issue of whether, as Boyle put it to Stubbe, Greatrakes like "our Saviour could communicate the power of working miracles to others at his pleasure."[64]

Because of these puzzles, when the natural philosophers started working with Greatrakes at Lincoln's Inn, at Boyle's lodgings in Stoke Newington, and at his sister Lady Ranelagh's house in Pall Mall in April and May 1666, they deliberately—often violently—tried whether they could reproduce his cures and master the fumes he emitted. On 10 April Sydenham enthusiastically sent Boyle a crippled scrivener to be treated by Greatrakes, while Boyle and his kinswomen ordered their victim to demonstrate just how much he was suffering. A week later John Wilkins was there when Greatrakes tried to paralyze the fingers of a woman who suffered pains in her hands. Wilkins "took her fingers between my hands endeavouring by friction to recover them to some warmth and limberness, and having continued this friction for some while, I let her hand goe, and would have persuaded her that she was cured." The woman insisted she was still in pain— Wilkins shoved a pin into the woman's hand without response, until Greatrakes stroked the affected area and restored feeling in the hand.[65] On 3 April one Robert Furnace, a Clerkenwell tinker previously treated for lameness at Saint Bartholomew's Hospital, visited Greatrakes and had his pain driven out of him through his foot. Three days later he came back to be treated again in Boyle's presence, "his paines . . . being not quite removed." Boyle then took Greatrakes's glove "and turning ye inside outward stroak'd therewith the affected Arme and (as I ghesse) withn a minute of an hower drove ye paine as ye Tinker told me into his wrist where he sayd it much afflicted him." Boyle then carefully followed Greatrakes's technique. Using the healer's glove, he drove the pain out of Furnace's side, and down into "ye extreame parts of his Toes." Boyle, calmly judi-

64. Boyle, *Some Considerations Touching the Usefulnesse,* part 2, 225–33; [Oldenburg], "Some Observations of the Effects of Touch." Boyle's inquiries in Maddison, *Robert Boyle,* 124–26, and Boyle to Stubbe, 9 March 1666, in Boyle, *Works,* 1:lxxviii.

65. British Library MSS. add. 4293 fol. 53 (Sydenham and Boyle, 10 April 1666); for Wilkins see Greatrakes, *Brief Account,* 56–57, 73.

cious, recorded that the tinker "appeared to be more tormented than ever and expressed an Impatience yt at my delaying to stroake yt I might inquire into wt manner of paine he felt and how intense." Keen to see whether his own palm had the same effects as that of Greatrakes, Boyle eventually "quite stroaked away" the tinker's ailment.[66]

The natural philosophers' capacity both to interpret and reproduce his cures did not damage, and helped reinforce, the carefully defined spiritual message which they held the Irishman taught. The matter needed care. Greatrakes insisted that "several instances seemed to me to be Possessions by dumb Devils, deaf Devils and talking Devils," thus reaffirming the widespread culture of possession and exorcism and claiming the "extraordinary Gift of God." This was not quite what Boyle wanted to hear from his regenerate and sanative informant in an age which saw enthusiasts as dangerous or immoral. He recorded an Eastertide discussion with Greatrakes: "he thinks most Epileptick persons to be Daemoniacks notwithstanding what I could say to the contrary."[67] When Greatrakes set out to produce "matters of fact . . . so clear that I need not use any further arguments to gain belief" in the first half of May, before his return to Ireland, he had to rely on the virtuosos for support. Wilkins, Cudworth, Rust, and Boyle all sent him signed statements. More reckoned Boyle's testimony "is likely to doe [Greatrakes] more creditt then any body." Greatrakes's cures were judged matters of fact, according to Rust, because they were examined by the Royal Society, "whom we may suppose as unlikely to be deceived by a contrived Imposture as any persons extant." Greatrakes commented that Boyle's "repute and testimony to the world will be so powerful . . . that truth may find belief; God have glory; and his poor instrument be justified before men."[68] This instrument's meaning hinged on the configuration of the body of the philosophers. They calibrated the capacities of Vermaasen, Coga, and Greatrakes against their own. Regeneration was established through these practical judgments and shows. Against these idealized bodies—the blind seer, the sheepish fool, the sanative healer—others proffered instead the fleshly bodies of charlatanry, asinine pretension, and carnivalesque satire. Greatrakes himself reckoned it was "the sober party" who were

66. Greatrakes, *Brief Account,* 48–49, 63–64, 71, and Boyle's notes (6 April 1666) in British Library MSS, add. 4293 fol. 52.

67. Greatrakes, *Brief Account,* 33–34; Boyle's note, British Library MSS, add. 4293 fol. 51.

68. Greatrakes to Conway, 24 April 1666, in Conway, *Correspondence* 272; Greatrakes, *Brief Account,* 41, 3; More to Lady Conway, 28 April 1666, Conway, *Correspondence,* 273; Rust to Glanvill, in Glanvill, *Blow at Modern Sadducism,* 89. Chronology of Greatrakes's work in May 1666 is described in Steneck, "Greatrakes the Stroker," 168. See Duffy, "Valentine Greatrakes," 273: "the reactions to his cures were dictated as much by fear of democracy or anarchy, of popery or religious enthusiasm, of the upsetting of the Restoration social, political and religious order, as by the intellectual demands of science or reason."

"most of them believers and my champions," and that the "large and full testimonials" of the virtuosos were the chief way of countering "the mouths" of the wits with the facts of sobriety.[69]

No doubt, as Marc Bloch and Keith Thomas have argued of the rituals of royal healing, early modern spectacles of touch, regeneration, and experiment hinged on practitioners' special status. The status of the body of the king's subjects, the natural philosophers, was also connected to and displayed in the body techniques which they used. Their capacity to judge other bodies on the basis of their own persons and senses was as political as that lauded by Stuart apologists on behalf of the Merry Monarch's sanative virtue. The specific claim that spirits could be managed experimentally in the eye, bloodstream, or hand was intimately bound up with the power of the experimenters' own bodies. Much has recently been made of the role of courtesy, especially of self-disciplined manners, in the emergence of early modern natural philosophy: experimenters handled instruments, it is claimed, just as the courteous body was carefully managed. This chapter has complemented that account with attention to the patriarchal and traditional resources that allowed natural philosophers to judge others by representing their own idiosyncrasies. In his brilliant sketch of the social meaning of body techniques, Marcel Mauss observed that "there is no technique and no transmission in the absence of a tradition. You will see that in this it is no different from a magical, religious or symbolic action." In early modern natural philosophy there was an especially close relation between the instrumental use of the body and the traditions of magical, religious, and symbolic action. This is why they could be described by some as asses, by others as angels.[70]

69. Greatrakes to Conway, 24 April 1666, in Conway, *Correspondence*, 272.
70. Bloch, *Rois Thaumaturges*, 428–29; Thomas, *Religion and the Decline of Magic*, 244; Biagioli, "Tacit Knowledge," 78; Mauss, "Body Techniques," 104.

■ ACKNOWLEDGMENTS ■

I thank the British Library and Library of the Royal Society for permission to use material in their possession.

■ REFERENCES ■

A., T. [Thomas Allen]. Χειρεξοκη: *The Excellency or Handywork of the Royal Hand*. London, 1665.
Albert, Jean-Pierre. *Odeurs de Sainteté: La Mythologie Chrétienne des Aromates*. Paris: EHESS, 1990.
Ashcraft, Richard. *Revolutionary Politics and Locke's Two Treatises of Government*. Princeton: Princeton University Press, 1986.
Beacher, Lionel. *Wonders if not Miracles*. London, 1666.

Biagioli, Mario. "Tacit Knowledge, Courtliness, and the Scientist's Body." In *Choreographing History,* ed. Susan Leigh Foster, 69–81. Bloomington: Indiana University Press, 1995.

Birch, Thomas. *The History of the Royal Society of London.* 4 vols. 1756–57. Reprint, Brussels: Culture et Civilization, 1967.

Bloch, Marc. *Les Rois Thaumaturges.* Paris: Galllimard, 1924.

Boyle, Robert. *Experiments and Considerations Touching Colours.* London, 1664.

———. *New Experiments Physico-Mechanicall Touching the Spring of the Air.* Oxford, 1660.

———. *Some Considerations Touching the Usefulnesse of Experimental Natural Philosophy.* Oxford, 1663.

———. "Tryals Proposed by Mr Boyle to Dr Lower, to be made by him, for the Improvement of Transfusing Blood out of one live Animal into Another." *Philosophical Transactions* 1 (1666): 385–88. It appears in manuscript as "Tryals proposed by Robert Boyle to Dr Lower," Royal Society RBC 2.299–301.

———. *Works.* 6 vols. London, 1772.

Brown, Harcourt. "Jean Denis and the Transfusion of Blood." *Isis* 39 (1948): 15–29.

Browne, John. *Adenochoiradelogia.* London, 1684.

Cavendish, Margaret. *Observations upon Experimental Philosophy; to which is Added, The Description of a New World called the Blazing World.* London, 1666.

Classen, Constance, David Howes, and Anthony Synnott. *Aroma: The Cultural History of Smell.* London: Routledge, 1994.

Cohen, Alfred. "Prophecy and Madness: Women Visionaries during the Puritan Revolution." *Journal of Psychohistory* 11 (1984): 411–30.

[Coker, Matthew]. *A Short and Plain Narrative of Mat. Coker.* London, 1654.

[———]. *A Whip of Small Cords to Scourge Antichrist.* London, 1654.

Conway, Anne. *The Correspondence of Anne, Viscountess Conway, Henry More and Their Friends, 1642–1684.* Ed. Marjorie Hope Nicolson. London: H. Milford, 1930.

Cook, Harold J. "Physicians and the New Philosophy: Henry Stubbe and the Virtuosi-Physicians." In French and Wear, 246–71.

Cope, Jackson I. *Joseph Glanvill, Anglican Apologist.* Saint Louis: Washington University Press, 1956.

Crawfurd, Raymond. *The King's Evil.* Oxford: Clarendon Press, 1911.

Cunningham, Andrew. "Thomas Sydenham: Epidemics, Experiment and the Good Old Cause." In French and Wear, 164–90.

Dailey, Barbara Ritter. "Visitation of Sarah Wight." *Church History* 55 (1986): 438–55.

Davis, Audrey B. *Circulation Physiology and Medical Chemistry in England 1650–1680.* Lawrence, Kans.: Coronado Press, 1973.

Denis, J. "A Letter Concerning a New Way of Curing Sundry Diseases by Transfusion of Blood" [London, 1667]. Reproduced in Farr, 154–62.

Dennis, Michael. "Graphic Understanding: Instruments and Interpretation in Robert Hooke's *Micrographia.*" *Science in Context* 3 (1989): 309–64.

Descartes, René. *Oeuvres,* ed. C. Adam and P. Tannery. 12 vols. Paris: J. Vrin, 1964–76.

Dewhurst, Kenneth. *John Locke: Physician and Philosopher.* London: Wellcome Historical Medical Library, 1963.

Duffy, Eamon. "Valentine Greatrakes, the Irish Stroker: Miracle, Science and Orthodoxy in Restoration England." *Studies in Church History* 17 (1981): 251–73.

Edwards, John. *A Compleat History of all the Dispensations and Methods of Religion.* London, 1699.

Farr, A. D. "The First Human Blood Transfusion." *Medical History* 24 (1980): 143–62.

Fisch, Harold. "The Scientist as Priest: A Note on Robert Boyle's Natural Theology." *Isis* 44 (1953): 252–65.

Frank, Robert G., Jr. *Harvey and the Oxford Physiologists: Scientific Ideas and Social Interaction.* Berkeley and Los Angeles: University of California Press, 1980.

French, Roger. "Surgery and Scrophula." In *Medical Theory, Surgical Practice: Studies in the History of Surgery,* ed. Christopher Lawrence, 85–100. London: Routledge, 1992.

French, Roger, and Andrew Wear, eds. *The Medical Revolution of the Seventeenth Century.* Cambridge: Cambridge University Press, 1989.

Fuller, Thomas. *The Appeal of Injured Innocence.* London, 1659.

[Gadrois, Claude]. "A Relation of Some Trials . . . Lately Made in France." *Philosophical Transactions* 2 (1667): 559–64.

Gent, Lucy, and Nigel Llewellyn, eds. *Renaissance Bodies: The Human Figure in English Culture c. 1540–1660.* London: Reaktion Books, 1990.

Glanvill, Joseph. *A Blow at Modern Sadducism in Some Philosophical Considerations about Witchcraft.* London, 1668.

———. *Plus Ultra: or the Progress and Advancement of Knowledge since the Days of Aristotle.* London, 1668.

———. *The Vanity of Dogmatizing or Confidence in Opinions Manifested.* London, 1661.

Greatrakes, Valentine. *A Brief Account of Mr Valentine Greatraks and Divers of the Strange Cures by Him Lately Performed.* London, 1666.

Hall, A. Rupert, and Marie Boas Hall. "The First Human Blood Transfusion: Priority Disputes." *Medical History* 24 (1980): 461–65.

Hall, Marie Boas. *Promoting Experimental Learning: Experiment and the Royal Society 1660–1727.* Cambridge: Cambridge University Press, 1991.

Hammond, Paul. "The King's Two Bodies: Representations of Charles II." In *Culture, Politics and Society 1660–1800,* ed. Jeremy Black and Jeremy Gregory, 13–48. Manchester: Manchester University Press, 1991.

Harvey, William. *Exercitationes Generatione Animalium.* London, 1651.

Harwood, Jonathan, ed. *The Early Essays and Ethics of Robert Boyle.* Carbondale: Southern Illinois University Press, 1991.

Henry, John. "Boyle and Cosmical Qualities." In Hunter, 119–38.

Hill, Christopher. *The Experience of Defeat.* New York: Viking Penguin, 1984.

———. *Intellectual Origins of the English Revolution.* London: Panther Books, 1972.

———. *Milton and the English Revolution.* London: Faber, 1977.

———. *The World Turned Upside Down: Radical Ideas during the English Revolution.* New York: Viking Press, 1972.

Hobbes, Thomas. *Leviathan.* London, 1651.

Hooke, Robert. *Micrographia.* London, 1665.

Hunter, Michael. *Establishing the New Science: The Experience of the Early Royal Society.* Woodbridge: Boydell Press, 1989.

———, ed. *Robert Boyle Reconsidered.* Cambridge: Cambridge University Press, 1994.

Hutton, Sarah. "Edward Stillingfleet, Henry More, and the Decline of *Moses Atticus.*" In *Philosophy, Science, and Religion in England 1640–1700,* ed. Richard Kroll, Richard Ashcraft, and Perez Zagorin, 68–84. Cambridge: Cambridge University Press, 1992.

Huygens, Christiaan. *Oeuvres complètes.* 22 vols. The Hague: Nijhoff, 1888–1950.

Jacob, James R. *Henry Stubbe, Radical Protestantism and the Early Enlightenment.* Cambridge: Cambridge University Press, 1983.

———. *Robert Boyle and the English Revolution.* New York: Burt Franklin, 1977.

Johnston, Nathaniel. *The Excellency of Monarchical Government.* London, 1686.

Jones, H. W. "Mid–Seventeenth Century Science: Some Polemics." *Osiris* 19 (1950): 254–74.

Jones, Richard Foster. *Ancients and Moderns: The Rise of the Scientific Movement in Seventeenth-Century England.* Saint Louis: Washington University Press, 1961.

Jose, Nicholas. *Ideas of the Restoration in English Literature 1660–1671.* London: Macmillan, 1984.

Kaplan, Barbara Beigun. *"Divulging of Useful Truths in Physick": The Medical Agenda of Robert Boyle.* Baltimore: Johns Hopkins University Press, 1993.

———. "Greatrakes the Stroker: The Interpretations of His Contemporaries." *Isis* 73 (1982): 178–85.

Kerby-Miller, Charles, ed. *Memoirs of the Extraordinary, Life, Works and Discoveries of Martinus Scriblerus.* New York: Oxford University Press, 1988.

[King, Edmund]. "An Account of the Experiment of Transfusion practised upon a Man in London." *Philosophical Transactions* 2 (1667): 557–59 (the original version, which differs in some respects, is in Royal Society Classified Papers 12 (1) no. 16).

Knoefel, Peter, ed. *Francesco Redi on Vipers.* Leiden: Brill, 1988.

Korshin, Paul, ed. *Studies in Change and Revolution.* London: Scolar Press, 1972.

Lloyd, David. *Wonders no Miracles.* London, 1666.

Locke, John. *An Essay Concerning Human Understanding.* 28th ed. London: Tegg, 1838.

———. *Some Thoughts about Education.* Ed. John W. Yolton and Jean S. Yolton. Oxford: Clarendon Press, 1989.

[Lower, Richard]. "The Method Observed in Transfusing the Bloud out of one Animal into Another." *Philosophical Transactions* 1 (1666): 353–58.

[———?]. *Tractatus de Corde.* London, 1669.

Macintosh, J. J. "Locke and Boyle on Miracles and God's Existence." In Hunter, 193–214.

Mack, Phyllis. *Visionary Women: Ecstatic Prophecy in Seventeenth-Century England.* Berkeley and Los Angeles: University of California Press, 1992.

Maddison, R. E. W. *The Life of the Honourable Robert Boyle F.R.S.* London: Taylor and Francis, 1969.

Maehle, Andreas Holger. "Literary Responses to Animal Experimentation in Seventeenth- and Eighteenth-Century Britain." *Medical History* 34 (1990): 27–51.

Malloch, Archibald. *Finch and Baines.* Cambridge: Cambridge University Press, 1917.

Maluf, N. S. R. "History of Blood Transfusion." *Journal of the History of Medicine* 9 (1954): 59–107.

Mauss, Marcel. "Body Techniques." In *Sociology and Psychology: Essays,* 97–123. London: Routledge, 1979.

McKeon, Michael. *Politics and Poetry in Restoration England.* Cambridge: Harvard University Press, 1975.

More, Henry. *Enthusiasmus Triumphatus.* In *A Collection of Several Philosophical Writings.* 4th ed. London, 1713.

Newton, Isaac. *Mathematical Principles of Natural Philosophy.* 1729. Reprint, Berkeley and Los Angeles: University of California Press, 1934.

Nicolson, Marjorie Hope. *Pepys' Diary and the New Science.* Charlottesville: University Press of Virginia, 1965.

Nuttall, Geoffrey. *James Nayler: A Fresh Approach.* London: Friends' Historical Society, 1954.

Ogilby, John. *The Entertainment of His Most Excellent Majestie Charles II.* London, 1662.

[Oldenburg, Henry]. "An Account of an Easier and Safer Way of Transfusing Blood out of One Animal into Another." *Philosophical Transactions* 2 (1667): 449–52.

[———]. "An Account of More Tryals of Transfusion." *Philosophical Transactions* 2 (1667): 517–25.

[———]. "An Account of the Rise and Attempts of a Way to Conveigh Liquors immediatly into the Mass of Blood." *Philosophical Transactions* 1 (1665): 128–30.

[———]. "An Advertisement Concerning the Invention of the Transfusion of the Blood." *Philosophical Transactions* 2 (1667): 489–90.

———. *Correspondence of Henry Oldenburg.* 13 vols. Ed. A. Rupert Hall and Marie Boas Hall. Madison: University of Wisconsin Press; London: Mansell; London: Taylor and Francis, 1965–86.

[———]. "An Extract of a Letter of M. Denis . . . Touching the Transfusion of Blood." *Philosophical Trancactions* 2 (1667): 453.

[———]. "Some Observations on the Effects of Touch and Friction." *Philosophical Transactions* 1 (1666): 206–9.

Oster, Malcolm R. "The 'Beame of Divinity': Animal Suffering in the Early Thought of Robert Boyle." *British Journal for the History of Science* 22 (1989): 151–80.

Pepys, Samuel. *Diary of Samuel Pepys.* 11 vols. Ed. Robert Latham and William Matthews. London: Bell and Hyman, 1970–83.

Peumery, Jean-Jacques. *Les Origines de la Transfusion Sanguine.* Amsterdam: B.M. Israel, 1975.

Power, Henry. *Experimental Philosophy*. London, 1664.

Ray, John. *Correspondence of John Ray*. Ed. Edwin Lankester. London: Ray Society, 1848.

Reedy, Gerard. "Mystical Politics: The Imagery of Charles II's Coronation." In Korshin, 19–42.

Romanell, Patrick. *John Locke and Medicine*. Buffalo: Prometheus Press, 1984.

Rossi, Paolo. *Francis Bacon: From Magic to Science*. Chicago: University of Chicago Press, 1968.

Schaffer, Simon. "Godly Men and Mechanical Philosophers: Souls and Spirits in Restoration Natural Philosophy." *Science in Context* 1 (1987): 55–85.

Shadwell, Thomas. *The Virtuoso*. Ed. Marjorie Hope Nicolson and David Stuart Rhodes. London: Arnold, 1966.

Shapin, Steven. *A Social History of Truth: Civility and Science in Seventeenth-Century England*. Chicago: University of Chicago Press, 1994.

Shapin, Steven, and Simon Schaffer. *Leviathan and the Air-Pump: Hobbes, Boyle, and the Experimental Life*. Princeton: Princeton University Press, 1985.

Speak, Gill. "An Odd Kind of Melancholy: Reflections on the Glass Delusion in Europe (1440–1680)." *History of Psychiatry* 1 (1990): 191–206.

Spiller, Michael R. G. *"Concerning Natural Experimental Philosophie"*: *Meric Casaubon and the Royal Society*. The Hague: M. Nijhoff, 1980.

Sprat, Thomas. *History of the Royal Society*. London, 1667.

Stallybrass, Peter, and Allon White. *The Politics and Poetics of Transgression*. London: Methuen, 1986.

Steneck, Nicholas. "The Background to *Plus Ultra*." *British Journal for the History of Science* 14 (1981): 59–74.

———. "Greatrakes the Stroker: The Interpretations of Historians." *Isis* 73 (1982): 161–77.

Stubbe, Henry. *The Miraculous Conformist*. Oxford, 1666.

———. *The Plus Ultra Reduced to a Non Plus*. London, 1670.

Swift, Jonathan. "Travels into Several Remote Nations of the World by Lemuel Gulliver." In *Selected Writings*, ed. John Hayward, 1–292. London: Nonesuch, 1968.

Thomas, Keith. *Religion and the Decline of Magic*. Harmondsworth: Penguin, 1972.

Tribby, Jay. "Cooking (with) Clio and Cleo: Eloquence and Experiment in Seventeenth-Century Florence." *Journal of the History of Ideas* 52 (1991): 417–39.

Venn, John, and John Archibald Venn. *Alumni Cantabrigienses*. 6 vols. Cambridge: Cambridge University Press, 1922–54.

Weber, Max. *The Theory of Social and Economic Organization*. Ed. Talcott Parsons. New York: Free Press, 1964.

Williams, Tamsyn. "Magnetic Figures: Polemical Prints of the English Revolution." In Gent and Llewellyn, 86–110.

Wilson, Catherine. *The Invisible World: Early Modern Philosophy and the Invention of the Microscope*. Princeton: Princeton University Press, 1995.

FOUR

ISAAC NEWTON

Lucatello Professor of Mathematics

ROB ILIFFE

In very hot countries as Italy in summer two or 3 doses cure. In Holland five or six. Gentlemen & delicate <active> bodies are <much> more easily cured than labourers.
——Newton, "The method how to use the tincture of sol"[1]

He was turning Grey, I think, at Thirty, and when my Father observed yt to him as ye Effect of his deep attention of Mind, He would jest wth ye Experimts he made so often wth QuickSilver, as if from thence he took so soon that Colour.
——Nicholas Wickins to Robert Smith, 16 January 1728[2]

In early modern culture, the philosopher or scholar was frequently "melancholic," an appellation which was a central term in a multiplicity of theological, literary, and medical discourses. Robert Boyle described his youthful delicate and solitary self as such, while contemporaries argued that Isaac Newton corresponded to this type. Without using the precise term, Nicholas Wickins referred to his father John's initial meeting with Newton in the early 1660s when "he retired one day into ye Walks, where he found Mr Newton solitary and dejected." And in what is probably the definitive description of the relatively young Newton, offered to John Sharp in August 1680, Henry More remarked that Newton's countenance was "ordinarily melancholy and thoughtfull, but then mighty lightsome and chearfull."[3] In addition to the many anecdotes concerning Newton of the *Principia* period, Godfrey Kneller's magnificent portrait of 1689 (fig. 4.1) is unmistakably that of the black-cloaked melancholic.

1. Angled brackets represent additions to text; struck-through words indicate deletions. Trinity College Cambridge MSS, R.16.448A, fol. 1r.
2. King's College, Cambridge Library, Keynes MSS, 137. Wickins was the son of Newton's roommate at Trinity.
3. Wickins quoted ibid. Conway, *Correspondence*, 478–79; for melancholy and solitariness, see Beier, *Sufferers and Healers*, 121; Babb, *The Elizabethan Malady*, 21, 25–28, 30–32, 36–37, 47, 63, 96–100, 182–83; for the significance of bodily presentations as pointing to the "internal" person see Bryson, "The Rhetoric of Status"; Pelling, "Appearance and Reality"; Beier, *Sufferers and Healers*, 249–50; for the constructedness of the division between internal and external, see Frank, "Sociology of the Body," 47.

FIGURE 4.1. Isaac Newton at forty-six, in a portrait by Godfrey Kneller, 1689. Courtesy Lord Portsmouth and the Portsmouth Estates.

With his gray hair and brooding demeanor, the second Lucasian Professor of Mathematics is undoubtedly the most famous and perfect epitome of the early modern scholarly persona described by Shapin in this book (chapter 1). However, the pictorial and verbal representation of Newton as the archetype of such self-neglecting individuals should not distract our attention from other evidence that is sometimes at odds with this conventional depiction. Indeed, I argue here that although he has been portrayed as a disembodied mind in communion with natural or divine truth, the soul of an unhealthy self that engaged in no bodily exercise, Newton employed a well-known classical regimen designed to support intellectual endeavor. I look at the accounts of this medical and ethical practice in the work of Marsilio Ficino and Robert Burton and then examine the bodily conduct of various other fellows and masters of Trinity College, Cambridge. Finally, I pay particular attention to Isaac Newton, as that college's greatest product. Not only were there at least *two* narratives relating to Newton's presentation of corporeal self, but these were combined with an intensely ascetic regimen that was remarked upon by the two men who knew him most intimately in the late seventeenth century. I relate his "breakdown" in 1693 to his private views regarding the bodily practices and sexual conduct of early Christians, and conclude by showing how his status as genius was thereafter attached to discourses of disembodiment.[4]

■ *The Sick Scholar* ■

IN THE LATE FIFTEENTH century, Marsilio Ficino composed *Three Books on Life,* a work in which he offered himself up as "the first to attend as a physician to sick and invalid scholars." The physician was, with "a prudent father and a thoroughly accredited teacher," one of the three earthly Muses (of nine in all) needed to help the intellectual reach the Muses' high temple. Echoing Juvenal's *Mens sana in corpore sano,* Ficino pointed to five particular enemies of the intellectual, namely phlegm, black bile, sexual intercourse, gluttony, and sleeping in the morning. Phlegm "dulls and suffocates the intelligence," while an overabundance of black bile "vexes the mind with continual care and frequent absurdities and unsettles the judgment." Mental agitation made the brain dry and cold, and the subsequent need to draw the more subtle spirits from the blood left the latter dense, dry, and black. Too little exercise failed to allow the dense vapors to be exhaled, compounding the problem. Ficino cited Avicenna to the effect that by draining the scholar of his most subtle spirits, too much sexual intercourse was a debilitating evil that weakened the brain and ruined the stomach and heart, while excessive indulgence in wine filled the head with "humours and very bad fumes." For complex reasons, lack of sleep contributed to the scholar's unhealthy condition. The fancy was "distracted and upset by many long and contrary imagina-

4. For the place of the mind, see Shapin, "'Mind Is Its Own Place'"; for asceticism see Harpham, *Ascetic Imperative;* Martin, Gutman, and Hutton, *Technologies of the Self;* for bodily technologies of self see also Bynum, *Fragmentation and Redemption;* Foucault, *Discipline and Punish.*

tions, cogitations, and cares while it is awake," and only "during the quiet of the night [was] that agitation finally calmed and put to rest."[5]

To properly care for the scholarly self, Ficino offered a series of ascetic practices and dietetic precautions. Such self-concern had to start from the break of dawn or even earlier, and Aristotle had wisely urged that philosophers rise before daybreak, but although he concurred with this, Ficino counseled that scholars should "carefully avoid early morning indigestion by taking a quick and moderate dinner." A massage was advisable before one even vacated the bed and, first thing, at least half an hour was to be spent on *expurgatio*. Following an hour's study, he advised, "relax for a little while your mind's intentness, and meantime, comb your head carefully and moderately with an ivory comb, drawing the comb back from forehead to neck about forty times. Then rub your neck with a rather rough cloth."[6] Exercising twice daily (though not strenuously) on an almost empty stomach would keep phlegm at bay, while techniques such as the meticulous cleansing of the skin, the avoidance of excessively cold foods, and the occupation of a dwelling "high and far away from heavy and cloudy air" were all invaluable for preventing the buildup of clogging vapors. To avoid an excess of black bile a host of foods were ruled out, as were "dryness and everything that is black; anger, fear, pity, sorrow, idleness, solitude, and whatever offends the sight, smell and hearing, and most of all, darkness." Light wine, raw eggs, the meat of birds, young cocks, and four footed sucklings, but especially the infusion of "gold or silver, particularly red-hot, and their leaves, in drinks or even in soup," were highly conducive to health. A detailed regimen for the stomach and a whole series of herbal medicines directed toward intellectually useful parts of the body such as the eyes, the tongue, and the stomach, as well as to occupational hazards such as sleeplessness and forgetfulness, completed Ficino's scholarly physic.[7]

The work of Richard Napier and Robert Burton in the early seventeenth century shows how the diseases of scholarship remained a significant problem in theology, medicine, and literature, and Napier in particular relied upon Ficino's analysis. By the middle of the sixteenth century, melancholy in England was associated with both idleness and mental overexertion, "the one causing the blood to be thicke through setting, and the other, by spending the brain overmuch, and

---

5. Ficino, *Three Books*, 107, 113, 123, 125; Klibansky, Panofsky, and Saxl, *Saturn and Melancholy*, 258–62, 266–67; for ancient conceptions of the relations between melancholy and creativity, see ibid., 18–19, 71–73, 114–15; for other aspects of melancholy, see Schleiner, *Melancholy, Genius and Utopia*; Soufas, *Melancholy and the Secular Mind*. Ramazzini (*Diseases of Workers*, 379–81, 403) repeated Ficino's analysis of the physiological effects of too much study in the early eighteenth century, although he thought that some scholars should be allowed to study at night.

6. For an extended rumination on the significance of scholars' hair, see Ramazzini, *Diseases of Workers*, 395–99.

7. Ficino, *Three Books*, 129, 131, 133, 135–59.

drying it excessively." The author of these words, Timothy Bright, noted in his *A Treatise of Melancholie* (1586) that studies had "great force to procure melancholie: if they be vehement, and of difficult matters, and high misteries." In his *Anatomy of Melancholy* nearly four decades later, Burton devoted a whole section to this phenomenon and entitled it "Love of Learning, or overmuch Study. With a Digression of the Misery of Schollers, and why the Muses are Melancholy." [8]

Much of Burton's work was a compendium of quotations from both classical and modern sources; for example, he cited the example of Levinus Lemnius who had affirmed that many were "come to this malady by continuall study, and night-waking, and of all other men, Scollers are most subject to it." On these grounds, "Patritius therefore in the institution of Princes, would not have them to be great students." As Machiavelli had shown, "study weakens their bodies, dulls the spirits, abates their strength and courage; and good scholars are never good soldiers." Blending conventional medical and anecdotal lore, Burton's own account of the causes of scholarly melancholy blamed the fact that sufferers "live a sedentary, solitary life, *sibi & musis,* free from bodily exercise, and these ordinary disports which other men use: and many times if discontent and idlenesse concurre with it, which is too frequent, They are precipitated into this gulf on a sudden." Going into more detail than Ficino, Burton enrolled a battery of authors to prove this and he went on to mention that all these writers added that

> hard students are commonly troubled with gouts, catarrhes, rhumes, cacexia, bradiopepsia, bad eyes, stone and collick, crudities, oppilations, vertigo, windes, crampes, consumptions, and all such diseases as come by overmuch sitting; they are most part leane, dry, ill coloured, spend their fortunes, loose their wits, and many times their lives, and all through immoderate paines, and extraordinary studies.

Unable or unwilling to enter into a *vita activa,* scholars led an unhealthy, eremitic existence, with no concern for the mundane cares of daily life. Such a man was Archimedes, held by literary authorities to be so intent on his studies that he "never perceaved what was done about him." [9]

He recommended chess for idle melancholics, but if the melancholy came from too much study, "it may doe more harme than good; it is a game too troublesome for some men's braines, too full of anxiety, all out as bad as study, besides, it is a testy, cholericke game, and very offensive to him that looseth the

8. Bright, *Treatise of Melancholie,* 31, 243–45; see also Babb, *The Elizabethan Malady,* 21–26, 96–100; Burton, *Anatomy of Melancholy;* MacDonald, *Mystical Bedlam;* Rather, "Old and New Views"; Jobe, "Medical Theories of Melancholia"; for a good treatment of Burton's text, see Babb, *Sanity in Bedlam.*

9. Burton, *Anatomy of Melancholy,* 1:302–4, 306; Ramazzini (*Diseases of Workers,* 379) later pointed out that standing was also an occupational hazard for academics.

Mate." It was suitable for courtiers, idle gentlemen, and Muscovites (who lived where it was "much used"), but "not altogether so convenient for such as are Students. The like I may say of *Cl. Bruxers* Philosophy game, *Dr Fulkes Metromachia*, and his *Ouranomachia,* with the rest of those intricate Astrologicall and Geometricall fictions, for such especially as are mathematically given; and the rest of those curious games." Scholars and those who were "*fracti animis,* troubled in minde," needed to be refreshed, and for this he recommended "*Dancing, Singing, Mumming, Stage-plaies . . .* if opportunely & soberly used"; there was "no better Physicke for a melancholy man then change of ayre and variety of laces, to travell abroad and see fashions." For others he commended study itself but for the student such a remedy "addes fuell to the fire, and nothing can be more pernicious; let him take heed he doe not overstretch his wits, and make a *Skeleton* of himself." For full health, both body and mind must be exercised, though in the case of students, they "(as Plutarch observes) have no care of the body."[10]

Another physician, Richard Napier, left a detailed account of his methods for curing scholars whose brains overheated either because they were too idle or because they had studied too much or too hard. All such individuals were suffering from a form of melancholy, a condition that affected one-fifth of all cases Napier saw, and if it were not treated immediately, it was thought that this condition in scholars led to madness. For melancholy, commonly conceived as being caused by fumes from the spleen that traveled up to becloud the brain, Napier (who also drew from the *Anatomy of Melancholy*) mixed traditional herbal remedies, medicines recently available from the New World, and metallic Paracelsian concoctions. According to Michael MacDonald, he became "an avid medical alchemist" and used his skill to prepare a number of what he himself called "chemical medicines" to be employed as laxatives and vomits to relieve the pressure in the head. To cure melancholy, he used a classical potion containing aloes, colocynth, and black hellebore—all laxatives—and a powerful Arabic purge. For example, purges, vomits, and bleeding were used on the luckless Daniel Georg, who was described in Napier's records as "very mopish and lightheaded" on the grounds that he was "crazed with studying." On top of this, he also recommended the conventional nonmedicinal cures for melancholy and overstudying, such as bright colored dressing, jovial company and surroundings, and the avoidance of solitude and of abstruse and gloomy contemplations.[11]

10. Burton, *Anatomy of Melancholy,* 2:81, 82, 64, 84–90, 95. William Fulke's Ουρανομαχια was published in 1571, his Μετρομαχια in 1578.

11. MacDonald, *Mystical Bedlam,* 117, 152–54, 186–88, 128. Napier recorded twenty-seven men or women suffering from too much study; ibid., 186. For a detailed and comprehensive account of all the cures available to the physician in the early seventeenth century, see Burton, *Anatomy of Melancholy,* vol. 2.

A detailed example of what happened when the scholar's health broke down in connection with his work regimen is visible in the description by his brother Roger of the ill-fated John North, master of Trinity College, Cambridge, between 1677 and 1683. Roger recorded that John was badly served from the very beginning by "a non-natural gravity, which in youth is seldom a good sign," a view that was supported by the "grave disposition" exhibited in a very early portrait by Thomas Blemwell (fig. 4.2). Roger described Blemwell as an accomplished artist who drew John "in his red cloak *alla naturalle,* just as he wore it," the Cavalier master of John's school dressing the boys in red because "scarlet was commonly called the king's colour." Roger appealed to this portrait, the only one John ever allowed to be painted, "for demonstration of what I have alleged concerning his grave disposition. The countenance is modest and composed, and copied from pure nature, wherein nothing is owing to the painter, for it was very like him." John was apparently "much observed for this amiable gravity, and after he grew up to man's estate, he retained a florid youthfulness in his countenance, of which more will be observed afterwards," although Roger later referred to what he called "the bizarre posture and habit expressed in that picture." Such a constitution presaged "an imbecility of body or mind, or both," but, in the case of John, "his lay wholly in the former, for his mental capacity was vigorous, as none more." His body inclined to the "effeminate" and he was always aware of this "weakness, and during the whole course of his life laboured to conquer it." Nor did he ever lose an opportunity to "improve himself," and

he kept himself bent with perpetual thinking and study, which manifestly impaired himself. Even conversation, which used to relieve others, was to him an incentive of thought. He was sensible of this, but did not affect any expedients of relief to his mind. I have heard him say that he believed if Sir Isaac Newton had not wrought with his hands in making experiments, he had killed himself with study; a man may so engage his mind as almost to forget he hath a body, which must be waited upon and served. The doctor could oversee in himself what in others plainly appeared to him.

Unfortunately, John North had no such manual exercise available to him and being "abstemious in extremity proved of ill consequence to his health."[12]

12. North, *General Preface and Life,* 98, 100–103, 108, 111. (The words "afterwards" and "the bizarre" together with a linking phrase are missing from this edition. I am grateful to Peter Millard for this information.) Ward, *Step to Stir-Birch-Fair,* 12, recorded that Cambridge was a place where Black and Purple Gowns were strolling about Town, like Parsons in a Country Metropolis, during the Bishop's visitation; Some looking with as meagre Countenances, as if in search of the *Philosophers-Stone,* they had study'd themselves into a Hypochondriack

John's body epitomized the hazardous constitution of a scholar, and Roger noted that his "temperature of body and austere course of life were ill matched, and his complexion agreed with neither." His "face was always tincted with a fresh colour, and his looks vegete and sanguine, and as some used to jest, his features were scandalous, as showing rather a *madame en travestie* than a bookworm." He suffered from the classic complaints of sedentary scholars, such as rheums and gravel, and he used to inspect his urine every morning to see whether he had voided red gravel, which symptom was supposed to be evidence that he was not suffering from the stone. Nevertheless, his main problem was work discipline:

> It is certain that he was overmuch addicted to thinking, or else he performed it with more labour and intenseness than other men ordinarily do, for in the end it will appear that he was a martyr of study. He scarce ever allowed himself any vacation; what he had was forced upon him [and] was the most intense and passionate thinker that ever lived and was sane.

In his struggle to curb his own excesses, "he performed so well that strangers seldom or never perceived his disorders," but to a man with this temperament, the appointment to the mastership of Trinity was deleterious in the extreme. No more could he engage in the "frequent, easy, fine and pleasant conversation" that had so far staved off a worse fate, but he was now thrust into "an anxious, solitary and pensive course of life, which, with his austere way of ordering himself, drew upon him a most deplorable sickness and that proved the decadence of all his powers of body and mind." [13]

John had none of the physical helps to indefatigable study that had been the possession of Isaac Barrow, the previous master. North was "frail and infirm and of a nature that needed recruits, and, to reinstate its forces, some measure of indulgence." He dined with other fellows but did not imbibe their quantities of wine; Roger reckoned that had he supped even a little alcoholic juice "it would have produced freedoms and dispersed those cloudy formalities which fell between a superiour and inferiours, unless the nerve is cut with the glass and humour hath a free play." The position of a master in a Cambridge college was utterly unsuitable for someone like this: "all are upon their guard where he is, and very few if any were thought capable of a true and familiar friendship, that is,

---

Melancholly; other's seemingly so profoundly thoughtful, as if in pursuance of *Agrippa's* Notions they were studying how to raise Sparagrass from Rams Horns.

For Locke's view that "melancholick persons" were subject to "affections of ye spleene to Hypochondriacall affections particularly quartan agues. &c," see Romanell, *John Locke*, 102–3.

13. North, *General Preface*, 139, 141–42, 144–45.

FIGURE 4.2. John North as a boy, by Thomas Blemwell, early 1660s. Courtesy Dr. Tom North.

clear of all design or project." Nevertheless, despite the unwelcome distractions of new fellows imposed by letters mandate, John kept up his rigorous work discipline and "nothing but a sense of duty could have made him so swerve from the interests of his health." Many friends told him to let himself go,

> to indulge a little, go abroad and be free with a glass of wine with good company in his college as he used to be free with them, that his self-denials would endanger his health, and the like. To which sort of discourses I have heard him return a tradition of Bishop Wren, who when he was told he must not keep Lent, his body would not bear it, 'Will it not?' said he, 'then it is no body for me.'[14]

Inevitably, John North's body could not take the strain, and a stroke was brought on by the necessity of having to discipline some loutish Trinity undergraduates. In his last years, enfeebled by his condition, he would recruit his spirits by drinking glass after glass of the strongest sherry, and surprisingly he recovered most of his previous power of thought. After his premature death in 1683, his brother was left to wonder whether a wife might not have preserved him longer.[15]

■ *Working Lives* ■

INDUSTRY AS THE END of the scholar's calling was stressed in the sermons of Isaac Barrow, Newton's predecessor as Lucasian Professor of Mathematics. Newton attended Barrow's lectures on optics and mathematics and probably heard Barrow's sermons on the spiritual value of hard work. In these, Barrow advised that truth lay "lodged deep in the bowels of things, and under a knotty complication of various matters; so that we must dig to come at it, and labour in unfolding it." Idleness was the "nursery of sins" and the "general trap, whereby every tempter assayeth to catch our soul . . . in places where there is least work, the worst sins do prevail." Keeping "our hearts from vain thoughts and evil desires," Barrow asked, what "vigilancy of mind, what intention of spirit, what force of resolution, what command and care over ouselves doth it require? . . ." Of the idle drone he posed the question: "what is he but an unnatural excrescence, sucking nutriment" from the common weal, "without yielding ornament or use?" The gentleman-scholar was particularly prone to this danger, and without industry,

> his time will lie upon his hands, as a pestering incumbrance. His mind will be infested with various distractions and distempers; vain and sad thoughts,

14. Ibid., 145, 150.
15. Ibid., 151–59.

foul lusts, and unquiet passions will spring up therein, as weeds in a neglected soil. His body will languish and become destitute of health, of vigour, of activity, for want of due exercise.

These "angelical operations of the soul" would allow Barrow's boys to escape
from the swinish commonfolk, from their "erroneous conceits" and from their
"perverse affections." Without the work of the soul, the gentleman-scholar "cannot continue like himself"; with it, he was elevated above "those brutish men
(beasts of the people) who blindly follow their motions of their sensual appetite,
or the suggestions of their fancy, or their mistaken prejudices." [16]

The Trinity at which Newton arrived in the summer of 1661 was in the grip
of a return to prewar discipline, and the nonconformist John Ray (who vacated
his Trinity Fellowship in 1662) told a correspondent in September 1660 that
High Churchmen "have brought all things here as they were in 1641: viz; services morning and evening, surplice, Sundays and holydayes, and their eves, organs, bowing, going bare, fasting nights." A series of "Rules to be Observed by
Young Pupils and Schollers in the University" was penned by the Trinity fellow
James Duport at about the time of Newton's arrival, and they offer an excellent
indication of the sets of regulations facing the young men entering Restoration
Trinity College for the first time. Although warned to "be more carefull to
trimme your soules then bodyes," the scholar was exhorted to be mindful of the
correct college dress code, to behave properly, and to "omit no acts nor exercise,
that either the statutes require, or your tutor appoints you . . . be diligent at lectures, acts, and exercises, both in the university and the college." If in trouble, he
was told "go to your tutor as to your oracle upon all occasions, for advice, and
directions, as also for resolution of any scruple, or doubt, or difficulty in religion
or learning." He was also told to "take heed of picking your nose, or putting your
hat or hand to your face, or any such odd, uncouth, or unseemly gesture." [17]

Whatever his conduct as a young man (and there is evidence that he spent
some time on the college bowling green), Newton told his future biographers
that he had been an autodidact who as a very young man had proceeded by almost plodding hard work to his mastery over mathematical and scientific texts.
He had discovered the law of universal gravitation "by thinking upon it continually." One set of narratives about this early life linked his extraordinary labors to
a fit and powerful body. Conduitt recorded how Newton had physically worsted
the school bully by smashing his face against the side of the church and his cunning use of the wind to outjump his schoolmates in a long-jump contest, while,

16. Barrow, *Theological Works*, 3:360, 380, 363, 413, 428–29, 427, 437–38.
17. Ray to Courthorpe, 26 September 1660, in Ray, *Further Correspondence*, 18; Trinity College
Cambridge MSS, O.10A.33, cited in Curtis, *Oxford and Cambridge*, 114–15.

after a wealth of stories concerning Newton's mechanical acumen, Conduitt concluded that "Sr Isaac had the mechanicks hands as well as the head of a Philosopher." He noted that on entering Trinity his hero "went at once upon Descartes *Geometry* & made himself master of it by dint of genius and application without going thrô the usual steps or the assistance of any other person," and these tales of his work habits were woven back into anecdotes about Newton's surprising physical strength and his manual dexterity:

> Happy was it then Sr I. thus exercised at once his body & his mind & hands as well as his thoughts—As the operations of the soul depends upon the condition of the organs of the body, & they are lively & vigorous or weak & faint according to the condition of those mechanical organs the machines by wch they Act, the Philosophical productions of the mind are in a great & advanced measure supported by ready operations of the hands, & Sr I. had never carried improvements of his intellectual discoveries so far if by an early habit & constant exercises he had not acquired a manual dexterity to invent & perform those experiments his invention & sagacity contrived.

Newton admitted to Conduitt that, although he had made his own tools, "the only help I had said he in those operations was my next chamber fellow who was stronger than I and used to help me on with my kettle." John Wickins himself "had several furnaces in his own chambers for chymical experiments."[18]

His amanuensis of the *Principia* period, Humphrey Newton, gave a detailed account of his work discipline and linked it to the bodily presentations that became so widely known. Unlike the accounts of his physical prowess which Newton himself reported to Conduitt, the stories related by Humphrey and others concentrated on Newton's *neglect* of the corporeal self. The portrait coalesced of a being completely undeterred by the loss of food or sleep, and Wickins, his roommate in these early years, apparently told his son that Newton frequently worked throughout the night. Newton himself told Conduitt that he "sate up so often in the year 1664 to observe a comet that appeared then, that he found himself much disordered and learned from thence to go to bed betimes." Like Wickins (who had passed the same story on to his son), Humphrey recalled Newton's extraordinary eating habits: "So intent, so serious upon his Studies, yt he eat very sparingly, nay oftentimes he forgot to eat at all" and could not resist comment-

---

18. Westfall, *Never at Rest*, 105 n. 1; Keynes MSS, 130 (9); 129 (A) fols. 2v–3r; 137; 130 (10), 1–2. In 1682, Newton recommended Edward Paget for the office of mathematical lecturer at Christ's Hospital, confirming that Paget was skilled in numerous branches of mathematics and, "wch is ye surest character of a true Mathematicall Genius, learned these of his owne inclination, & by his owne industry without a teacher." He told Bentley in 1692 that "if I have done ye publick any service this way 'tis due to nothing but industry & a patient thought," a statement that was remembered by its recipient and passed on to John Conduitt in the 1720s; Newton, *Correspondence*, 2:375, 3:233.

ing on Newton's sleeping patterns: "I believe he grutch'd yt short Time he spent in eating & sleeping." He scarcely slept

> especially at Spring & Fall of ye Leaf, at wch times he used to employ about 6 weeks in his Elaboratory, the Fire scarcely going out either Night or Day, he siting up one night, as I did another, till he had finished his Chymical experiments, in ye Performances of wch he was ye most accurate, strict, exact: what his Aim might be, I was not able to penetrate into, but his Pains, his Diligence at those sett Times, made me think, he aim'd at something beyond ye reach of humane Art & Industry.

Humphrey also explained Newton's absences from Church by the fact that he studied so hard "yt He scarcely knew ye Hour of Prayer. . . . Thinking all Hours lost, yt was not spent in his studyes, to wch he kept so close, yt he seldom left his Chamber, unless at Term Time," a sentiment that, as we have seen, was shared by John North. Roger North noted that Newton was "famous in the college for his insensate behaviour in the college hall." He "would stand at the hearth, and warm his hands at midsummer; and if one changed his meal for a dry bone, he would be satisfied, thinking he had dined, and the like." Humphrey and William Stukeley repeated almost identical anecdotes. As Shapin suggests in chapter 1 of this book, these have been remarkably stable commonplaces about scholars for over two millennia.[19]

By the time of Newton's death, such stories were widely circulated, and Nicholas Wickins reported that these narratives were "what ye world has so often heard of Sr Isaac." Westfall has pointed out that Stukeley's tales sound remarkably like those of Humphrey, and it is known that they conferred when both resided in Grantham in 1727–28. Clearly, the most significant source of these stories was Humphrey, since apart from John Wickins, he was the man closest to Newton in his private life. His portrait of Newton, absent-mindedly making his way around Trinity "wth Shooes down at Heels, Stockins unty'd, surplice on, & his Head scarcely comb'd," was classical. What he actually knew of his master's work is uncertain, but one of Wickins's reminiscences (reported by Stukeley) echoes well-known classic topoi. Thus Wickins said that when Newton walked in the fellows' garden, "if some new gravel happen'd to be laid on the walks, it was sure to be drawn over and over with a bit of stick, in Sir Isaac's diagrams;

---

19. Keynes MSS, 130 (4) (notes by Conduitt, mid-1720s); ibid., 135; B.L. Add. MSS, 32,516, fols. 18r–v. Ramazzini (*Diseases of Workers*, 393) noted that mathematicians had to keep their mind "detached from the senses and have hardly any dealings with the body; hence they are nearly all dull, listless, lethargic, and never quite at home in the ordinary affairs of men." The entire body of professors of mathematics was especially prone to decay: "For while the light given by the spirits is engaged in the innermost cavities of the brain, whatever is outside must be encompassed by darkness and languish accordingly."

which the Fellows would cautiously spare by walking beside them." Compare this with Plutarch's account of Archimedes, who

> possessed so lofty a spirit, so profound a soul, and such a wealth of philo-sophical enquiry [that] he had acquired through his inventions a name and reputation for divine rather than human intelligence . . . some attribute this to the natural endowments of the man, others think that it was the result of exceeding labour that everything done by him appeared to have been done without labour and with ease . . . there is no need to disbelieve the stories told about him—how, continually bewitched by some familiar siren dwell-ing with him, he forgot his food and neglected the care of his body; and how, when he was dragged by force, as often happened, to the place for bathing and anointing, he would draw geometrical figures in the hearths, and draw lines with his finger in the oil with which his body was anointed.

Yet Newton did *not* neglect his body. While one narrative of his person stressed his disembodied scholarly bearing, another, as we have seen, emphasized just how well fitted and disciplined was his physique for engaging in manual exercise. Such attention to his corporeal self stretched to the production of detailed me-dicinal preparations.[20]

■ *Self-Composition* ■

MANUEL NOTES OF NEWTON that his longevity was "an index of his stam-ina, and there were no protracted physical illnesses that we know of, even though he was somewhat hypochondriac." Despite this, there is no reason to think that Newton's behavior differed very much from that exhibited by other scholars of the period. As a young man, he kept a notebook in which he entered under the heading "Of Diseases" a number of recipes taken from Francis Gregory's *Nomen-clatura Brevis* of 1654, and Manuel is correct to stress the importance of Newton's immersion in the culture of medicinal recipes for his later chemical and alchemi-cal work. This was recognized even during Newton's own lifetime, and (possibly on the basis of Newton's own testimony) Stukeley recorded that "Children are always imitators, and perhaps his being brought up in an apothecary's house might give him a turn towards the study of nature: and indoubtedly it gave him a love for herbarizing." It was at the abode of a Mr. Clark (a Grantham apothe-cary) that Newton constructed the mechanical contraptions for which he was lo-cally famous, and it was in the notebook that he had begun using just after he left the apothecary's shop in late 1659 that he wrote the first surviving "letter" of his

20. Westfall, *Never at Rest*, 191, 361; Keynes MSS, 135; Stukeley, *Memoirs*, 61; Plutarch, *Marcel-lus*, xvii, 4–6, cited in Thomas, *Selections*, 2:31–32. Humphrey (Keynes MSS, 135) recorded that Newton "was very curious in his garden, which was never out of order, in which he would at some seldom times take a short walk or two, not enduring to see a weed in it."

correspondence. Although this was transliterated into a universal language with which he was then experimenting, it smacks of his confident self-righteousness and may well have been sent as an actual letter to a "Loving friend":

> It is commonly reported yt you are sick. Truly I am sorry for yt. But I am much more sorry yt you got your sicknesse (for yt they say too) by drinking too much. I ernestly desire you first to repent of your having beene drunk & yn to seek to recover your health. And if it please God yt you ever bee well againe yn have a care to live healthfully & soberly for time to come.[21]

Such concern for self and others was carried into his life at Trinity and involved the use of potions and concoctions which we know he applied to himself. A number of his homemade recipes are extant, testimony to the care with which he looked after his own health while in college. One of these, "a Medicine to cleare the eye-sight," referred to lumps of "yellow matter" in Hungary that were found on the sides of mines and were harvested in the middle of summer. Another, "Lucatello's Balsam," was of particular significance. This was a universal medicine capable of being taken both internally and externally, and the list of illnesses that it purported to treat included measles, plague, and small pox, for which he recommended "a ¼ of an ounce in a little broth take it warme & sweate after it." It also cured gangrene, burns, and bruises, although for rabies and poisoning it had to be taken internally as well as externally. Readers can try the recipes themselves (see appendix). John Evelyn used Lucatello's Balsam when his two-year-old son Richard choked on a bone. While the boy was struggling, a maid fainted and, there being "no chirurgeon near," Evelyn held the child's "head down [and] incite[d] it to vomit. It pleased God, that on the sudden effort & as it were struggling his last for life, he cast forth bone." Once the drama was over, Evelyn "gave the child some Lucatellus Balsame, for his throat was much excoriated."[22]

Newton gulped down a substantial quantity of the balsam when he believed he had consumption and seemed to take it at regular intervals. It is significant that this was one of the facts that Nicholas Wickins remembered his father telling

21. Manuel, *Portrait of Isaac Newton*, 4, 69–70; Stukeley, *Memoirs*, 54–55; Newton, *Correspondence*, 1:1. For Newton's stay in Grantham, see Westfall, *Never at Rest*, 56–64. Clark was stepfather to the two Storer boys, one of whom Newton recorded striking in another early notebook; they were the nephews of Humphrey Babington, fellow of Trinity and the most likely patron of Newton in his early years at Cambridge.

22. Evelyn, *Diary*, 355–56. See Beier, *Sufferers and Healers*, esp. 4–5, 10–15, 139–50, 167–73, 200–201. The Royal College of Physicians published the *London Pharmacopoeia* from 1618 onward, although there were other commonly available texts such as *The Englishman's Doctor. Or the School of Salerne* (London 1607), and Levins, *A Right Profitable Book of All Diseases*; for these and others, see O'Hara-May, *Elizabethan Dyetary*; Porter, *Popularization of Medicine*; and Slack, "Mirrors of Health"; given the extent of illiteracy, however, texts were unlikely to have been the most important means of disseminating such practical information.

him; John Wickins apparently said of Newton that "when he had compos'd Himself, He would now and then melt in Quantity abt a Qr of a Pint & so drink it." That he (and his son) recalled this habit at all testifies to the role the balsam played in Newton's private life, and this ostensibly peculiar reminiscence about Newton's Trinity routine is evidence of the careful attention Newton paid to himself, in contrast to other stories of his bodily and sartorial abuse.[23]

Perhaps the most remarkable aspect of Newton's medicinal regime was the extent to which his advice was valued by others. He remained in contact with Clark the apothecary up to the end of the 1680s and, in December 1691, Humphrey Newton (now returned to Grantham) reminded him that when he was working for Newton, Clark had visited Trinity and left some pills whose recipe could no longer be found. Humphrey asked him to "impart" to him the mixture, "If You have a knowledge of ye Receipt, or of the chiefest of ye Ingredients, wch, if I misremember not, You told me was a Gum." Later, Humphrey recalled that Newton was "only once disorder'd with Paines at ye Stomach, which confin'd Him for some dayes to his Bed"; these he "bore wth a great deal of Patience & Magnanimity, seemingly indifferent either to live or dye." Newton took pity, and "seeing me much concern'd at his Illness, bid me not trouble my self, for if said he, I dye, I shall leave you an Estate, wch he then mention'd."[24]

In 1695, Newton prescribed a plaster to be applied to the chest for the wife of his step-brother Benjamin Smith, who replied that

> the fomentation [was] applyed as soone as wee could possible; The effects have pved very successfull; for the swelling is verry much abated, and the blacknes quite gone. Although at certaine times shee hath still a paine in her brest; after every bathing, shee put the same plaster to her Brest againe, you sent, I could not perswade her to take the Sowes, for since her being wth child almost every thing goes against her Stomach; but shee is resolved to try.

His skills were likewise to the fore in what was one of the traumatic events of his life, the death of his mother. She had herself been tending Benjamin in early 1679 when she caught a fever. Newton left Cambridge and immediately mixed up medicines and applied poultices to her body with his own hands. Conduitt recorded that he

> attended her with a true filial piety, sate up whole nights with her, gave her all her Physick himself, dressed all her blisters with his own hands, and made use of that manual dexterity for wch he was so remarkable to lessen

23. Keynes MSS, 137; Newton, *Correspondence*, 7:368.
24. Newton, *Correspondence*, 3:278–79; Humphrey to Conduitt, 17 January 1727/28 (Keynes MSS, 135).

the pain wch always attends the dressing, the torturing remedy usually applied in that distemper with as much readiness as he ever had employed it in the most delightfull experiments.

Manuel points out that she left her body to be buried as Isaac "shall think fit," and he wrapped it up in white wool. Such care also extended to his onetime protégé, the young Swiss mathematician Fatio de Duillier, in the early 1690s, and Newton perhaps lessened his attention to his young friend only when he himself desperately needed help.[25]

■ *Being Embroiled* ■

NEWTON'S SO-CALLED breakdown of 1693 has notoriously attracted comment since Biot drew attention to it in the early nineteenth century. Responding to what he took to be the Frenchman's impertinence, the Scottish scientist and biographer of Newton David Brewster "felt it to be a sacred duty to the memory of that great man . . . to enquire into the nature and history of that indisposition which seems to have been so much misrepresented and misapplied." Nevertheless, the crisis of the early to mid-1670s paralleled the later problems in a number of ways. As Newton's work gradually appeared in the *Philosophical Transactions* in early 1672, it was attacked on all sides by individuals such as Robert Hooke and Franciscus Linus. As early as May 1672 he indicated to John Collins that such publicity had caused him to lose his "serene former liberty," and he repeatedly indicated to people like Robert Hooke and Henry Oldenburg that he wished to withdraw from the relatively public life to which he had just been exposed. By November 1676 the dispute with the Jesuits had taken its toll, and he told Collins, "I wish I could retract what has been done, but by that, I have learn't what's to my convenience, wch is to let what I write ly by till I am out of ye way." Similarly, he told Oldenburg ten days later that he had enslaved himself to philosophy, and if he could get clear of "Mr. Linus's business I will resolutely bid adew to it eternally, excepting what I do for my privat satisfaction or leave to come out after me." Newton was strategically threatening the death of the philosophical author if not of himself.[26]

On that occasion, Newton's health was spared by his almost total withdrawal

25. Newton, *Correspondence,* 7:187; Keynes MSS, 130 (2), 12, and 130 (8); Manuel, *Portrait of Isaac Newton,* 27. "Sowes" refers to sowthistles, and is almost certainly not a mistake for "sowens" (a kind of porridge), as argued in Westfall, *Never at Rest,* 503 n. 112. Culpeper said they were "cooling and somewhat binding and are very fit to cool a hot stomach and ease the pain thereof"; they were also recommended to nursing mothers. Cited in Grieve, *Modern Herbal,* 756–57.

26. Biot, *Life of Sir Isaac Newton;* Brewster, *Memoirs of Sir Isaac Newton,* 2:133; Newton, *Correspondence,* 1:161, 172, 262, 345; 2:179, 182–83. The best modern account of the "black year" is Manuel, *Portrait of Isaac Newton,* 213–25; but see also Westfall, *Never at Rest,* 531–41, Spargo and Pounds, "Newton's 'Derangement of the Intellect,'" and Keynes, "The Personality."

into the solitude of Trinity, where he remained for over a decade. By the spring of 1687, his public defense of the Protestant integrity of Sidney Sussex College, Cambridge, and the publication of the *Principia Mathematica* made him so famous that there could never again be a retreat into complete solitude. After the Glorious Revolution, which took place in December 1688–January 1689, his public opposition to the Catholic James II endeared him to the new regime of William of Orange and the question of preferment forced him to seek out the patronage of the politically powerful, such as William Sacheverell, Samuel Pepys, John Hampden, Charles Montague, and John Locke. Yet although he was elected to the position of M.P. for the university in the Convention Parliament of January 1688/89, this was not enough to gain him the provostship at King's College, Cambridge, for which he, Hampden, Fatio, and Christiaan Huygens lobbied the king on 10 July 1689. In February 1689/90, Fatio told him that Hampden had lost favor with William but that he himself had tried to get Locke to speak on Newton's behalf to Lord Monmouth, a man who had the ear of the king.[27]

This obviously had some effect, since on 28 October 1690 Newton told Locke, "I am extremely much obliged to my Ld & Lady Monmoth for their kind remembrance of me & whether their designe succeed or not must ever think my self obliged to be their humble servant," and two weeks later wrote that "their favour is such that I can never sufficiently acknowledge it." Westfall not unreasonably links this to a potential position at the mint that was mentioned in a letter from a legal expert called Henry Starkey as early as May 1690. Newton's prospects soon diminished. After he had lost his parliamentary seat he made at least two journeys to London in the autumn of 1691 seeking a public position there, but he rejected the possibility of the mastership of Charterhouse in December, complaining to Locke that "the confinement to ye London air & a formal way of life is what I am not fond of."[28]

On the last day of 1691, just at the point when David Gregory reported that Newton was about to publish a treatise on fluxions, Robert Boyle died. This drew him back to London for the funeral; within a week he was to lose perhaps his closest colleague at Trinity, Humphrey Babington. On 9 January 1691/92 (five days after Babington's death) he was engaged in a discussion with John Evelyn, Pepys, and Thomas Gale on the topic of the forthcoming Boyle lectures. On 26 January, he told Locke that "being fully convinced that Mr Montacue upon an old grudge wch I thought had been worn out is false to me, I have done wth him & intend to sit still unless my Ld Monmouth be still me friend." A footnote

27. Westfall, *Never at Rest*, 480–83, 496–500, and Newton *Correspondence*, 3:390–91, for Fatio's letter to Newton of 24 February 1689/90.
28. Newton, *Correspondence*, 3:79, 82, 184; Westfall, *Never at Rest*, 497.

about Locke's possible possession of Boyle's "process about ye red earth & [mercury]" indicated one of two other areas that were then troubling Newton; the other, perhaps more important, concerned his anti-Trinitarian discourse on "Two Notable Corruptions in Scripture," which he had sent to Locke for publication but now wished to suppress. This was confirmed in another letter of 16 February, in which he told Locke that he was glad Monmouth was still his "friend" but intended "not to give his Lordship & you any further trouble. My inclination is to sit still." Montague, who had attempted to found a philosophical society at Cambridge with Newton in 1685, and whom Newton had described as his "intimate friend" in a letter to Edmond Halley of February 1686/87, received no mention.[29]

In his letter to Locke, Newton apologized for "pressing" into the company of Monmouth when they had last met, but he claimed he would "not have done it but that Mr Paulin prest me into the room." It was Robert Pawling who had told Newton of a request that Locke had made for Boyle's red earth, while a letter of 16 January from Pawling to Locke sheds some light on the situation between Newton, Monmouth, and Montague. Pawling related how he had on that day come across Mr. Boyle's "man," who promised Locke the earth if he could distinguish it "amongst so many sorts as they have." He went on to describe his two meetings with Newton, which were evidence to the bewildered Pawling that Newton was in a sorry state:

> he is in Suffolk street up two pairs of stairs in a pittifull room. he would have been glad to have seen you. he enquired of me concerning somewhat whereof I could give no account. It seems somewhat was said to him by a great man when he supped at my house, he was puzzled to know whether it was a reflection on him or designed for good advice. you seemed to take notice of it afterwards, but how you apprehended it he knew not. I blamed him that he was not so free with you as to know your mind more fully, in that he was well assured of your friendship to him. whatever twas I assured him you never spake to me of it. he intends to consult you about it when he sees you. I understand by him that some great man in whom he hath put great confidence, he is fully satisfied, hath been false to him. I enquired not into it, so am in the dark. it may be you may pick something out of this.

From the subsequent letters from Newton to Locke it seems that the first "great man" was Monmouth, while the second was Montague. This incident must have made Newton's behavior of 1693 more comprehensible to Locke, though back at their first meeting in 1692, the Oxford-trained Pawling was distressed by the

---

29. Newton, *Correspondence,* 3:192–93, 195; 2:464. Montague, the grandson of the first earl of Manchester, was elevated to the title of earl of Halifax in 1700.

professor's sad condition: "I could not but pitty the Gentleman when he brake the business to me at first a bashfull young man could not be more concernd the first time he acquaints his Mistress of his affection to her, etc. than he was." Linking Newton's behavior to the conventional bearing of the philosopher, he concluded: "tis a sad thing to be a meer, though a great, scholar." [30]

Once Newton learned of the suppression of his theological tract, his mood lightened, and in May 1692 he even asked Locke to stay with him in Cambridge. Intellectual matters now ranged over almost the totality of his interests, with theology and alchemy—in particular Boyle's mid-1670s recipe for the incalescence of gold with mercury—being the chief subjects of his correspondence with Locke. However, there was precious little other communication with the outside world aside from his exchanges with Fatio over the latter's health and his recommendations to Richard Bentley for the text of his forthcoming Boyle lectures. In May 1693, Fatio indicated that Locke wished them all to stay at the Masham estate at Oates in Essex, and, for whatever reason, Newton vacated Trinity on the thirtieth of that month, only days after the death of his close friend John Spencer. He continued to record alchemical experiments in his notebook up to June 1693, when he again visited London, almost certainly in connection with the medical-alchemical interests of Fatio. There were no other contacts of which we are aware until the notorious series of letters of autumn. [31]

Newton's letter to Pepys of 13 September 1693, written while its author was in a state of total discomposure, broke the silence. As he had done with Pawling nearly two years earlier, Newton accused Pepys's friend John Millington of trying to "press" him into seeing Pepys the next time he was in London. Newton was made to agree to this, but only "before I considered what I did, for I am extremely troubled at the embroilment I am in, and have neither ate nor slept well this twelve month, nor have my former consistency of mind." "Pressing" was the

30. Ibid., 3:195; Locke, *Correspondence*, 4:364–66.

31. Newton, *Correspondence*, 3:214, 216–19. Newton drafted a letter to Otto Mencke on 30 May; cf. ibid., 3:270–71. Westfall, *Never at Rest*, 533, points out that Newton referred to Fatio's 4 May letter in his alchemical manuscript "Praxis"; Spargo and Pounds, "Newton's 'Derangement of the Intellect,'" 21, shows that Newton was well aware of the dangers of the steams of sulfur, antimony, and arsenic, which were "able to make those stagger or perhaps strike ym down that wthout a competent wariness unlute ye vessels wherein they have been distilled or sublimed; of wch there have been divers sad examples": CUL Add. MSS, 3975, 98. For other examples of Newton's connections between alchemy and medicine, see Keynes MSS 67, 58 fol. 8v (for a "febrifuge"). See in particular the marginal notation to his 1669 copy of *Secrets Reveal'd*, which speaks of a process involving mercury and gold that gives rise to a "sweet white oyle wch is good for mitigating the pain of wounds, & in wch Phers have placed their great secrets for medicine: especially if it be made wth the virgins milk of the ⊙ [i.e., gold] ffor then it is a red fragrant . . . sweet oyle or balsam permiscible in all things, the highest medicine"; cf. Duveen, *Bibliotheca Alchemia*, plate XII, a facsimile of Newton's notes in *Secrets Reveal'd*, 90–91.

term Newton had used about his Jesuit enemies in the mid-1670s, in letters of 28 November 1676 and 5 March 1677/78, and a final letter on the affair to John Aubrey in mid-1678. If the tensions involved in a solitary scholar seeking public office were becoming too much to bear, then Samuel Pepys, implicated in the nonexistent Popish Plot over a decade earlier, may have epitomized for Newton the moral and religious enemies that were now returning with vengeance to torment him. Newton told him that he had "never designed to get anything by your interest, nor by King James' favour, but am now sensible that I must withdraw from your acquaintance."[32]

In his letter to Locke of three days later, scribbled "At the Bull in Shoreditch," he apologized for thinking that Locke had "endeavoured to embroil me wth weomen & by other means" as well as for saying that it would be better if Locke had died when Newton was told Locke was sickly. He craved forgiveness for saying that he had "struck at ye root of morality in a principle you laid down in your book of Ideas & designed to pursue in another book," and for taking Locke "for a Hobbist." He concluded his self-destructive damage limitation exercise by apologizing for "saying or thinking that there was a designe to sell me an office, or to embroil me," and signed it "your most humble & most unfortunate servant." Ostensibly directed against his friends, there was nothing that Newton said in these letters, about women, the selling of offices, and rank atheism, that could have been more damaging to *Newton,* and whatever the proximate cause of their appearance, the demons of women and Catholicism were Newtonian and had haunted him for some time.[33]

After these letters, some major work had to be done to rebuild Newton's sanity and reputation. Pepys told Millington at the end of September that he feared Newton's behavior "should arise from that which of all mankind I should least dread from him and most lament for,—I mean a discomposure in head, or mind, or both," and Millington took it upon himself to see Newton in Huntingdon, whereupon an unprompted Newton admitted that he had written Pepys a "very odd letter, at which he was much concerned." Millington linked his problem and the strange letter to Newton's inability to find a post in London, telling Pepys

> that it was in a distemper that much seized his head, and that kept him
> awake for above five nights together, which upon occasion he desired I

32. Newton, *Correspondence,* 3:279; 2:184, 260, 262–63, 266. The Buttery Book suggests that he left Trinity sometime in the middle of the week between 8 and 15 September 1693, probably on 10 or 11, and returned not long after 22 September; cf. Edleston, *Correspondence,* lxxxv, xc.

33. Newton, *Correspondence,* 3:280. For Locke's fear of being labeled a Hobbist, see Dunn, *Political Thought of John Locke,* 81–82, esp. 82 n. 4.

would represent to you, and beg your pardon, he being very much ashamed he should be so rude to a person from whom he hath so great an honour. He is now very well, and, though I fear he is under some small degree of melancholy, yet I think there is no reason to suspect it hath at all touched his understanding, and I hope never will; and so I am sure all ought to wish that love learning or the honour of our nation, *which it is a sign how much it is looked after, when such a person as Mr. Newton lies so neglected by those in power.*

Locke also sought to convince Newton of his continuing friendship and that "I truly love & esteem you & yt I have still the same goodwill for you as if noe thing of this had happened." He offered to meet Newton anywhere and urged him to point out the places in his work "yt gave occasion to that censure that by explaining my self better I may avoid being mistaken by others, or unawares doing ye least prejudice to truth or virtue."[34]

Newton's own contribution to this refashioning process was the molding together of pieces of information from the past year to tell a plausible story. Some time earlier, he had drafted a letter—almost certainly to Fatio—in which he had explained a Latin passage of an alchemical recipe. In it he mentioned having seen a recent work by John Craige and told the prospective recipient that if his "friend should go into Flanders," or if anything happened so as to prevent him from working, he should spend the winter with Newton. He concluded with the comment that a fortnight earlier "I was taken ill of a distemper wch has been here very common, but am now pretty well again." Months later, in mid-September 1693, he told Pepys that he was extremely troubled by an "embroilment," and that he had "neither ate or slept well this twelve month." Three days after this, he twice mentioned being "embroiled" to Locke, but by mid-October, the narrative of his troubles made reference only to the disease. In a letter to Locke he claimed that the problem was purely physiological:

> The last winter by sleeping too often by my fire I got an ill habit of sleeping & a distemper wch this summer has been epidemical put me further out of order, so that when I wrote to you I had not slept an hour a night for a fortnight together & for 5 nights together not a wink. I remember I wrote to you but what I said of your book I remember not.

If Locke wished to send him a copy of the offending passage he would attempt to explain it. The story has some degree of consistency with other passages in his

34. Newton, *Correspondence*, 3:281, 283–84.

letters—such as the catching of a "distemper," the yearlong malaise, and the five-day loss of sleep—although it would clearly be presumptuous to offer a totally consistent etiology of Newton's condition.[35]

The day after writing the mid-October letter to Locke, he wrote another to Leibniz in which he remarked that he did his "best to avoid philosophical and mathematical correspondences," but had become afraid that their friendship "might be diminished by silence." This was, he wrote, because "I value friends more highly than mathematical discoveries . . . my aim in these pages has been to give proof that I am your most sincere friend and that I value your friendship very highly." In spite of this, the epistle did contain mathematics and represented a rapid return to intellectual work. As in a letter to Otto Mencke of 22 November, Newton ascribed his delay in replying to Leibniz to the fact that his letter had "got mislaid among my papers" and that he had not been able to "lay hands on it" until the middle of October. Nevertheless, the reference to "sincere friendship" in Newton's letter to Leibniz eerily and precisely echoed the language of Locke's to Newton of 5 October, as did his request to Leibniz to let him know if there was "anything that deserves censure" in his writing. Perhaps, Locke's letter had become the template for Newton's reconstruction of himself.[36]

News of Newton's troubles soon spread abroad. Three days after he had told his brother that Newton had been suffering from a kind of madness for a year and a half, Christiaan Huygens reported in his journal on 29 May 1694 that someone called Colm had told him that Newton had been overworking and had been the victim of a fire that had destroyed a great deal of his philosophical material. Newton had betrayed his "not being in his right mind" in a conversation with the archbishop of Canterbury: "At this his friends took him in hand; he was kept at home, and, whether he liked it or not, treatment applied whereby he has now recovered his sanity so far as now again to begin to understand his own book of the Mathematical Principles of Philosophy." To Leibniz, Huygens repeated his claim that Newton had been the victim of a "frenesie" for the past eighteen months and added that his friends had cured him "by means of remedies, and

35. Ibid., 7:367; 3:280. The editors of the Newton *Correspondence* (4:265) propose that the draft is that of the letter of 2 May 1693, to which Fatio referred two days later. The mention of a "friend" going abroad certainly links it to Fatio's potential medicinal partner (described in the letters of 4 and 18 May), although the reference to the forthcoming "winter" makes its placement in May problematic. Craige's work, *Tractatus Mathematicus de Figurarum Curvilinearum Quadraturis* came out in 1693, so the letter is unlikely to refer to the winter of 1692–93, although this would indeed be the case if the book appeared early in 1692/93 and the "distempers" described in this and the letter to Locke were one and the same.

36. Newton, *Correspondence*, 4:285–87; Manuel, *Portrait of Isaac Newton*, 223.

keeping him shut up." All this, he added, the English were desperately trying to cover up—a feat at which they succeeded for over a century.[37]

▪ *The Way to Chastity* ▪

IN 1692 AND 1693, Newton was unable to make the leap into the new social identity that was required of him in London. Just as in 1676–77, when he had strategically withdrawn from the scientific republic of letters, he threatened to utterly efface himself when confronted with the problems of public exposure. Nevertheless, in September 1693 he went all the way and almost literally extinguished himself in a dingy London pub. Surrounded by enemies like Montague, who had been false to him, Locke, the libertine and Hobbist, and Pepys, the crypto-papist and seller of offices, there was no escape. This was why, as he told Pepys, he had to withdraw from all his "friends." As I show in this section these men, now also present in the form of the early Christians of the third and fourth centuries about which he was writing in private, stormed the barricades which divided Newton from his Others. The only way to retreat was to disappear completely.

The "breakdown" of September was only a point in a long drawn-out process that began with his defense of Sidney Sussex and the publication of the *Principia*. Neither the mercury hypothesis nor Manuel's reference to his relationship with Fatio are sufficient to capture the extremely rich set of narratives that tormented him over a two-year period from the end of 1691. When Newton's well-being was assaulted and his health taken apart, his ascetic self was also assailed. Recall his lament that he had not eaten or slept properly for a year, and (yet again) his complaint about Locke's efforts to "embroil [him] with weomen." Newton's history of the corrupt bodily practices of the early church provided the link between these discourses of women and religion. In his secluded body work, he proposed a strict means of avoiding these twin corruptions and wrote of the despair and madness experienced when one succumbed to temptation or perversion.

In his graduate notebook of the mid-1660s (within which he developed his early ideas on light and colors), Newton wrote down a series of excerpts from authors like More, Descartes, Charleton, and Hobbes. He inserted various original optical and visual experiments of his own, and followed these self-experiments with a series of reflections on the vagaries of the imagination or "fancy." The imagination was held to be the typical source of creativity and disorder in the individual, and Burton joined others in assuming that the imagination threatened mental and physical health when it broke the bounds of reason. Melancholy men had overactive imaginations since it was the "medium deferens

37. Manuel, *Portrait of Isaac Newton*, 222–24.

[instrument] of passions, by whose means they work and many times produce prodigious effects . . . as the phantasy is more or less intended and remitted, and their humours disposed, so do perturbations move, more or less, and take deeper impression." [38]

In describing the power of the imagination Newton specified a set of guidelines for leading the life of a healthy scholar and for producing the optimal body for the experimental natural philosopher:

> We can fancie ye thing wee see in a right posture wth ye heeles upward. Phantasie is helped by good aire fasting moderate wine but spoiled by ~~not~~ drunkenesse, Gluttony, too much study, (whence & from extreame passion cometh madnesse) dizzinesse, commotions of ye spirits . . . Meditation heates ye ~~oun~~ braine in some to distraction in others to an akeing & dizzinesse. The boyling blood of youth puts ye spirits upon too much motion or else causet too many spirits, but could [i.e., old] age makes ye brain either two dry to move roundly through or else is defective of spirits yet their memory is bad . . . A man by heitning his fansy & immagination may bind anothers to thinke what hee thinks as in ye story of ye Oxford Scollar in Glanvill Van of Dogmatizing.

This was immediately followed by the report of an experiment in which he looked at the sun and tried to reproduce various images of it by shutting his eyes. Later he concluded after some even more terrifying experiments on his eyes that from such tests "may be gathered yt ye tenderest sight argues ye clearest fantasie of things visible & hence something of ye nature of madnesse." [39]

Consideration of the nature of dreams and madness was common currency among writers, and Protestant cultures had a rich understanding of the relationship between imagination and melancholy, sex and dreams. The imagination permanently tempted the godly man toward sin, while it was well known that idleness allowed the untrustworthy imagination to run riot in the undisciplined minds of the immoral and slothful rabble. Specifically, Protestants criticized Roman Catholicism for allowing priests to manipulate the bodies and imaginations of the vulgar in order to increase their power over them. The Jesuits compounded this felony since their founder Loyola believed that it was only by embracing and then grappling with temptation and sin in the imagination that one could gain the strength of will to avoid them and choose God. Protestants suspected that this was an excuse to indulge in immoral sexual practices; sensuality, carnal lust, and enslavement to the imagination, it was believed, were the bed-

38. McGuire and Tamny, *Certain Philosophical Questions*, 377; Burton, *Anatomy of Melancholy*, 1:257–58.

39. McGuire and Tamny, *Certain Philosophical Questions*, 394, 444.

rocks of a Catholic faith that was skilfully managed by monks and Jesuits. The Protestant had to avoid the lure of the fancy by avoiding sinful thoughts.[40]

In a remarkable manuscript that was probably composed in the early 1690s, Newton drew from his earlier experiments on the extent of the fancy's dominion and related in detail the sexual pitfalls of those who failed to construct a strict work regimen and who played with the fire of their imaginations. For Newton, few if any of the Catholic histories of the Church were reliable, and most had been composed to justify the monkish way of life: "it has been ye constant practise of ye Catholicks to cry out fraud at every thing wch makes against 'em before they know whether it be so or not, & thereby have corrupted all history." The monks themselves were much worse, and they always corrupted texts and doctrines "by pretending (wthout proof) that hereticks had corrupted them before." Nevertheless, he granted the truth of some of the early stories of the monks' promotion of the "magical sign of the Cross" and the spread of exorcisms and witchcraft, by which means they came to have power over the ordinary people. By promising such power to the clergy, monkery under the promotion of Saint Anthony "overflowed the world." The monks were hardly paragons of virtue, and Newton noted that they were "ye most unchast & superstitious part of mankind as well in this first age as in all following ages":

> it was a general notion amongst them that after any man became a Monk he found himself more tempted by the Devil to lust then before & those who went furthest into ye wilderness & profest Monkery most strictly were most tempted, <the Devil (as they imagined) tempting them most when it was to divert them from the best purpose>. So that to turn a Monk was to run into such temptation as Christ has taught us to pray that God would not lead us into. For lust ~~by a violent prohibition~~ <being forcibly restrained> & by struggling wth it is always inflamed. The way to be chast is <not to contend & struggle wth unchast thoughts but to decline them> keep the mind imployed about other things: for he that's always thinking of chastity will be always thinking of weomen, & ~~struggle~~ <every contest> wth unchast thoughts will leave such impressions upon the mind as <shall> make those thoughts to return more frequently.[41]

Newton linked this ascetic discipline to his working life as a mechanically minded scholar. Paralleling the Protestant commonplace that idle hands and minds were led inexorably into sin, he was surely describing himself in a letter of

40. For the commonplace of Protestant accusations of sexual malpractice against Catholics, see Lake, "Anti-popery," 74–75, 99.

41. From the William Andrews Clark Library, UCLA, entitled "Paradoxical Questions Concerning Athanasius" (hereafter WACL).

1694 which contrasted the "Vulgar Mechanick" with the man who could "reason nimbly and judiciously about figure, force and motion," and who was "never at rest till he gets over every rub." There is evidence that Newton subjected himself to an intense bodily regimen when he worked on his optical researches in the late 1660s, and Shapin has noted George Cheyne's claim that "to quicken his faculties and fix his attention, [Newton] confined himself to a small quantity of bread, during all the time, with a little sack and water, of which, without any regulation, he took as he found a craving or failure of spirits."[42] Given the habits of the early nuns and monks, ascetic practices of *any* kind were not to be taken to excess, and Conduitt reported that Newton "had a patience and perseverance in any study he was pursuing [and] was very temperate & sober in his diet tho without ever observing any rules or strict regimen." Nevertheless, the existence of the man who was never at rest was "a life wch was one contrived series of labour." The Conduitts joined with Humphrey in relating Newton's self-discipline to that of the ancients. Catharine Conduitt recorded that "Sir I. resembled Socrates in keeping between the medium of luxury & penury [&] was grave & chearfull," while Humphrey told John Conduitt that in his rooms Newton "walk'd so very much, yt you might have thought him to be educated at Athens among ye Aristotelian sect," adding that he was "always thinking with Bishop Sanderson, Temperance to be ye best Physick."[43] In his own account of Newton's ascetic practices William Stukeley observed,

> Sir Isaac's moral character [was] ever eminently good, and never impeach'd in any one instance. Somewhat of it depends on a good state of the body and a government over passions. Sir Isaac by his great prudence and naturally good constitution, had preserv'd his health to old age, far beyond what one could have expected in one so intirely immers'd in solitude, inactivity, meditation and study; in an incredible expence of mind; and that thro' a long series of years.[44]

To make, as Newton put it, "a body fit for use," he recommended the golden mean, such as "fasting duly." This was one of the main moral virtues, though it had its "vitious extreames like all the rest," and similar regimens were commonly recommended at Cambridge. In 1656, for example, Henry More praised the virtue of temperance which he defined as "a measurable Abstinence from all

42. Cheyne, *Natural Method of Curing Diseases,* 81; see Shapin, chapter 1 in this volume.
43. Newton, *Correspondence,* 3:359–60; Keynes MSS, 129 (B) 11 and 130 (2); Edleston, *Correspondence,* xli–xlii; Keynes MSS, 130 (7), typescript p. 15; Humphrey to Conduitt, 14 February 1727/28, Keynes MSS, 135. For the commonplace anti-Jesuit accusation involved here, see Stachniewski, *Persecutory Imagination,* 36–37.
44. Stukeley, *Memoirs,* 67.

hot or heightening meats or drinks, as also from all venerous pleasures and tac-
tuall delights of the body, from all softness and effeminacy." In addition to this
he praised "moderate exercise of Body, and seasonable taking of the fresh aire,
and due and discreet use of Devotion, whereby the Blood is ventilated and
purged from dark oppressing vapours; which a temperate diet, if not fasting,
must accompany." In the instances of enthusiasm that More had described, it
was plain "that the more hidden and lurking fumes of *Lust* had tainted the Phan-
sies of those Pretenders to *Prophecie* and *Inspiration*." In the light of such ascetic
practices, Newton reworked his old notes on the fancy and added in remarks on
the temptations of women:

> such a moderate degree of fasting as best suits with every mans body so as
> ~~to~~ wthout unfitting it for use to keep down lust, is the due mean of intem-
> perance. . . . To pamper ye body enflames lust & makes it lesse active & fit
> for use. <And on the other hand> to macerate it <by fasting & watching>
> beyond measure does ye same thing. It does not only render ye body feeble
> & unfit for use, but also enflames it <& invigorates> lustful thoughts: <~~for~~
> ~~<at length< it invigorates ye imagination & brings>~~ <the want of sleep &
> due refreshment disorders the imagination & <at length> brings> men to a
> sort of distraction <& madnesse> so as to make them ~~think they~~ have vi-
> sions of weomen conversing with 'em ~~& sitting upon their knees~~ & think
> they really see <& touch> them & hear them talk.[45]

Newton noted that he had "not met with <more uncleannesse &> greater
arguments of unchast <minds> in any sort of people then in the lives of the first
Monks." If one considered Newton's warnings, "what else means their doctrine
that its better to contend with & vanquish unchast thoughts then not to have
them, their frequent visions of naked weomen, their digging up the bodies of
dead weomen ~~their lust~~ wth wch they burned in lust, their lusting even after passive sodomy, & their relating these & other such histories wthout blushing?"
Such postures betrayed their filthy and idolatrous superstition:

> ffor what else is their placing so much of religion in bodily exercises, or in
> extravagant fasting, in torturing their bodies by irons, ulcers, sordid habi-
> tations & ye like, in standing daily upon a pillar or in some other strange
> posture to be seen of men; in using ye signe of ye cross, <dead mens bones,
> holy water, consecrated oyle &> other consecrated things to do super-
> natural operations.[46]

45. More, *Enthusiasmus Triumphatus*, 37; WACL; and see Shapin's account of More and Anne
Conway in this volume (chapter 1).
46. WACL.

■ *Losing Possession of One's Self* ■

IN THE DONNISH CULTURE of which Newton was a part, no women usually entered college walls except for bedders and it is precisely to this intense gathering of narrow bachelordom that we should look if we wish to put Newton's beliefs in their relevant context. In this academic, male society, taking the virtuous path of sexual and marital renunciation was one of the last vestiges of monkish life that remained acceptable in Protestant England, though the cloistered life of the scholar was also associated with ill-health, madness, perversity, and the detested "solitary vice." Protestants who saw scholarship as a religious vocation necessarily guarded their souls and bodies against the twin evils of females and papists. Granted the evidence of his histories of the early Church, it is scarcely surprising that Newton linked the menace of Catholicism to the threat posed by Locke's supposed attempt to "embroil" him with women, and it is a crucial document for understanding his situation in the early 1690s.

The scholar had traditionally renounced the *vita activa* for the seclusion of the eremitic cell and his distracted appearance was the trademark he bore wherever he went. The social identity that Newton now sought was in complete opposition to such a life, and he lacked the repertoires to manufacture a simultaneously withdrawn and yet highly public scholarly identity. His inability to resolve the ensuing tension unleashed the dual enemies of lust and irreligion that had perverted and then corrupted the wrongdoers who preyed on the heroic early anti-Trinitarians. For a brief period, no *punctum* of Newton's scholarly self remained, nor any distinction between the public demons of his cosmopolitan friends and the intimately personal existence of the evil monks of the fourth century. In this sense, Newton literally suffered a breakdown.[47]

Not until the early part of the next century, with his acquisition of the presidency of the Royal Society, the publication of his *Opticks,* and the award of his knighthood, was his identity as an intensely private man working in the most public of employments successfully established and secured. The price he paid for his new identity was the complete loss of self in 1693 and a period of consolidation first as warden and then master of the Royal Mint. Nevertheless, the prize of immortality granted to this ambitious man was adequate recompense for his pains. Contemporaries soon spoke of him as wholly Other, and already in the 1690s the French mathematician the Marquis de l'Hôpital supposedly asked John Arbuthnot "every particular about Sr I. even to the colour of his hair [and] said does he eat & drink & sleep. is he like other men?" With serene implausibil-

47. For important analyses of the Christian extinction of self and body, see Brown, *Body and Society,* and in particular Harpham, *The Ascetic Imperative,* part 1.

ity, Arbuthnot replied that the great man "conversed chearfully with his friends assumed nothing & put himself on a level with all mankind." [48]

For many years the question of Newton's sanity was hardly raised. Henry Pemberton's claim in the preface to his 1728 exposition of Newton's philosophy that the octogenarian Newton *did* understand his own writings, "contrary to what I had frequently heard in discourse from many persons," possibly points more to contemporary narratives concerning the unintelligibility of the *Principia* than to its author's sanity. In his fine analysis of Newton's reputation in Britain after 1760, Yeo has drawn attention to the difficulties faced by contemporaries in fitting the person of the great constructor of philosophical method into the typology containing the category of "genius." Although Newton's scholarly quirks of dress and regimen were extremely well known and were conventionally linked to some of the accepted cultural stereotypes of madness, by the middle of the eighteenth century Conduitt and other allies had succeeded in elevating his character to the position of the most rational and virtuous man ever produced by Albion. As Yeo points out, the two most significant texts written on "genius" after Young's *Essay on Originality* of 1759 (William Duff's *An Essay on Original Genius* of 1767 and Alexander Gerard's *An Essay on Genius* of 1774)—somewhat ironically, given Newton's own distrust of the imagination—praised the imaginative component of the "Original Philosophic Genius" of which Newton was the paragon. They did not associate genius with insanity, although they did argue that excessive use of the imagination would be deleterious to the constitution, and, as Roy Porter has argued, the Georgian medical profession came to concur that the genius was sane and healthy. In the next few years, despite occasional discordant voices such as William Blake's, Newton's saintly virtue and superhuman intellect were elevated still further, and by the first decades of the next century, it was once again de rigueur to argue that he was somehow wholly Other than the adoring millions that worshipped his name. [49]

This changed in 1829 with Henry Brougham's translation of Biot's 1821 article in the *Biographie Universelle*, after which it could not be denied that Newton had suffered a mental illness of sorts in 1693. Biot drew upon the letters by Christiaan Huygens describing Newton's belated recovery of his understanding and, citing eighteenth-century lore about his legendary work habits and the destruction by fire of an early book on light and colors, he argued that the great man "had become insane . . . his mind being affected either by excess of exertion or through grief at seeing the results of his efforts destroyed." After his derange-

---

48. Keynes MSS, 130 (5), 2–3.
49. Yeo, "Genius, Method and Morality," 262; Porter, "Bedlam and Parnassus." I am indebted to Yeo's paper for the material in this and the next paragraph.

ment, the mentally shattered Newton turned to theology. Like Biot, Brewster denied that genius could be reduced to the "plodding drudgery of inductive discipline," but he pointed to the "weakness of [Newton's] imaginative powers" (which presumably might have led him astray) and argued that his genius could not in any way be related to his mental illness. However, while trying to correct the negative impression given by Biot, he was forced to struggle with manuscript material that seemed to prove that his hero was an alchemist and an anti-Trinitarian as well as with other information recently discovered from the Flamsteed archive that suggested that Newton was far from being the white knight beloved of Enlightenment and Romantic legend.[50]

The incontrovertible references to Newton's black year that would prove such a godsend to late Victorian commentators on the links between genius and insanity such as J. F. Nisbet and Francis Galton remained uncomfortable to those seeking to construct a proper place for Newton in the moral discourse of 1830s Britain. Where Biot used the example of Newton to draw attention to the existential condition of all mankind—"Such is the frightful condition of man, Genius and madness may exist in his mind side by side simultaneously"—the British either downplayed the severity of the illness or incorporated the madness story into a new moral narrative about the trials of genius. Baden Powell and William Whewell (who, as Warwick shows in chapter 8 of this volume, viewed such discipline as ideal for the moral training of Cambridge students) both pointed to the heroic demands of such disembodiment. Whewell claimed that "often, lost in meditation, he knew not what he did, and his mind appeared to have quite forgotten its connection with the body," while Powell echoed this thought: "The truth is that the intellect which had most deeply sounded and explored the mysteries of external nature was at times perplexed and obscured by the mysteries and infirmities of its own constitution, and in embracing the system of the universe Newton at times lost possession of himself."[51]

As disembodied and now immortal mind, wandering alone on the seas of thought, he had shed his mere corporeal self to occupy the high temple of the Muses. In his proposed biography, Conduitt recast the Arbuthnot anecdote and groped blindly for words to describe his idol: "When wee consider his talents his virtues. Even wee that knew him can hardly think of him without a sort of superstition wch demands all our reason to check—nor forbear saying with the Marquis de l'Hospital was Newton a man?"[52]

50. Brewster, *The Life of Sir Isaac Newton*, 336, 224–25; Yeo, "Genius, Method and Morality," 275.

51. Whewell, *History of the Inductive Sciences*, 2:141, and Powell, *Historical View*, 534 (cited in Yeo, "Genius, Method and Morality," 276); Becker, *Mad Genius*.

52. Keynes MSS, 130 (16).

■ Appendix ■

1. To make Lucatellos Balsome[53]

Venus Turpentine 1ts put ~~it~~ into it a pint of ye best Damask rose water beat ym together till it looke white yn take 4 ounces of Bee's wax sliced red sanders[54] halfe an ounce oyle Olive of ye best a pint <½ an ounce of oyle of St Johns wort>, & ½ a pint of Sack. Set it <(ye sack)> on ye fire in a new Pipkin ad to it ye oyle & wax. let it stand on a soft fire where it must not boyle but melt till it be scalding hot. yn take it off. when it is cold take out ye cake & scrape of ye dirt from ye bottom & take out ye Sack wipe ye Pipkin put in ye cake againe, set it on ye fier let them melt together yn put in also ye Turpentine & sanders let ym not boyle but bee well melted & mixed together take it of & stir it now & yn till it bee cold. If you would have it to take inwardly add to it when it is of from ye fier ½ an ounce of Pouder of Scuchineal[55] & a little naturall Balsoms.[56]

ffor ye Measell Plague or Small Pox a ¼ of an ounce in a little broth take it warme & sweate after it, & against poiyson & ye biting of a mad dog. for ye last you must dip lint & lay upon ye wound besides taking it inwardly. There are other vertues of it. <ffor greene wounds, sore breasts, new scalds or C[ ]es anoynt ye place greived. & so for bruises.>

A Medicine to cleare the eye-sight[57]

In ye gold Mines in Hungary ten or twenty fathoms deep or deeper upon ye sides of ye Mines there grow lumps of a yellow matter wch if <one> day it be as big as a horse bean ~~in another migh~~ the next day it will be as big as a nutt & so it increases until it come to its full growth. <It has no name> In the months of June & July it becomes ripe & fit for use & then it is <hard like a stone &> covered over wth a perfect yellow golden colour shining like the best gold but within it is of ye colour of a yellow earth or stone wthout shining. Before & after it is ripe it is spongy & unfit for use. ffor after it is ripe it wasts & loseth its <spirits &> colour. This stony concrete must be gathered in <& when it is hardest & shines ~~most like perfect gold~~ wth ye best golden colour wch happens> in the months of June & July. And then it must be beaten in a Mortar & powdered very fine & made into a past with eye-bright water.

53. Stanford University Library MSS, M132/2/5. Printed in Taylor, *Catalogue*, 7.
54. "Sanders-wood" is a synonym for sandalwood.
55. I.e., cochineal.
56. Balsams are the most significant sweet-smelling oils, assumed to be of great use in dealing with nervous complaints and supposed to be particularly useful for driving away melancholy and helping to seal open wounds.
57. Trinity College Cambridge MSS, R.16.38.439A, fol. 1r.

■ Acknowledgments ■

For comments on previous drafts on this paper I would like to thank David Harley, Simon Schaffer, Andrew Warwick, Moti Feingold, and Roy Porter. For permission to publish from their collections I am grateful to the Master and Fellows of Trinity College, Cambridge, and The William Andrews Clark Memorial Library, University of California, Los Angeles. The Keynes manuscripts are cited by kind permission of the Provost and Fellows of King's College, Cambridge.

■ References ■

Babb, Lawrence. *The Elizabethan Malady: A Study of Melancholia in English Literature from 1580 to 1642.* East Lansing: Michigan State University Press, 1951.

———. *Sanity in Bedlam: A Study of Robert Burton's Anatomy of Melancholy.* East Lansing: Michigan State University Press, 1959.

Barrow, Isaac. *The Theological Works.* 9 vols. Ed. A. Napier. Cambridge: Cambridge University Press, 1859.

Becker, George. *The Mad Genius Controversy: A Study in the Sociology of Deviance.* Beverly Hills, Calif.: Sage, 1978.

Beier, Lucinda. *Sufferers and Healers: The Experience of Illness in Seventeenth-Century England.* London: Routledge and Kegan Paul, 1987.

Biot, Jean-Baptiste. *Life of Sir Isaac Newton.* Trans. H. Brougham. Library of Useful Knowledge. London: Baldwin and Craddock, 1829.

Brewster, David. *The Life of Sir Isaac Newton.* Edinburgh: John Murray, 1831.

———. *Memoirs of Sir Isaac Newton.* 2 vols. Edinburgh: T. Constable and Co., 1855.

Bright, Timothy. *A Treatise of Melancholie.* London, 1586.

Brown, Peter. *The Body and Society: Men, Women and Sexual Renunciation in Early Christianity.* London: Faber and Faber, 1988.

Bryson, Anna. "The Rhetoric of Status: Gesture, Demeanour and the Image of the Gentleman in Sixteenth- and Seventeenth-Century England." In *Renaissance Bodies: The Human Figure in English Culture c. 1540–1660,* ed. Lucy Gent and Nigel Llewellyn, 136–53. London: Reaktion Books, 1990.

Burton, Robert. *The Anatomy of Melancholy.* 3 vols. Ed. Thomas C. Faulkner, Nicholas K. Kiessling, and Rhonda L. Blair. 1628. Reprint, Oxford: Clarendon, 1989.

Bynum, Caroline Walker. *Fragmentation and Redemption: Essays on Gender and the Human Body in Medieval Religion.* New York: Zone Books, 1991.

Cheyne, George. *Natural Method of Curing Diseases of the Body and Disorders of the Mind.* London, 1742.

Conway, Anne. *The Correspondence of Anne, Viscountess Conway, Henry More and Their Friends, 1642–1684.* Ed. Marjorie Hope Nicolson. New Haven: Yale University Press, 1930.

Curtis, Mark. *Oxford and Cambridge in Transition, 1558–1642.* Oxford: Clarendon Press, 1959.

Dunn, John. *The Political Thought of John Locke: An Historical Account of the Argument of the "Two Treatises of Government."* Cambridge: Cambridge University Press, 1969.

Duveen, Denis. *Bibliotheca Alchemica et Chemica.* London: E. Weil, 1949.

Edleston, J., ed. *The Correspondence of Sir Isaac Newton and Professor Cotes.* London: F. Cass, 1850.

Evelyn, John. *Diary: Now First Printed in Full from the Manuscripts Belonging to Mr. John Evelyn.* Ed. E. S. de Beer. Oxford: Clarendon Press, 1955.

Ficino, Marsilio. *Three Books on Life* [1489]. Ed. and trans. Carol V. Kaske and John R. Clark. Binghamton, N.Y.: Renaissance Society of America.

Foucault, Michel. *Discipline and Punish: The Birth of the Prison.* Trans. Alan Sheridan. Harmondsworth: Penguin, 1979.

Frank, Arthur W. "For a Sociology of the Body: An Analytical Review." In *The Body: Social Process and Cultural Theory,* ed. Mike Featherstone, Mike Hepworth, and Bryan S. Turner, 36–102. London: Sage, 1991.

Grieve, Maude. *A Modern Herbal: The Medicinal, Culinary, Cosmetic and Economic Properties, Cultivation and Folklore of Herbs, Grasses, Fungi, Shrubs and Trees with Their Modern Scientific Uses.* London: Cresset Press, 1992.

Harpham, Geoffrey Galt. *The Ascetic Imperative in Culture and Criticism.* Chicago: University of Chicago Press, 1987.

Jobe, Thomas H. "Medical Theories of Melancholia in the Seventeenth and Early Eighteenth Centuries." *Clio Medica* 11 (1967): 217–31.

Keynes, Milo. "The Personality of Isaac Newton." *Notes and Records of the Royal Society* 34 (1995): 11–32.

Klibansky, Raymond, Erwin Panofsky, and Fritz Saxl. *Saturn and Melancholy: Studies in the History of Natural Philosophy, Religion, and Art.* London: Thomas Nelson, 1964.

Lake, Peter. "Anti-Popery: The Structure of a Prejudice." In *Conflict in Early Stuart England: Studies in Religion and Politics 1603–1642,* ed. Richard Cust and Ann Hughes, 72–106. London: Longman, 1989.

Levins, P. *A Right Profitable Book of all Diseases, called the Path-way to Health.* London, 1632.

Locke, John. *The Correspondence of John Locke.* Ed. E. S. de Beer. 8 vols. Oxford: Clarendon Press, 1976.

MacDonald, Michael. *Mystical Bedlam: Madness, Anxiety, and Healing in Seventeenth-Century England.* Cambridge: Cambridge University Press, 1981.

Manuel, Frank. *A Portrait of Isaac Newton.* Cambridge: Belknap Press, 1968.

Martin, Luther H., Huck Gutman, and Patrick H. Hutton, eds. *Technologies of the Self: A Seminar with Michel Foucault.* Amherst: University of Massachusetts Press, 1988.

McGuire, James E., and Martin Tamny. *"Certain Philosophical Questions": Newton's Trinity Notebook.* Cambridge: Cambridge University Press, 1983.

More, Henry. *Enthusiasmus Triumphatus; or, a Brief Discourse of the Nature, Causes, Kinds and Cure of Enthusiasm.* London, 1656.

Newton, Isaac. *The Correspondence of Sir Isaac Newton.* 7 vols. Ed. Herbert Westren Turnbull et al. Cambridge: Cambridge University Press, 1959–77.

North, Roger. *General Preface and Life of Dr. John North.* Ed. Peter Millard. London: University of Toronto Press, 1984.

O'Hara-May, Jane. *Elizabethan Dyetary of Health.* Lawrence, Kans.: Coronado Press, 1977.

Pelling, Margaret. "Appearance and Reality: Barber-Surgeons, the Body and Disease." In *The Making of the Metropolis: London 1500–1700,* ed. A. L. Beier and Roger Finlay, 60–81. London: Longman, 1985.

Porter, Roy. "Bedlam and Parnassus: Mad People's Writings in Georgian England." In *One Culture: Essays in Science and Literature,* ed. George Levine, 258–86. Madison: University of Wisconsin Press, 1987.

———, ed. *The Popularization of Medicine, 1650–1850.* London: Routledge, 1992.

Powell, Baden. *An Historical View of the Progress of the Physical and Mathematical Sciences.* London: Longman, 1834.

Ramazzini, Bernardino. *Diseases of Workers: The Latin Text of 1713 Revised, with Translation and Notes,* ed. Wilmer Cave Wright. Chicago: University of Chicago Press, 1940.

Rather, Lelland J. "Old and New Views of the Emotions and Bodily Changes." *Clio Medica* 1 (1965): 1–25.

Ray, John. *Further Correspondence of John Ray.* Ed. Robert William Gunther. London: Ray Society Publications, 1991.

Romanell, Patrick. *John Locke and Medicine: A New Key to Locke.* Buffalo: Prometheus Books, 1984.

Schleiner, Winfried. *Melancholy, Genius and Utopia in the Renaissance.* Wiesbaden: Harrossowitz, 1991.

Shapin, Steven. "'The Mind Is Its Own Place': Science and Solitude in Seventeenth-Century England." *Science in Context* 4 (1991): 191–218.

Slack, Paul. "Mirrors of Health and Treasures of Poor Men: The Use of the Vernacular Medical Literature of Tudor England." In *Health, Medicine and Mortality in the Sixteenth Century,* ed. Charles Webster, 237–73. Cambridge: Cambridge University Press, 1979.

Soufas, Terese Scott. *Melancholy and the Secular Mind in Spanish Golden Age Literature.* Columbia: University of Missouri Press, 1990.

Spargo, Peter E., and C. A. Pounds. "Newton's 'Derangement of the Intellect': New Light on an Old Problem." *Notes and Records of the Royal Society* 34 (1979): 11–32.

Stachniewski, John. *The Persecutory Imagination: English Puritanism and the Literature of Religious Despair.* Oxford: Clarendon Press, 1991.

Stukeley, William. *Memoirs of Sir Isaac Newton's Life.* Ed. A. Hastings White. 1752. Reprint, London: Taylor and Francis, 1936.

Taylor, John. *Catalogue of the Newton Papers Sold by Order of Viscount Lymington.* London: Sotheby and Co., 1936.

Thomas, Ivor. *Selections Illustrating the History of Greek Mathematics.* 2 vols. London: William Heinemann Ltd., 1939–41.

Ward, Nicholas. *A Step to Stir-Birch-Fair: with Remarks on the University of Cambridge.* London, 1700.

Westfall, Richard S. *Never at Rest: A Biography of Isaac Newton.* Cambridge: Cambridge University Press, 1980.

Whewell, William. *The History of the Inductive Sciences.* 3 vols. London: J. W. Parker, 1837.

Yeo, Richard. "Genius, Method, and Morality: Images of Newton in Britain, 1760–1860." *Science in Context* 2 (1988): 257–84.

■ FIVE ■

MEDICAL MINDS, SURGICAL BODIES

Corporeality and the Doctors

CHRISTOPHER LAWRENCE

Butchers, William had observed, categorising even then, tend to be well-fleshed men, outward-looking and with strong opinions.
—A. S. Byatt, *Angels and Insects*

■ *The Body of Physic* ■

IN THE 1930S the English surgeon, Sir William Heneage Ogilvie, observed that his choice of surgery as a career "was the inevitable one of a young man who has the normal combative instincts of the healthy male." Surgery, he continued,

> appeals to the craftsman who likes using his hands, to the artist whose mind works on visual images, to the romantic who enjoys the drama of life particularly when it affords him the opportunity to play a decisive role, to the extrovert . . . the footballer, the mountaineer, the yachtsman, are drawn instinctively towards the surgical side of practice.

The surgeon he noted is "a doer rather than a thinker." Ogilvie particularly appreciated the American surgeon, whom he took to be "courageous, independent, self-sufficient, distrustful of tradition, a pioneer by instinct, able to make quick decisions and to act upon them." By contrast "the prizewinner, the editor of the hospital journal, the debater, the naturalist, tend to find their vocation in [internal] medicine."[1] It is still tribal lore among medical professionals in Europe and North America that it is possible to distinguish physicians and surgeons generically, identifying them by a characteristic combination of corporeal and mental qualities.[2] The physician is stereotypically lean, aquiline, bookish, inscrutable, solitary, and given to deep musing on medical problems. The surgeon, by contrast, is muscular, bluff, practical, theatrical, gregarious, and ever ready for dramatic intervention (figs. 5.1 and 5.2).

1. Ogilvie, *Surgery,* 3, 36–37, 38, 66, 37.
2. For reasons of historical accuracy I use the British distinction between physicians and surgeons. This approximates the American distinction between internist and surgeon.

156

Auf meine Diagnose können Sie Gift nehmen!

FIGURE 5.1. A physician. Characteristically he is lean and bespectacled, most viewers no doubt associating this attribute with scholarship. The caption means, "On my diagnosis you can take poison!" Color process print by Carl Josef, ca. 1930. Courtesy Wellcome Institute Library, London. V11945.

FIGURE 5.2. A surgeon. Characteristically he is a stocky figure, the image conjuring up associations with down-to-earth matters, notably butchery. The spectacles on this figure are more likely to be associated by the viewer with the close attention to visual detail required by the surgeon's work than with scholarship. Color process print by Carl Josef, ca. 1930. Courtesy Wellcome Institute Library, London. V11950.

Each of these groups of characteristics can be considered as constituting what Shapin has called repertoires.[3] Whether these repertoires have any important cultural employment in modern medicine (they well might) will not be addressed here. In the past, however, they were put to significant use. The constituents of these repertoires, alone and in various combinations, were frequently deployed, both positively and negatively, by surgeons, physicians, and others to enhance their authority or devalue that of others. These repertoires were compiled over a long period; the scholarly nature of the physician, for instance, was constructed in the middle ages. By the middle of the seventeenth century most of the other constituents of these repertoires were available, including an obvious gendered element. Here I trace the employment of these repertoires by British (and to a limited extent Continental European and North American) physicians, surgeons, and others from the early modern period to the recent past. A particularly crucial moment in the use of these repertoires occurred in the nineteenth century, when surgeons turned what had, even by then, become a caricature of themselves to their distinct advantage.

In 1675 Mary Trye published her *Medicatrix or the Woman-Physician,* a vindication of her father, the chemist and courtier, Thomas O'Dowde. In this work Trye observed, "The Body of Man . . . is the Subject of Physick. To preserve the health thereof . . . and heal it, when it is Sick, is the chief and great End of Physick. In these few words is the Body of Physick contained."[4] That physic, in the sense of all the healing arts, comprised a body was a commonplace in the early modern period and, indeed, this is a usage that can be traced to the present.[5] But exactly which healing practices and practitioners were legitimate components of that body and exactly how their hierarchical relationships should be constituted have long been subjects of bitter dispute. It is well known that in the seventeenth century and long after, orthodox physicians considered themselves the head of physic. In 1668 Everard Maynwaringe noted, "To make good Medicines, and be well skilled in *Pharmacy,* is one good part of a Physician [but] . . . Here is a Body but no Head."[6] Physicians, in considering themselves this head, regarded surgeons and apothecaries as the hands of healing. Practice without theory, noted

3. Shapin, "'A Scholar and a Gentleman'"; idem, *A Social History of Truth.*

4. Trye, *Medicatrix,* 78.

5. *Physic* was sometimes used to mean internal medicine only. For example, John Colbatch, brought up as an apothecary, recalled how he was "induced to look into the Body of Physick itself." *Physico Medical Essay,* preface. More often it comprehended the other branches, "Surgerie beynge . . . one of the chiefest parts of phisike," as John Securis noted in 1566, *A Detection and Querimonie,* EIII. See also Sprackling, *Medela Ignorantiae,* 121.

6. Maynwaringe, *Medicus Absolutus,* 45.

one physician, could no more produce regular success than "Limning from a hand altogether rude, though arm'd with the Pencils and Colours of *Vandike*."[7] The manual trades, physicians declared, were the instruments of medicine. Thus in 1676 the physician Charles Goodall observed that the College of Physicians was "fully resolved, . . . to encourage and protect those two necessary instruments of physic, the surgeons and the apothecaries."[8] The analogue of this in nature could be found described in medical theory, where the voluntary muscles were designated as the instruments of the soul.[9] Such a hierarchy privileged mind over body, art over artifice, and master over servant.

■ *Physicians as Scholars* ■

IF, THEN, ORTHODOX physicians designated themselves as the head of medicine, what was it about the head that gave it the right and ability to govern the manual arts of healing? The answer not surprisingly was medical learning. Learning, all agreed, was naturally confined to man's "upper sphere" where "are planted all the Faculties, viz. his Sense, Will, Reason and Understanding."[10] The faculty (as in a university faculty) was a generic term frequently applied to physicians, by themselves and others. The scholarly role cultivated by physicians was thus deployed to regulate surgeons, apothecaries, and unlicensed practitioners and to attract an elite clientele. Medical learning in this period, of course, did not simply mean possessing special knowledge; it meant, physicians said, *being* a philosopher. That Hippocrates and, following him, Galen had decreed that the finest physicians were also philosophers was an observation physicians ceaselessly reiterated. "*Hippocrates,*" noted James Primrose, in 1651 "saies, that a physician which is a Philosopher, is God-like."[11] In 1683 Walter Harris observed, "Where the Philosopher Ends, there the Physician does Begin."[12] Even in the recent past, when doctors have defined their role in terms of expertise, and even as philosophy was given its more modern specialized connotation, physicians have still drawn on philosophical identity. In 1858 Sir Russell Reynolds was insisting that medicine must "gain the advantages derivable from the prevalent philosophical system of the day."[13]

Creating physic as the head of healing by tying it to learning had been cen-

7. Castle, *Chymical Galenist*, 18. A similar trope appears in Maynwaringe, *Medicus Absolutus*, 47.
8. Cited in Cook, *The Decline of the Old Medical Regime*, 204.
9. *"But whence comes this great and constant Variety of Motions? Is it not from these Instruments of Motion we call the* Muscles?"*: William Cockburn, prefatorial letter to Browne, *Myographia.*
10. Preface, ibid.
11. Primrose, *Popular Errours*, 2.
12. Harris, *Pharmacologia*, 36. This dictum, in Latin, was a Renaissance coining. There are many variants and its origins are obscure. See Schmitt, "Aristotle among the Physicians."
13. Reynolds, *Essays and Addresses*, 22.

tral to Thomas Linacre's creation of the College of Physicians in 1518. As Webster observes, "In view of his varied associations with Corpus Christi College [Oxford], it is not surprising that Linacre organized the College of Physicians in such a way as to suggest an affinity between the academic and professional bodies."[14] The college's entrance examinations, for example, were modeled on those of the university and its charter of 1518 decreed that physic could be practiced only by "those persons that be profound, sad [i.e., serious], and discreet, groundedly learned, and deeply studied in physic."[15]

How Linacre presented such a corporeal self to the world we do not know, for there is no known contemporary portrait or description of him. There are, however, many paintings, drawings, and engravings that physicians have long venerated as representing the *real* Linacre. The most widely disseminated of these images is the Windsor Castle portrait to which Linacre's name became attached in the middle of the eighteenth century (fig. 5.3). "The Windsor portrait," says one authority, "cannot represent Linacre."[16] It shows a lean figure in academic robes that physicians have long taken to embody the profundity, seriousness, and discretion of a man "groundedly learned and deeply studied in physic."

That physic required deep learning and that the best physicians were the most deeply learned is a refrain that spans from the seventeenth through the twentieth century. However, physicians have also insisted that mere depth of learning was insufficient to make a good doctor; what also counted was breadth. As James Makittrick put it in 1772, "no science requires so extensive a knowledge of what is called general learning, than physic."[17] A similar sentiment was voiced by William Osler, eventually Regius Professor of Medicine at Oxford, in the early twentieth century: "In no profession does culture count for so much as in medicine"[18] (fig. 5.4).

■ *The Problems of Scholarly Identity* ■

PHYSICIANS, HOWEVER, were not merely scholars, they also aspired to practice among the gentry, aristocracy, and royalty. In an important paper Cook has argued that, in the sixteenth and early seventeenth centuries, physicians presented themselves to their clients not principally as experts in the cure of disease but as counselors. This role was recognized by the upper orders, who consulted physicians for advice about how to live their lives. A university education, physicians agreed, gave a man a disciplined and sound moral character and was the

14. Webster, "Thomas Linacre," 219.
15. Munk, *Roll,* 1:8.
16. Hill, "An Iconography of Thomas Linacre," 357.
17. Makittrick, *Commentaries on the Principles,* xvii.
18. Osler, *Student Life,* 113.

FIGURE 5.3. Thomas Linacre (1460?–1524). Stipple engraving by H. Cook after W. Miller, 1810, at the Royal College of Physicians, copy of a painting by the School of Q. Metsys at Windsor. Courtesy Wellcome Institute Library, London. V357B.

foundation of good judgment. The physician's "gravity, learning and discretion," said Justice Walmesly at the beginning of the seventeenth century, meant "he may be trusted with anything." [19] Cook concludes that "what the physicians did and who they claimed to be were intimately related." [20] To this day the force of grav-

19. Cited in Cook "Good Advice and Little Medicine," 11.
20. Ibid., 13.

FIGURE 5.4. Sir William Osler (1849–1919). Osler was by no means an insubstantial individual, and he was venerated as the supreme diagnostician and is seen here embodying the gravity of the academic physician. Photographic reproduction of a section of the painting "Four Doctors" by J. S. Sargent. Original at Johns Hopkins University. Courtesy Wellcome Institute Library, London. V4381B.

ity has remained an important constituent of the physician's self-presentation. Nonetheless, Cook further argues that, for a variety of reasons, notably increasing commercialism and competition, from about the middle of the seventeenth century onward "the idea that practitioners engaged in medicine—the application of remedies to diseases—gained influence not only among the public at large but among the physicians themselves."[21] Physicians increasingly presented themselves, like their competitors, as sources of advice about how to achieve instrumental effects.[22] But in doing so the physician risked becoming indistinguishable from other practitioners. To distinguish himself "the physician now strove to become a polished ornament to society."[23] There was a move in the late seventeenth century from gravity to gentility.

It is not clear to what extent physicians in the Elizabethan age were regarded, or regarded themselves, as gentlemen. A common, although often contested, opinion was that to be a gentleman one must have an independent income from land. But by the end of the seventeenth century gentility had definitely been linked to learning. In 1677 one author observed "in these days he is a Gentleman, who is commonly so taken." He could be one "whosoever studieth the Laws of this Realm, whosoever studieth in the University, who professeth Liberal Sciences, and to be short, who can live without Manual Labour, and will bear the port, charge and countenance of a Gentleman, he shall be called Master, and shall be taken for a Gentleman."[24] One of the imperatives, that physicians style themselves and present themselves as gentlemen, came, of course, from the most worldly of places, the court. As Sir George Clark put it, "From the granting of the charter onwards, the main footing of the College in the world of power was that of the royal physicians with their daily access to ministers, favourites and the sovereign." William Butts, physician to Henry VIII was, as Clark proudly notes, "first in the long line of English medical knights."[25] By the late seventeenth century, college members were definitely designating themselves gentlemen. Edward Baynard, a fellow left off the college's "catalogue" printed in 1693, complained "I am a gentleman, and no footman."[26] Thomas O'Dowde, attacking the college's physicians in 1665, referred to them as "Gentlemen and Students in Physick."[27] Likewise, the physician John Twysden charged the college critic, Marchamont Nedham, with behavior "beneath a Gentleman, unworthy a Scholar."[28] But such self-styling, of course, was a resource open to all and sundry. Thus Thomas

21. Ibid., 21.
22. This transformation was not confined to physicians.
23. Cook, "Good Advice and Little Medicine," 30.
24. Logan, in Guillim, *A Display of Heraldry*, 154–55. This formulation was itself a paraphrase of an Elizabethan stipulation: see Shapin, *A Social History of Truth*, 57.
25. Clark, *History of the Royal College of Physicians*, 1:76, 72.
26. Cited in Cook, *Decline of the Old Medical Regime*, 226. In fact very few physicians at mid-century could genuinely claim to have gentle blood: Axtell, "Education and Status," 147; Prest, *The Professions*, 8.
27. O'Dowde, *Poor Man's Physician*, 40.
28. Twysden, *Medicina Veterum Vindicata*, 207. On Nedham see Cook, *Decline of the Old Medical Regime*, 145–47.

Trigge, author of an astrological calendar, styled himself "Gent, student in Physic and Astrology."[29]

Scholars and gentlemen, however, have distinct repertoires available to them. Seventeenth-century scholars and gentlemen variously drew on a number of attributes to identify themselves, each other, and their social places. Scholarship was often associated with solitude as opposed to the civic life, with otherworldliness as opposed to worldliness, with the cultivation of self rather than others, with knowledge for its own sake rather than as a basis for action, with self-neglect rather than self-respect, and, most pertinent to the issues here, with disembodied asceticism. In various ways physicians have long associated themselves with the disembodied knowledge—seeking tradition described in this volume by Shapin (chapter 1). This placement of the physician as scholar had its naturalistic confirmation in medical theory where wit and perspicacity were associated with delicate and tender constitutions.[30] But from the gentlemanly perspective the scholar's role was widely regarded as unsociable, boorish, and pedantic. By the late seventeenth century the physician who was accounted too much of a scholar might well find himself losing his gentlemanly clientele.

■ *Medical Practice* ■

EARLY MODERN PHYSICIANS had to balance not only the scholarly and genteel repertoires, there was yet a further factor to be juggled: practice. After the Restoration observers grappling with the elusive definition of a gentleman agreed that, whatever else he did, he did not support himself by manual labor. But how were physicians to *practice* medicine while still being scholars and also gentlemen? In the world of practical physic, physicians had to present themselves as men of skill and action. How were they to distinguish themselves from those other practitioners—surgeons and apothecaries—who composed the meaner instruments of the body of physic? Too much cultivation of the practical could jeopardize their genteel aspirations and could also identify them with their competitors, who in turn could designate *themselves* as the real men of action. That physicians did indeed have problems in joining these things is clear from their critics' perceptions of their vulnerability. For example, when physicians professed themselves as learned, as philosophers, the custodians of scholarship, they could find themselves dismissed by their competitors as impractical. Marchamont Nedham, for example, declared that "a Doctor bred up in the Contemplative Philosophy of the Schools, may be a Scholar and a very fine Gentleman; but

29. Cook, *Decline of the Old Medical Regime,* 39.
30. Shapin, chapter 1, this volume; idem, "'A Scholar and a Gentleman.'" On constitution and wit, idem, *A Social History of Truth,* 76, 154.

what is that to the Curing of a Disease . . . ? This is to be expected rather from one that is qualified for the work by acquaintance with Mechanick and *Experimental Philosophy.*"[31]

What, then, was the nature of this practice that, physicians declared, required and combined the learning of a philosopher and the bearing of a gentleman? Practice, physicians agreed, consisted in the exercise of what they called skill, but skill was not a manual activity. It was constituted by the ability, as one physician put it in 1670, "to exercise . . . a piercing Judgement." This could be done, he said, only by having a "large Comprehension of . . . subtle and numerous natures and things."[32] Such discriminatory power, as Schaffer shows in chapter 4 of this volume, was often associated with disembodiment. Seventeenth-century natural philosophers regarded themselves as highly reliable reporters of their own sensations and, for this reason, acutely perceptive of goings-on in other bodies. Judgment also required, as Walter Harris put it in 1683, "diligent *Observations* from *Experience* and *Practice*" or else the physician, qua practitioner, would be overcome "even by *Ideots* and *Women.*"[33] A store of experience, coupled with scholarly knowledge, gave the physician the ability to reason out the complexities of a case and tailor therapy to individual needs, rather than applying, as empirics and apothecaries were said to do, a remedy as a "universal appointment to all bodys, all constitutions, and all tempers."[34] Practical knowledge of this sort, John Freind noted in 1726, was "incommunicable."[35] This was a stipulation used by William Bowman to his students nearly 124 years later.[36] The concept remains a refrain in clinical medicine to the present day. If there is a long-term continuity in medicine it is this: until the middle of the nineteenth century, and, with significant provisos, well beyond, the physician by his own account was a philosopher and a gentleman, utilizing skills gained by supplementing lectures and reading with extensive experience.[37]

Medical historians frequently say that before the nineteenth century physicians spurned manual activity because of its association with trade. In fact, in the seventeenth century, many physicians began to discern and value a manual aspect to their profession. Competitors asserted the value of hands-on experience and, as noted above, doctors felt increasingly compelled to adopt those com-

31. [Nedham], *Medela Medicinae,* 255.
32. Goddard, *A Discourse,* 12.
33. Harris, *Pharmacologia,* 36–37.
34. Twysden, *Medicina Veterum Vindicata,* 23.
35. Freind, *The History of Physick,* 1:215.
36. Bowman, *An Introductory Address,* 19.
37. Such a view is ubiquitous in the writings of Sir William Osler. Osler credited the Hippocratic school with "the conception and realization of medicine as the profession of a cultivated gentleman": *Student Life,* 44.

petitors' strategies. Apothecaries, for example, whose work physicians designated as being solely the compounding of medicines, were increasingly treating the sick. This is significant, not only because of the obvious things it says about economic competition, but because of the light it throws on the physicians' crisis of identity. In diagnosing and prescribing, apothecaries were usurping one of the physicians' roles; that is, they might *seem* to the world as though they *were* physicians. Since skill in medicine, as defined by physicians, had only a slight manual dimension—feeling the pulse, for example—how could the public judge whether or not apothecaries were *reasoning* like physicians? As the physician Christopher Merret put it, because apothecaries have acquired "a little smattering of diseases . . . and by applying the terms of Art . . . they made people believe they had acquired some skill in the Art."[38] Apothecaries were in fact turning the world upside down. It was not simply that the irregular practice of physic took meat from the physician's table. The place of the head of medicine was, for the physician, woven into his bodily constitution and into the social fabric by virtue of its equation with learning and, increasingly, with gentility. Irregular practice threatened the social order by confounding natural place, elevating hand over head. Upstart apothecaries practicing physic disrupted the boundaries between trade and profession, master and servant, gentleman and laborer. Merret observed that a "manifest sign of . . . [the apothecaries'] endeavour to usurp our Practice is, their absurd [habit of] calling the sick their *Patients,* for 'tis most certain that in all reason and language the *Physician* and *Patient* only have relation to each other, but not to the Apothecary, who is but a Tradesman, and manual Operator." "When Dispensatories were first made," he added, "the *Apothecaries* were really *Physicians* Servants."[39] Such an inversion, another physician said, makes physic *"prey to Hostlers, Coblers and Tinkers"* and will bring *"Taylors to invade the Bar and Iuglers the Pulpit."*[40]

To remedy this situation some physicians held that they themselves should dispense. But how this was to be done without tainting physic with the odor of trade needed careful management. As Jonathan Goddard noted, "how can it be honorable for a Physician to sell Medicaments, may be a question." He answered that if it was for the benefit of physic at large, it was proper, because, he said, stretching somewhat the identity of physic and gentility, it was no more dishonorable for a physician to dispense "than to persons of Honour, and great Estates to sell their Corn, Cattel, [or] Wooll."[41] In fact, the longer-term response of

38. [Merret], *A Short View of the Frauds,* 51.

39. Ibid., 26, 40.

40. C. T. [Clarke?], *Some Papers Writ,* 13. On C. T.'s probable identity as Clarke, see Cook, *Decline of the Old Medical Regime,* 66.

41. Goddard, *Discourse,* 37–38.

many elite clinicians was to resign themselves to the reality that apothecaries were practicing and to retreat from any hint of manual activity. Take, for instance, Sir Henry Halford talking of midwifery in 1834:

> I think it is considered rather as a manual operation, and we should be very sorry to throw anything like a discredit upon the men who have been educated at the Universities, who had taken time to acquire the improvement of their minds in literary and scientific acquirements by mixing it up with this manual labour. I think it would rather disparage the highest grade of the profession.[42]

Ironically, Halford's opinion was given on the eve of the wholesale transformation of bedside medicine into a hands-on activity.

■ *Crisis of Identity* ■

THE PROBLEM FOR late seventeenth-century physicians was this: they did not have estates; they did not work with their hands; their activities were both intellectual and practical; their occupation enjoyed only the loosest of legal priveleges: how were they to create a specific and valued identity, both individually and corporately? The highly fluid structure of seventeenth-century society meant that having and displaying identity was a pervasive concern.[43] For the physician, beset on all sides by competitors, the presentation of self posed acute problems. How was the public to recognize such a man? What constituted his identity and his entitlement to authority?

There was not, for example, any special behavior or ethical code by which physicians identified themselves. As Fissell has pointed out for the eighteenth century, and it certainly seems true for the earlier period, what governed the behavior of physicians were the same rules that regulated relations between equals, superiors and inferiors, masters and servants, or men and women in ordinary life. The guide to such conduct was gentlemanly manners.[44] As the sixteenth-century author John Securis put it, the physician must of course have study and learning, but in addition, "he must be well disposed: for there is nothyng that getteth a man better estimation and authoritie then to bee endued with an honest lyfe and good maners."[45] Such prescriptions hardly gave physicians a distinct identity.

---

42. Cited in Richardson, *Death, Dissection and the Destitute,* 34.
43. Stone, "Social Mobility."
44. Fissell, "Innocent and Honorable Bribes." Physicians did appeal to the Hippocratic oath and the works of Galen as justifications for their conduct, but these resources largely seem to have been used in the context of legitimating themselves as the head of physic: see Wear, "Medical Ethics."
45. Securis, *A Detection and Querimonie,* AIIII.

Until well into the nineteenth century physicians, in fact, found it easier to define their identity by what they were not than by what they were. For example, they were not empirics. Empirics, physicians agreed, displayed all the qualities not found in the ideal physician: loquacity, hastiness, and a fondness for slander and self-advertisement. Another thing a physician could not be until late into the Victorian age was a woman. John Cotta in 1612 found women unequipped for medical practice since God had not made them "commissioners in the sessions of learned reason and understanding." They were given too much to "babling" and were unhelpful to the sick.[46]

It seems then that it was much easier for physicians to describe how they should not appear and behave than to give anything other than the most bland account of themselves. One physician, writing around 1680, noted that the physician's "Morality is sober, grave, continent, pitiful to, and careful of his Patients." "He is continent wise and prudent, mild and modest" and so on. He does not envy physicians "who gain great Estates and Glory, by outward shews," for instance, those who display "a modish Habit, a . . . well-furnished House, a Coach and Horses, a Velvet Coat, a fine Periwig, a curiously-headed Cane, a smooth and starched Behaviour, flattering Language."[47] Similarly, Campbell in his *London Tradesman* of 1747 declared "I would have the Physician, a Man endowed with Sagacity, Learning and Honesty. . . . He must, besides a solid discerning Judgment, be possessed of a tenacious Memory."[48] By the end of the seventeenth century, however, as physicians increasingly courted gentlemanly status, "outward shews" *were* becoming the mark of quite distinguished practitioners and, as Cook observes, "A number of physicians of the College were gaining reputations not by emphasizing their academic learning but by underscoring their special views of disease or their witty and worldly conversation. . . . Samuel Garth and Richard Blackmore acquired reputations as wits and poets and consulted with patients in coffeehouses."[49]

If Cook's argument is correct and physicians were ceasing to be counselors, the reasons would seem to relate to the physician's re-presentation of himself as a gentleman, as someone very worldly. This was part of the broader attempt to link gentility to scholarship and in turn involved dispensing with the image of learning as totally grave.[50] The grave physician had been an object of satire since the Elizabethan age and remained so, but the satire of ostentatious display of

46. Cotta, *A Short Discoverie*, 25, 28.
47. [Merret?], *The Character of a Compleat Physician*, 6, 7.
48. Campbell, *London Tradesman*, 43–44.
49. Cook, *Decline of the Old Medical Regime*, 215.
50. Cook, "Good Advice and Little Medicine."

wealth among physicians seems to have been on the increase during the seventeenth century.[51] Although the topos of the solitary, learned, lean physician persisted in the eighteenth century, clubbability, claret, and the doctor were also frequently associated.

Yet if a genteel status was what, in the eighteenth century, physicians sought to achieve, not everyone was ready to accord it to them, and their pretensions were frequently mocked. But something else was going on. While acknowledging that their behavior should be governed by gentlemanly codes of conduct, doctors also self-consciously began to develop a code of ethics specific to medicine, detailing their relations with patients and with other practitioners. This was a complex business and by about 1800 both systems—the gentlemanly code and a specific ethical code—seem to have been in operation. Thus some medical authors assumed that older gentlemanly and Christian conventions would be sufficient for conducting medical business, but others began writing specific guides for medical conduct.[52] Physicians were both aspiring to be acknowledged as gentlemen and also creating a role for themselves as scientific experts. In doing so they developed specific professional codes of practice. They thus exposed themselves doubly to satire. They were mocked for aspiring to be gentlemen and also for assuming manners at the bedside that were not "natural" but were those of the self-interested professional. The creation of medical ethics and the rise of self-advertisement among physicians are noteworthy, since, in the satire they engendered, the historian can discover the corporeal and intellectual qualities physicians said they possessed.

Thus, although in 1772 James Makittrick observed that the physician must have "Humanity," "Prudence," "Decency of manners," and "Candour," he added that other, more visible qualities often identified the doctor: "Some peevish men have . . . alledged, that success in the profession [depends on] . . . sauntering in coffee-houses, or tippling in clubs;—by the size of their wig . . . [by] effusions of unmeaning jargon . . . by forming dirty connexions with nurses and ladies women."[53] As Roy Porter has shown, throughout the "long eighteenth

51. "*Gravity*," wrote Shaftesbury in "A Letter Concerning Enthusiasm," (1707), was "the very essence of imposture" (cited in Agnew, *Worlds Apart,* 162). Martinus Scriblerus personified the absurd, grave physician. See Kerby-Miller, ed., *Memoirs of Martinus Scriblerus.* On the debates about gentility, learning, profession, and trade among medical men at this time, see Guerrini, "'A Club of Little Villains.'" On gravity as a disguise for sexual predation, see Porter, "A Touch of Danger," 218. On the satire of gravity earlier in the century, see Hattori, "Performing Cures." I am grateful to Dr. Hattori for the point that satire of display seems to have increased later.

52. Fissell argues that the publication of Lord Chesterfield's *Letters to his Son* (1774) was an important turning point, but this seems late: Fissell, "Innocent and Honorable Bribes"; see also Porter, "Thomas Gisborne."

53. Makittrick, *Commentaries on the Principles,* xxxvii, xxxviii, xxxix–xl.

century" it was such "outward shews" that permitted many quacks to ape orthodox physicians.[54]

Satirizing the outward shews of physicians in 1813, the pseudonymous
satirist Peter MacFlogg'em reckoned all physicians to be shallow pretenders. He
noted that a great physician should prove himself a patron of the fine arts and
have a "*museum* furnished with fossils, minerals, stuffed birds, beasts, snails,
beetles, anatomical preparations, lusus naturae, and every other sort of natural
curiosity." Furthermore, the physician "should make it a maxim, never if possible to visit a patient, when he is absolutely so *drunk* that he can neither *walk*
nor *stand*." He was also quite explicit about the doctor's physiognomical and
corporeal presentation:

> a physician about to engage in the practice of medicine . . . should undergo
> a *complete metamorphosis* . . . If he ever possessed any thing like *ingenuous-*
> *ness, hilarity,* or good *humour,* these must all be speedily exchanged for a
> character the very *reverse* . . . by attempting to invest himself with every *ap-*
> *pearance* of sanctity, by assuming a *solemn* visage, and *formality* of deport-
> ment he should on no account ever be seen to smile.

At the bedside in a doubtful case the physician must indicate his judgment by "a
long face, and grave shake of the head."[55] Early in the twentieth century, William
Osler recommended, in all earnestness, that physicians adopt such an expression:

> The physician who shows in his face the slightest alteration, expressive of
> anxiety or fear, has not his medullary centres under the highest control,
> and is liable to disaster at any moment. I have spoken of this to you on
> many occasions, and have urged you to educate your nerve centres so that
> not the slightest dilator or contractor influence shall pass to the vessels of
> your face under any professional trial.[56]

■ *Picturing the Physician* ■

THE GRAVE, LEAN, AND LEARNED physician, who is a gentleman and philosopher, has a continuous iconographic history from the early modern period
to the present (figs. 5.5 and 5.6). Writing to his protégé studying at the Edinburgh medical school in 1792, the Bath physician Anthony Fothergill observed:

> Hence perhaps it is that the Scotch people, by dint of meagre diet become
> as keen as the northern blast, and excel all other nations in deep specula-
> tion. . . . If you but just kept from starving so much the better. You'll return,

54. Porter, *Health for Sale.*
55. MacFlogg'em, *Aesculapian Secrets Revealed,* 33, 53, 21, 63.
56. Osler, *Student Life,* 37.

FIGURE 5.5. Anthony Askew (1722–74), physician to Saint Bartholomew's Hospital. The book, of course, signifies learning. Mezzotint by T. Hodgetts after a painting by A. Ramsay in Emmanuel College Cambridge. Courtesy Wellcome Institute Library, London. V227B.

we expect, as thin as a rat, and with famine in your countenance but then you'll return sharp set and with your senses as keen as a razor.[57]

The association of slightness of build and intellectual excellence was maintained in medical theory. In 1807 Sir John Sinclair observed that "Brilliant faculties are

57. Fothergill, "Letters," 18 April 1792.

FIGURE 5.6. Sir Henry Wentworth Acland (1815–1900), Regius Professor of Medicine at Oxford. Again, like Osler, not a slight man but the cerebral gesture denotes the physician's association with scholarship. Lithograph by W. Rothstein. Courtesy Wellcome Institute Library, London. V34.

seldom accompanied with great strength of body . . . by far the greatest proportion of men, distinguished by their talents, are of puny, irritable frames and, consequently, not so likely to live long, as their more robust . . . brethren." [58] Physicians, therefore, had to be careful how they used this resource. As Iliffe (chapter 4 of this volume) and Browne (chapter 7) show, in the instances of Newton and Darwin sickness could be designated as constitutive of intellectual excellence; as Winter (chapter 6) demonstrates, this could also be so for women, although with more difficulty. This was scarcely a promising strategy for physicians to pursue. Indeed the early eighteenth-century chronicler of the diseases of

58. Sinclair, The Code of Health, 1:79–80.

FIGURE 5.7. Holmes here is being confronted by Watson, who is in a conventional nineteenth- and twentieth-century diagnostic pose—at his desk. In this instance, of course, the bluff Watson, although a doctor, is the stocky figure since he is confronted with the ultimate rational machine. See also figs. 5.10 and 5.11. From Arthur Conan Doyle, *The Memoirs of Sherlock Holmes* (London, Newnes, 1894), 258. Courtesy Wellcome Institute Library, London. L15282B.

workers, Bernardini Ramazzini, prided himself on how infrequently physicians became sick.[59]

The archetypal angular diagnostician is, of course, Sherlock Holmes, created by Arthur Conan Doyle and modeled on the Edinburgh diagnostician Joseph Bell[60] (fig. 5.7). A self-image, however, is not always the same thing as a public image and the doctor has a long history as a figure in caricature. But it is noteworthy, especially in examples from the golden age of caricature—the late eighteenth and early nineteenth centuries—how successful the physician had been

59. Ramazzini, *Diseases of Workers,* 391. I am grateful to Rob Iliffe for pointing this out to me.

60. Bell was actually a surgeon, but by this time surgeons had begun to borrow from the physicians' repertoire (see below). In any case what Doyle admired in Bell was his shrewd reasoning in difficult cases: Edwards, *The Quest for Sherlock Holmes.*

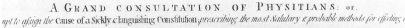

FIGURE 5.8. This is a political satire incorporating a demand for government reform. The physicians surrounding the bloated patient are all relatively slight and their depiction uses conventions appropriate for gentlemanly posture. Etching with line engraving by J. Eynon, 23 July 1756. Courtesy Wellcome Institute Library, London. V10894B

in transmitting his self-image to the public. As Brewer has shown, eighteenth-century caricaturists developed into a fine art the depiction of social place through such things as physiognomy, costume, and body language.[61] Plebeians have coarse features, they ogle and leer, they stand or sit pigeon-toed, they stoop and assume awkward postures. Gentlemen by contrast have out-turned feet and elegantly placed hands. Doctors, even when satirized, were presented as lean gentlemen rather than chunky plebeians (figs. 5.8 and 5.9). The physician retained his comparatively skeletal body in visual art throughout the nineteenth century, for example, in the cartoons of *Punch* (fig. 5.10). Neither has twentieth-century visual art dispensed with the convention (fig. 5.11). Repertoires, however, are not static. In representing the nineteenth-century doctor's impeccable understatement, *Punch*'s artists relied not only on the trope of the lean physician but on the sartorial style and postures of Beau Brummel (fig. 5.12).

61. Brewer, *The Common People*.

FIGURE 5.9. A medical satire on George IV's opposition to Catholic emancipation and Wellington's support of it. Colored etching by T. Jones, 1829. Courtesy Wellcome Institute Library, London. V11341B.

FIGURE 5.10. Stereotypical depiction of the physician as reasoner into the causes of disease. *Punch*, 1 April 1925, 346. Compare figs. 5.7 and 5.11. Courtesy Wellcome Institute Library, London. V11521.

FIGURE 5.11. "Sentence of Death" [ca. 1922?]. Compare figs. 5.7 and 5.10. Photogravure after J. Collier. Courtesy Wellcome Institute Library, London. V16100.

NOT A NICE WAY OF PUTTING IT.

She. "Oh, Dr. Pillsbury, I am so anxious about poor Mrs. Perkins. She is in your hands, is she not!"
Dr. Pillsbury. "She was; but I have left off attending her for the present."
She. "Oh that's good! She is out of danger then!"

FIGURE 5.12. Though the physician is represented as the conventionally lean figure, the joke at the physician's expense and the hint of dandyism are more likely to suggest to the reader the fashionable and possibly incompetent and predatory physician than the scholar. See also fig. 5.13. *Punch,* ca. 1900. Courtesy Wellcome Institute Library, London. V11458.

Thinness, however, is a resource that can be put to many uses. Shakespeare's Caesar, talking of Cassius, associates thinness with learning and a penetrating intellect, but also points to it as an indicator of coldness, reticence, and villainy: "I do not know the man I should avoid, so soon as that spare Cassius. He reads much; He is a great observer, and he looks quite through the deeds of men." [62] Narrowness, meanness, and smallness are the marks of deceit and cunning. The ingratiating, sinister, fashionable doctor is suitably angular (fig. 5.13). The dark side of the cerebral is nowhere more apparent than in the contrast between town and country physicians where refinement, urbanity, and slickness confront solid, honest rusticity (fig. 5.14). When doctors do appear as rotund, and they frequently do, it is usually in contexts in which they are concerned with material and corporeal things: money or seduction, for example (figs. 5.15 and 5.16).

62. *Julius Caesar*, 1.2.201.

THE FASHIONABLE PHYSICIAN.

He must have killed a great many people to get so rich.

MOLIERE.

FIGURE 5.13. "The Fashionable Physician." Wood engraving by J. Orrin Smith after J. K. Meadows, 1840. Compare fig. 5.12. Courtesy Wellcome Institute Library, London. V10930.

COUNTRY *and* TOWN
PHYSICIANS

FIGURE 5.14. "Country and Town Physicians." Etching, n.d. This is a thoroughly conventional representation of town-meets-country, and is just as likely to suggest downright honesty versus fashion than rusticity versus scholarship. Courtesy Wellcome Institute Library, London. V10891B.

Printed for & Sold by CARINGTON BOWLES,   No. 69 St. Paul's Church Yard, LONDON.

HOW MERRILY WE LIVE THAT DOCTOR'S BE
WE HUMBUG THE PUBLIC AND POCKET THE FEE.

FIGURE 5.15. "How merrily we live that Doctor's be. We humbug the Public and pocket the fee." There is no suggestion of scholarship in the plump doctor, and the lean figure suggests predation. Mezzotint after R. Deighton, 1793. Courtesy Wellcome Institute Library, London. V10904B.

FIGURE 5.16. "Medical Dispatch, or Doctor Doubledose killing two birds with one stone." Colored etching by T. Rowlandson, 1810. Courtesy Wellcome Institute Library, London. V11013.

■ *The Surgeon's Qualities* ■

THE PHYSICIAN THEN, since the early modern period, has presented himself as a lean, grave, learned gentleman, but what of the surgeon? Like physicians, surgeons struggled to identify themselves and to erect boundaries that would separate them from undesirable associations. Even elite surgeons were manual workers, users of instruments, and as such they had to differentiate themselves from the trades of, for example, barbering and, even worse, butchery. They had also to deal with the physicians themselves, who were emphasizing their liberal and learned image, and thus their superiority. In the sixteenth century English surgeons had displayed their learning. Physicians, apparently, were not threatened by this. Thus the physician John Securis: "Surgerie beynge counted as one of the chiefest parts of phisike . . . it were necessarie that the Surgion . . . were well learned in philosophie and phisicke."[63] But the surgeon also needed the resolve to operate and to exhibit physical skills that were not required by the physician. The surgeon, said William Bullein, himself a physician, in 1562, "must begin first in youth with good learning, and exercise in this noble art, he also must be clenly, nimble handed, sharpe sighted, pregnant witted, bolde spirited, clenly apparailed, pitefull harted, but not womenly affeccionated: to wepe or trimble, when he seeth broken bones."[64] That "womanly" qualities are inappropriate in a surgeon is a stipulation that has had a long, but by no means consistent, usage.[65]

The Elizabethan age did indeed see a flowering of English surgery, but there is evidence that the status of English surgeons fell in the following century.[66] By then physicians were designating surgery as merely a practice with a theory that rightly belonged to the learned head of the profession. James Primrose observed in 1651 that "now a daies [learning] is not desired in a Surgeon . . . Hence is it that whosoever have written any thing of Surgery . . . have been alwaies physicians."[67] Seventeenth-century English surgeons, however, gradually turned this designation into a Baconian virtue by emphasizing their practical, no-nonsense approach: "I am a Practiser, not an Academick," wrote Richard Wiseman, undoubtedly England's most famous surgeon of the day. Wiseman proclaimed his personal, hands-on approach to his art: "I continued the embrocation," "I

---

63. Securis, *A Detection and Querimonie,* Ellll.

64. Bullein, *Bulleins Bulwarke,* dialogue VIIr. For the earlier history, including a citation from the thirteenth-century surgeon Lanfranc to the effect that the surgeon should have a "strong body," see Schechter and Swan, "Historical Views of the Ideal Surgeon."

65. According to Jordanova, *Sexual Visions,* the surgical operation itself is a gendered action.

66. Lawrence, "Democratic, Divine and Heroic."

67. Primrose, *Popular Errours,* 39.

FIGURE 5.17. "Amputation." Compare the gross, robust surgeons and their postures with the physicians in fig. 5.8. Colored aquatint by T. Rowlandson, 1793. Courtesy Wellcome Institute Library, London. V11038.

chaffed the pained hip," "I let her blood," "I dressed her with an unguent."[68] This down-to-earth self-image was adopted by Wiseman's successors, for instance the naval surgeon James Handley, who, in 1710, observed that the surgeon, "ought to have . . . a competent Stock of Learning; to have an unshaken Courage, a steady Hand, a clear Sight, . . . and to be an *honest Man*."[69] In *The London Tradesman*, Campbell noted, "It is vulgarly said, that a Surgeon should have a Lion's Heart, a Hawk's Eye, and a Lady's Hand; Womanish Tenderness is very improper for a Surgeon."[70] The image of the robust surgeon was ready-made fodder for the caricaturist, who happily conflated country doctor, apothecary, and surgeon, whether high or low (figs. 5.17, 5.18, and 5.19). Even worse for surgeons, at least at first sight, is that they were also lumped with butchers. An association between meat eating and intellectual coarseness had been made

68. Wiseman, *Eight Chirurgical Treatises*, epistle, 17, 27.
69. Handley, *Colloquia Chyrurgica*, 1.
70. Campbell, *London Tradesman*, 48.

Breathing a vein.

FIGURE 5.18. "Breathing a vein." The surgeon or surgeon-apothecary here is, in his physique, posture, and dress the embodiment of weakness of intellect. Colored etching by J. Sneyd after J. Gillray, 1804. Courtesy Wellcome Institute Library, London. V11195B.

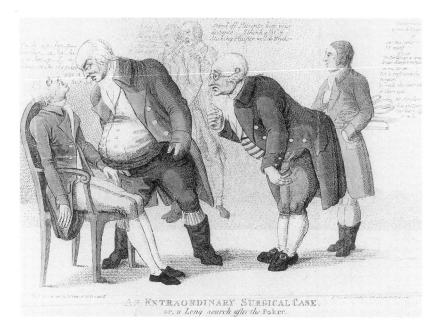

FIGURE 5.19. "An Extraordinary Surgical Case, or, a Long search after the Poker." Colored etching, 1800. Courtesy Wellcome Institute Library, London. V11041.

since antiquity and the view remained quite common. In *Twelfth Night* Sir Andrew Aguecheek observes, "I am a great eater of beef and I believe that does harm to my wit."[71] One eighteenth-century observer noted, "Vulgar and uninformed men, when pampered with a variety of animal food, are much more choleric, fierce and cruel in their tempers than those who live chiefly on vegetables." Likewise, the slaughtering of animals was viewed as predisposing to human cruelty. In Sir Thomas More's *Utopia* of 1516, slaves did the slaughtering. In Britain butchery was regarded by many as a thoroughly offensive and degrading trade. In Joshua Poole's *The English Parnassus* of 1677, butchers were described as "greasy, bloody, slaughtering, merciless, pitiless, cruel, rude, grim, harsh, stern, . . . surly." "The trade of a butcher," said Adam Smith, "is a brutal and an odious business."[72]

The association of surgeons with butchers goes back at least to the Middle Ages, when Henry de Mondeville declared that surgeons, like other specialists in flesh, must "boldly cut and destroy."[73] In 1665 George Thomson observed that

71. *Twelfth Night*, 1.3.90.
72. Cited in Thomas, *Man and the Natural World*, 287–300.
73. Cited in Pouchelle, *The Body and Surgery*, 76.

"we know . . . that Surgeon[s] are too forward to lop off parts, and butcherly to cut holes in the skin." [74] In 1787 Lord Herbert expostulated "Never for God's sake see a d——d D—ct—r" after he had done with the services of "butcher Pott." [75] In 1813 Peter MacFlogg'em recommended that the surgeon acquire "the unfeeling *brutality* of a butcher" and noted that, if in a head injury there seemed to be a deficit of brains, "You may procure enough of this article, to supply the melancholy defect, from some neighbouring *brother butcher.*" [76] It was widely supposed, incorrectly as it happens, that butchers and surgeons were excluded from jury service on account of a lack of feeling engendered by their trades. This error was repeated by Locke, Swift, and Mandeville.[77] Tyburn epitomized the connection of butchers and surgeons, for both occupations had close associations with the gallows. John Gay rhymed "Butchers! Whose hands are dy'd with blood's foul stain. And always foremost in the hangman's train." [78] Until 1832 surgeons in Britain took their anatomical subjects from the gallows. The Cato Street conspirators of 1820 were beheaded with a surgeon's knife.[79] Surgeons also used the language of butchers. For instance, in 1713 William Cheselden observed that the "COLON is the first of the great GUTS," the "PANCREAS, the *sweet Brede.*" [80] In caricature, surgeons almost uniformly appear as stout, relatively vulgar figures like their brother the butcher [81] (fig. 5.20).

■ *A Surgeon and a Gentleman?* ■

I HAVE USED the word *surgeon* thus far as though it designated a member of a single occupational group. But of course it did not. In the second half of the eighteenth century in London, Dublin, and Edinburgh, small numbers of highly skilled, learned practitioners, lumped with country doctors by the caricaturists, were endeavoring to elevate themselves and their craft *onto*, and indeed *above*,

74. Thomson, *Galeno-Pale*, 32. Thomson, a chemist, included physicians in this assessment.

75. Pott was the surgeon, Percival Pott. Quoted in Porter and Porter, *Patient's Progress*, 53.

76. MacFlogg'em, *Aesculapian Secrets Revealed*, 97, 111.

77. Stevenson, "On the Supposed Exclusion."

78. Gay, *Poetry and Prose*, 1:43. For the association of butchers, surgeons, and executioners see Pouchelle, *The Body and Surgery*, 75. In the eighteenth and nineteenth centuries they were joined by the "resurrection men." See Thomas Rowlandson's view of surgeons in Butterfield, "A Caricaturist."

79. Porter, *London: A Social History*, 154.

80. Cheselden, *Anatomy*, 103, 108. Susan Lawrence drew my attention to this quote in her paper "In the Details" for the symposium "Narrative and Medicine," at the Wellcome Institute, 2 June 1995.

81. All the medical orders were well aware of the stigmata of posture. For instance, a 1761 guidebook for apothecaries noted that one who has "the clumsiness, the walk, the air or the blunt rudeness of a plowman, can never be fit for this genteel profession"; cited in Fissell, "Innocent and Honorable Bribes," 24; see ibid., 25, for a quotation that indicates that in the eighteenth century the butcher could be cited as the embodiment of plebeian behavior.

FEMALE INFLUENCE: or, the DEVONS——E CANVAS.

FIGURE 5.20. "Female Influence; or, the Devons——e Canvas." 3 April 1794. A satire on the duchess of Devonshire canvassing for Fox. Note the butcher (recognizable by the apron and steel at his waist) in characteristic plebeian pose. Copyright British Museum. BMC6493.

that social plane occupied by physicians. They did this in two ways. First, they displayed surgery as a profession based on learning: the learning in this case being largely experimental science. One measure of this adoption of learning was the abandonment of the language of butchery for a technical terminology.[82] Second, and equally important, they began to present this scientifically principled art as something suitably practiced by gentlemen. Surgery was a significant site where scientific culture was adopted and used, not in opposition to gentility, as Shapin suggests happened among nonconformist circles, but to redefine gentlemanliness in professional terms.[83]

It is a striking feature of the representation of surgeons in the late eighteenth and nineteenth centuries that they are rarely depicted with their surgical instruments, that is, with the tools of their trade. They are, however, frequently represented as men of science and learning (figs. 5.21 and 5.22). By the twentieth century, neurosurgeons had plundered the physicians' repertoire so successfully

82. Again I am grateful to Susan Lawrence for this point; see note 80. On surgeons aspiring to be gentlemen in this period see her *Charitable Knowledge*.
83. Shapin, "'A Scholar and a Gentleman,'" 313.

FIGURE 5.21. John Hunter (1750–1809). Hunter is here, in his mission to raise the status of surgery, adopting the conventions used by physicians. He is a thin, cerebral man reading the books of scholars and the book of nature. Line engraving by W. Sharp, 1788, after a painting by Sir. J. Reynolds, 1786, at the Royal College of Surgeons. Courtesy Wellcome Institute Library, London. V3625.

FIGURE 5.22. Robert Liston (1794–1847), professor of surgery, University College, London. Mezzotint by C. Turner after himself, 1840. Courtesy Wellcome Institute Library, London. V3625.

that they could be regarded as the embodiment of penetrating intellect.[84] The representation by surgeons of themselves as men of science was part of that nineteenth-century restipulation of gentility in which, as Perkin observes, the professions, especially through Oxford and Cambridge, linked the idea of moral and intellectual superiority to a new concept of the gentleman, "no longer the aristocratic code of militaristic honour to be defended by duelling but the notion of a 'gentle man,' educated, courteous, well-spoken, and considerate of others."[85] As a guide to professional careers had expressed it in 1857, "The importance of the professions and the professional classes can hardly be overrated, they form the head of the great English middle class, maintain its tone of independence, keep up to the mark its standard of morality, and direct its intelligence."[86]

For physicians this move to consider themselves the quintessential embodiment of gentlemanliness, rather than simply as scholars aspiring to it, was not too difficult; they had long considered themselves gentlemen and, unencumbered as they were by the taint of manual activity, they could easily pass themselves off as learned, moral custodians. As the physician Patrick Black told the Saint Bartholomew's Hospital students in 1852, "Your profession *demands* from you that you shall possess what is called scientific knowledge, and it is expected from your acknowledged station in society that you should not be wanting in those accomplishments which distinguish the position of gentlemen."[87] The even greater professional triumph, however, was that of the surgeons, who were acknowledged by the Victorians as gentlemen. Who for instance would dare to say Joseph, Baron Lord Lister (fig. 5.23), was not a gentleman? In *Bleak House* Mr. Boythorn exclaims of Richard Jarndyce's intention to become a surgeon, "I rejoice to find a young gentleman of spirit and gallantry devoting himself to that noble profession!"[88] Nevertheless, as the perspicuous Anthony Trollope noticed, medical men's status fell in the shade of the other professions. Miss Marrable in *The Vicar of Bullhampton*, endorsed the church, law, and army and navy as suitable occupations for a gentleman but: "She would not absolutely say a physician was not a gentleman, or even a surgeon; but she would not allow to physic the same absolute privilege which, in her eyes, belonged to the law and the church."[89]

84. Speaking of the rift between Charles and Diana, the Prince and the Princess of Wales, a minister, Nicholas Soames, observed, "Clearly the marriage has irretrievably broken down. You don't need to be a brain surgeon to see that." *The Independent* (London), 22 November 1995. Quite how dealing with living brains confers on these specialists especially well-developed intellectual powers merits further investigation.

85. Perkin, *The Rise of Professional Society*, 84.

86. Cited ibid.

87. Black, *An Address*, 8.

88. Dickens, *Bleak House*, 219.

89. Cited in Peterson, *The Medical Profession*, 194.

FIGURE 5.23. Joseph Lister (1827–1912). Lister is represented as any other Victorian gentleman would have been. Photograph, possibly of a lithograph. Courtesy Wellcome Institute Library, London. V3618.

That surgeons had used science to make themselves into gentlemen was well recognized in the nineteenth century. John Hunter (fig. 5.21) was widely regarded in nineteenth-century Britain as the "founder of scientific surgery." James Paget noted,

> And now mark what he did for surgeons. Before his time they held inferior rank in the profession. . . . they were subject to the physicians, and very justly so, for the physicians were not only better learned in their own proper calling, but men of higher culture, educated gentlemen and the associates of gentlemen. From Hunter's time a marked change may be seen. Physicians worthily maintained their rank, as they do now, and surgeons rose to it. . . . Yes, more than any man that ever lived, Hunter helped to make us gentlemen. . . . Surely, that if we are to maintain the rank of gentlemen . . . It must be by the highest scientific culture to which we can attain.[90]

Arguably, it was this professional success of the surgeon that facilitated the introduction of instrumentation into bedside medicine, a possibility many gentlemen-physicians had once been reluctant to take up. Gentlemen, as the parallel case of geology confirms, could be tool-using creatures.

Nineteenth-century contemporaries recognized these gentlemanly aspirations of surgeons and had no compunction in mocking them: "Are you too fine a gentleman to think of contaminating your fingers by administering a clyster to a poor man?" wrote one satirist. The price of delicacy was effeminacy. The critic continued: "are your olfactory nerves so delicate, that you cannot avoid turning sick when dressing an old neglected ulcer? . . . If you cannot bear these things, put Surgery out of your head, and go and be apprentice to a man-milliner or perfumer."[91] Although, in general, qualities usually specified as feminine were designated as inappropriate in a surgeon, and although they were increasingly employed to stigmatize genteel aspirations, they could also be invoked to counter the ascribed coarseness of surgeons. John Flint South recalled that Henry Cline operating around this time "insisted on almost womanly tenderness in the necessary handling which his cases required."[92] In 1845 an anonymous author observed that surgery needed all the "milder virtues" and "gentleness," qualities

---

90. Paget, "The Hunterian Oration," 242.

91. Chamberlaine, *Tirocinium,* 51. Caricaturists were not impressed either. George Cruikshank's "The Examination of a Young Surgeon" of 1811 depicts the members of the court of examiners of the Royal College of Surgeons of England in the most plebeian of postures: Carty, "The Examination of a Young Surgeon," 214.

92. South, *Memorials,* 48.

usually associated with women. Elsewhere, however, he noted: "No man can know much of Anatomy, who is too finical or too lady-like to soil his delicate fingers."[93]

But in creating themselves as learned gentlemen-practitioners, surgeons did not simply appropriate the corporeal repertoire of the physician and appear as lean, grave, and learned. Far from it, they remodeled their traditional image. The robust practicality which surgeons attributed to themselves had been an object of satire in the eighteenth century and was easily caricatured as clumsiness and stupidity. As we have also seen, one means to denigrate the surgeon was to call him a butcher. But, as Brewer has observed, by the end of the eighteenth century the butcher, in some contexts, had come to represent common sense and the liberty-loving Englishman: "His work required skill, strength and lack of squeamishness; he was typically thought of in the eighteenth century as both bold and . . . extremely amorous" (fig. 5.24). Butchers were portrayed as heroes because they "presided over the preparation and sale of that most distinctive of English dishes, the roast beef of England." The butcher personified virility and was a "reminder of English prosperity."[94] Eighteenth-century caricaturists contrasted aristocratic effeminacy and the masculine, unaffected directness of the demotic butcher (fig. 5.25).

It was the sorts of qualities embodied in the butcher—a capacity for physical endurance, courage, solidity, and honesty—that were highly prized in the Victorian cult of manliness, a cult that, as Warwick (chapter 8 of this volume) demonstrates, was adopted by the most cerebral of creatures, Cambridge mathematicians. Butchers were hardly candidates for the Victorian manly ideal, whereas surgeons, increasingly accepted as gentlemen, were. In this context, as Johns has shown, French surgeons modeled themselves as the democratic heroes of everyday life. This modeling, she suggests, originated with the men who, in the Revolutionary years, created surgery as a medicine of the interior: "French surgeons put themselves forward . . . as the very embodiment of Enlightenment values and egalitarian opportunities taken seriously—as heroes of modern life."[95] By midcentury, British and American surgeons were also presenting themselves as the embodiment of heroism and manliness. In 1887 Lawson Tait said of his older contemporary Robert Liston that "we always spoke of him as a hero."[96]

93. Snipe [F. Garlick], *Remarks on Physicians*, 31.
94. Brewer, *The Common People*, 41–42.
95. Johns, *Thomas Eakins*, 58. Flaubert, in his *Dictionary of Platitudes*, observed that "Surgeons Have Hearts of Stone: Call Them Butchers": cited in Pouchelle, *The Body and Surgery*, 231 n. 29. See also the corpulent surgeon and hero of the hour, Canivet, in Flaubert's *Madame Bovary*.
96. Tait, "An Address," 167.

Io. Collet pinx.t   Le Francois a Londres.   C. White fecit.

The FRENCHMAN in LONDON.

Printed for Rob.t Sayer N.o 53. in Fleet Street, & Ino Smith, No 35. in Cheapside.

Published as the Act directs Nov.r 10. 1770.

FIGURE 5.24. "The Frenchman in London," 10 November 1770. The sturdy English butcher confronts the foppish Frenchman. For the butcher and sexuality see fig. 5.20 and Brewer, *The Common People and Politics 1760–1790s*, pl. 83. Copyright British Museum. BMC4477.

The PHYSICAL ERROR.

Butcher { Prepare Sir——You poison'd my wife——————and by Jupiter.
I'll Butcher you.

Doctor { Good Sir———— hear me ———— Indeed———— upon my Honor
it was only ?———— only only ———— A Physical Error.

Publish'd Jan 25 1782 by W.Humphrey N°227 Strand

FIGURE 5.25. "The Physical Error." A physician confronted by a butcher who claims "you poison'd my wife." It would not be difficult to mistake this picture for a caricature of the eighteenth-century physician and surgeon. The conflation of butcher and surgeon is complete. Etching by J. Kent after P. V. , 1782. Courtesy Wellcome Institute Library, London. V10959B.

American surgeons of the nineteenth century and beyond likened themselves to frontiersmen. As Frederick Dennis put it in 1904:

> There is no science that calls for greater fearlessness, courage, and nerve than that of surgery, none that demands more of self-reliance, principle, independence and the determination in the man. These were the charac-

FIGURE 5.26. "The Gross Clinic." The surgeon is elevated by Eakins far from the barbaric trade of the butcher, yet Gross remains firmly in the world of the here and now. Thomas Eakins, 1875. Oil on canvas, 78″ × 96″. Courtesy Jefferson Medical College of Thomas Jefferson University, Philadelphia, Pa.

teristics which were chiefly conspicuous in the early settlers of this country. And it is these old-time Puritan qualities, which, descending to them in succeeding generations, have passed into surgeons of America, giving them boldness in their art, and enabling them to win that success in surgery, which now commands the admiration of the civilized world.[97]

In Thomas Eakins's picture of the distinguished American surgeon Samuel Gross, the viewer could see the apotheosis of the surgeon as both solid democratic citizen and singular hero (fig. 5.26).

97. Dennis, *History,* 84.

What remains of these repertoires? In 1966 Sir Douglas Robb considered "the surgeon in the making . . . tends to be attracted to the real. . . . His understanding may not be very deep. . . . He is unhappy amid . . . the speculative." The surgeon is "more like the pilot of a jet aircraft."[98] More recently, a surgeon on a jumbo jet, on a flight from Hong Kong, performed an emergency, midair, lifesaving operation on a woman with a collapsed lung. The hero, Angus Wallace, was described by Professor Miles Irving of Manchester University as "a typical professor of surgery: ebullient, extrovert and confident. It was a brilliant piece of improvisation."[99] If the surgeon has achieved the status of hero, the unfortunate butcher has remained the symbol of stolidity. In one of his not so rare ad hominem references to a colleague, F. R. Leavis remarked "you won't learn anything from Dr Z. He looks like a pork butcher."[100]

98. Robb, "The Surgeon among Doctors." I have the impression, which I cannot detail, that surgeons have often compared themselves to pilots and even more often to captains of ships, notably sailing ships. See, for example, the opening quotation from Ogilvie, *Surgery.*
99. *The Times* (London), 24 May 1995.
100. Cited in MacKillop, *F. R. Leavis,* 12.

■ ACKNOWLEDGMENTS ■

I am grateful to Hal Cook and Roy Porter for their time spent talking about some of the themes developed here. Thanks to Caroline Overy for help with the research. For permission to quote from manuscript material I am grateful to the American Philosophical Society.

■ REFERENCES ■

Agnew, Jean-Christophe. *Worlds Apart: The Market and the Theater in Anglo-American Thought, 1550–1750.* Cambridge: Cambridge University Press, 1986.
Axtell, James L. "Education and Status in Stuart England: The London Physician." *History of Education Quarterly* 10 (1970): 141–59.
Black, Patrick. *An Address Delivered to the Students at St. Bartholomew's Hospital.* London, 1852.
Bowman, William. *An Introductory Address Delivered at King's College.* London, 1851.
Brewer, John. *The Common People and Politics 1760–1790s.* Cambridge: Chadwyck-Healey, 1986.
Browne, John. *Myographia Nova: Or a Graphical Description of all the Muscles in Humane Body.* London, 1697.
Bullein, William. *Bulleins Bulwarke of Defence Againste all Sicknes, Sornes, and Woundes.* London, 1562.
Butterfield, William C. "A Caricaturist of the Eighteenth Century Anatomists and Surgeons." *Surgery, Gynaecology and Obstetrics* 144 (1977): 587–92.
Campbell, R. *The London Tradesman.* 1747. Reprint, Newton Abbot: David and Charles, 1969.

Carty, Michael A. R. "The Examination of a Young Surgeon." *Journal of the History of Medicine* 36 (1981): 214–15.

Castle, George. *The Chymical Galenist.* London, 1667.

Chamberlaine, William. *Tirocinium Medicum.* London: for the author, 1812.

The Character of a Compleat Physician, or Naturalist. London, 1680.

Cheselden, William. *The Anatomy of the Humane Body.* London, 1713.

Clark, George. *A History of the Royal College of Physicians of London.* 2 vols. Oxford: Clarendon Press, 1964–66.

Colbatch, John. *A Physico Medical Essay.* London, 1696.

Cook, Harold J. *The Decline of the Old Medical Regime in Stuart London.* Ithaca: Cornell University Press, 1986.

———. "Good Advice and Little Medicine: The Professional Authority of Early Modern English Physicians." *Journal of British Studies* 33 (1994): 1–31.

Cotta, John. *A Short Discoverie of the Unobserved Dangers of Severall Sorts of Ignorant and Unconsiderate Practisers of Physicke.* London, 1612.

Dennis, F. S. *The History and Development of Surgery during the Past Century.* Philadelphia, 1905.

Dickens, Charles. *Bleak House.* 1853. Reprint, Harmondsworth: Penguin, 1971.

Drake, T. G. H. "The Medical Caricatures of Thomas Rowlandson." *Bulletin of the History of Medicine* 12 (1942): 323–25.

Edwards, Owen. *The Quest for Sherlock Holmes: A Biographical Study of Arthur Conan Doyle.* Mountrath, Ireland: Dolmen, 1983.

Fissell, Mary E. "Innocent and Honorable Bribes: Medical Manners in Eighteenth-Century Britain." In *The Codification of Medical Morality,* ed. Robert Baker, Dorothy Porter, and Roy Porter, 19–45. Dordrecht: Kluwer Academic Publishers, 1993.

Fothergill, Anthony. "Letters to John Woodforde." American Philosophical Society. BF823.

Freind, John. *The History of Physick: From the Time of Galen, to the Beginning of the Sixteenth Century.* 2 vols. London: J. Walthoe, 1725–9.

Gay, John. *Poetry and Prose.* 2 vols. Oxford: Clarendon Press, 1974.

Goddard, Jonathan. *A Discourse Setting Forth the Unhappy Condition of the Practice of Physick in London.* London, 1670.

Guerrini, Anita. "'A Club of Little Villains': Rhetoric, Professional Identity and Medical Pamphlet Wars." In Roberts and Porter, 226–45.

Guillim, John. *A Display of Heraldry. The fifth edition, much enlarged with great variety of bearings. To which is added A treatise of honour military and civil by Capt. John Logan . . .* London, 1679.

Handley, James. *Colloquia Chyrurgica.* 2d ed. London, 1710.

Harris, Walter. *Pharmacologia Anti-Empiric.* London, 1683.

Hattori, Natsu. "Performing Cures: Practice and the Interplay in Theatre and Medicine of the English Renaissance." D.Phil. thesis, University of Oxford, 1995.

Hill, Maureen. "An Iconography of Thomas Linacre." In Maddison, Pelling, and Webster, 354–74.

Johns, Elizabeth. *Thomas Eakins: The Heroism of Modern Life.* Princeton: Princeton University Press, 1983.

Jordanova, Ludmilla. *Sexual Visions: Images of Gender in Science and Medicine between the Eighteenth and Twentieth Centuries.* London: Harvester Wheatsheaf, 1989.

Kerby-Miller, Charles, ed. *Memoirs of the Extraordinary Life, Works, and Discoveries of Martinus Scriblerus.* New York: Russell and Russell, 1966.

Lawrence, Christopher. "Democratic, Divine and Heroic: The History and Historiography of Surgery." In *Medical Theory, Surgical Practice: Studies in the History of Surgery,* ed. Christopher Lawrence, 1–47. London: Routledge, 1992.

Lawrence, Susan C. *Charitable Knowledge: Hospital Pupils and Practitioners in Eighteenth-Century London.* Cambridge: Cambridge University Press, 1996.

MacFlogg'em, Peter [pseud.]. *Aesculapian Secrets Revealed.* London: C. Chapple, 1813.

MacKillop, Ian. *F. R. Leavis: A Life in Criticism.* London: Allen Lane, 1995.

Maddison, Francis, Margaret Pelling, and Charles Webster, eds. *Essays on the Life and Work of Thomas Linacre, c. 1460–1524.* Oxford: Clarendon Press, 1977.

Makittrick, James. *Commentaries on the Principles and Practice of Physic.* London, 1772.

Maynwaringe, Everard. *Medicus Absolutus.* London, 1668.

[Merret, Christopher]. *A Short View of the Frauds, and Abuses Committed by Apothecaries.* 2d ed. London, 1670.

Munk, William. *The Roll of the Royal College of Physicians of London.* 6 vols. London, 1878.

[Nedham, Marchamont]. *Medela Medicinae.* London, 1665.

O'Dowde, Thomas. *The Poor Man's Physician.* London, 1665.

Ogilvie, William Heneage. *Surgery, Orthodox and Heterodox.* Oxford: Blackwell Scientific, 1948.

Osler, William. *The Student Life.* Ed. Richard Verney. Edinburgh: E. and S. Livingstone, 1960.

Paget, James. "The Hunterian Oration." *Lancet* i (1877): 238–42.

Perkin, Harold. *The Rise of Professional Society: England since 1880.* London: Routledge, 1989.

Peterson, M. Jeanne. *The Medical Profession in Mid-Victorian London.* Berkeley and Los Angeles: University of California Press, 1978.

Porter, Dorothy, and Roy Porter. *Patient's Progress: Doctors and Doctoring in Eighteenth-Century England.* Oxford: Polity, 1989.

Porter, Roy. *Health for Sale: Quackery in England, 1660–1850.* Manchester: Manchester University Press, 1989.

———. *London: A Social History.* London: Hamish Hamilton, 1994.

———. "Thomas Gisborne: Physicians, Christians and Gentlemen." In Wear, Geyer-Kordesch, and French, 252–73.

———. "A Touch of Danger: The Man-Midwife as Sexual Predator." In *Sexual Underworlds of the Enlightenment,* ed. G. S. Rousseau and Roy Porter, 206–32. Manchester: Manchester University Press, 1987.

Pouchelle, Marie-Christine, 1990. *The Body and Surgery in the Middle Ages.* Trans. Rosemary Morris. New Brunswick, N.J.: Rutgers University Press.

Prest, Wilfrid, ed. *The Professions in Early Modern England.* London: Croom Helm, 1987.

Primrose, James. *Popular Errours. Or the Errours of the People in Physick.* London, 1651.

Ramazzini, Bernardino. *Diseases of Workers* [1713]. Trans. Wilmer Cave Wright. Chicago: University of Chicago Press, 1940.

Reynolds, J. Russell. *Essays and Addresses.* London: Macmillan, 1896.

Richardson, Ruth. *Death, Dissection and the Destitute.* London: Routledge and Kegan Paul, 1987.

Robb, Douglas. "The Surgeon among Doctors." *Surgery* 60 (1966): 948–49.

Roberts, Marie Mulvey, and Roy Porter, eds. *Literature and Medicine during the Eighteenth Century.* London: Routledge, 1993.

Schechter, David Chas, and Henry Swan. "Historical Views of the Ideal Surgeon." *Surgery* 50 (1961): 427–36.

Schmitt, Charles. "Aristotle among the Physicians." In *The Medical Renaissance of the Sixteenth Century,* ed. Andrew Wear, Iain Lonie, and Roger French, 1–15. Cambridge: Cambridge University Press, 1985.

Securis, John. *A Detection and Querimonie . . .* London, 1566.

Shapin, Steven. "'The Mind Is Its Own Place': Science and Solitude in Seventeenth-Century England." *Science in Context* 4 (1991): 191–218.

———. "'A Scholar and a Gentleman': The Problematic Identity of the Scientific Practitioner in Early Modern England." *History of Science* 29 (1991): 279–327.

———. *A Social History of Truth: Civility and Science in Seventeenth-Century England.* Chicago: University of Chicago Press, 1994.

Sinclair, John. *The Code of Health and Longevity.* 4 vols. Edinburgh: printed for A. Constable, 1807.

Snipe, Surgeon [F. Garlick]. *Remarks on Physicians, Surgeons, Druggists, and Quacks.* Halifax: H. Martin, 1845.

South, John Flint. *Memorials.* London: J. Murray, 1884.

Sprackling, Robert. *Medela Ignorantiae: Or a Just and Plain Vindication of Hippocrates and Galen.* London, 1665.

Stevenson, Lloyd. "On the Supposed Exclusion of Butchers and Surgeons from Jury Duty." *Journal of the History of Medicine* 9 (1954): 235–38.

Stone, Lawrence. "Social Mobility in England, 1500–1700: Conference Paper." *Past and Present* 33 (1966): 16–55.

T., C. *Some Papers Writ in the Year 1664.* London, 1670.

Tait, Lawson. "An Address on The Development of Surgery and The Germ Theory." *British Medical Journal* 2 (1887): 166–70.

Thomas, Keith. *Man and the Natural World: Changing Attitudes in England, 1500–1800.* London: Allen Lane, 1983.

Thomson, George. *Galeno-Pale.* London, 1665.

Trye, Mary. *Medicatrix or the Woman-Physician . . .* London, 1675.

Twysden, John. *Medicina Veterum Vindicata: Or an Answer to Book, entitled Medela Medicinae.* London, 1666.

Wear, Andrew. "Medical Ethics in Early Modern England." In Wear, Geyer-Kordesch, and French, 98–130.

Wear, Andrew, Johanna Geyer-Kordesch, and Roger French, eds. *Doctors and Ethics: The Earlier Historical Setting of Professional Ethics.* Amsterdam: Editions Rodopi B.V., 1993.

Webster, Charles. "Thomas Linacre and the Foundation of the College of Physicians." In Maddison, Pelling, and Webster, 198–222.

Wiseman, Richard. *Eight Chirurgical Treatises.* 4th ed. London, 1705.

■ SIX ■

A CALCULUS OF SUFFERING

Ada Lovelace and the Bodily Constraints on Women's
Knowledge in Early Victorian England

ALISON WINTER

■ *Introduction* ■

VICTORIAN INTELLECTUALS often presented themselves as sickly.[1] Ada
Lovelace, expositor of Charles Babbage's Analytic Engine and self-styled "high
priestess" of mathematics, was one of the sickliest. As Lord Byron's daughter, and
later countess of Lovelace, she was a prominent figure in aristocratic and intellectual circles throughout her short life. She was known for her mathematical
and scientific interests, which brought her into correspondence with Babbage,
Augustus De Morgan, Charles Wheatstone, William Carpenter, and other eminent figures in the 1840s. Early in her life these friends expected great intellectual
achievements of her. Lovelace's rich intellectual and social life is open to a number of interpretations. It is noteworthy, although I do not propose systematically
to address the issue here, that her various intellectual reflections were prompted
by her relations with specific men. Lovelace's only published writings were to be
a translation in 1843 of a paper by L. F. Menabrea with a set of "Notes" of her
own on Babbage's plans for his Analytic Engine. Her many other projects never
reached fruition. Instead, her personal life deteriorated, and her mathematical
skills were put to work at gambling, her Byronic passions turned from algebra to
extramarital affairs, and her health declined precipitously. In 1852 she died of a
uterine tumor, at the age of thirty-seven.

Lovelace's invalidism was by no means unusual. The frail, stoical, house-
bound, and self-monitoring existence of the invalid was one that many Victorian
ladies experienced at one time or another.[2] More significant, "life in the sickroom" replicated in extreme form certain characteristics otherwise definitive of
idealized Victorian femininity: here could be found a delicate creature, confined
to the domestic sphere, whose weak body left her dependent on others and simultaneously contributed to her enhanced spirituality.[3] Lovelace was unusual,
however, in the ambitions she voiced from this predicament. Intellectual achieve-

---

1. Oppenheim, *"Shattered Nerves,"* but see also Warwick, chapter 8 of this volume.
2. Bailin, *Sickroom in Victorian Fiction.*
3. Martineau, *Life in the Sick-Room.*

ments of any kind made a woman stand out in this culture. Lovelace was especially remarkable because her major field of study was one widely reckoned to be peculiarly inaccessible to a woman's mind: mathematics.

Throughout the late 1830s and 1840s, Lovelace sought to become an accomplished mathematician, summed up by the term "analyst"—a word that loosely evoked the notion not only of a master of mathematics but of a thinker of sufficient knowledge and mental power to dissect formal systems into their constituent parts and explain their interactions. As an analyst, Lovelace aspired both to contribute to existing mathematical knowledge and also, far more ambitiously, to study the bodily and mental dynamics involved in the production of such knowledge. All the more striking, then, was her claim that her menstrual, digestive, and nervous complaints and her mathematical powers were intimately connected. During the 1840s she explored their relations, claiming, at times, that her intermittent invalidism was a necessary condition for her intellectual ability.

This representation should not come as a complete surprise. Other essays in this volume demonstrate that sickliness has often been a prominent component in the self-representations of intellectuals and natural philosophers. They likewise reveal the use of gendered characteristics in representing the bodies of such individuals. Shapin (chapter 1) mentions how an understanding of women's relatively "colder" physiology provided both a justification for their lack of philosophical activity and an explanation for those rare exceptions to the rule; Iliffe (chapter 4) and Warwick (chapter 8) both note their actors' references to "effeminacy"; Lawrence (chapter 5) traces the ambition of surgeons to have "a lady's hand"; and Browne (chapter 7) illustrates how characteristics such as a flowing beard shaped Darwin's public identity.[4] Such arguments demand that we pay attention to the history of representations of gendered characteristics in relation to the development of knowledge. Claims about what masculine characteristics enabled often entailed views about what femininity obstructed. Conversely, stipulations about what feminine characteristics facilitated were not necessarily prescriptions that women could achieve success.[5]

Thus when Lovelace herself raised the question of how a Victorian *woman* could adapt to her own purposes conventions for portraying the intellectual, gender and sickness were crucial issues. Her answer had an audacity typical of her. She suggested not only that her body's complaints enhanced her mathematical powers, but that they provided her with a "laboratory" in which to explore the corporeal conditions for producing knowledge more generally. Her disorders

4. See in particular Shapin's discussion of Anne Conway in chapter 1 of this volume.
5. Shapin and Barnes, "Head and Hand"; Shapin and Schaffer, *Leviathan and the Air-Pump*; Shapin, *Social History of Truth*, 86–89 (for gender).

were, Lovelace stressed, evidence of a "frame so *susceptible* that it is an *experimental laboratory* always about me, & inseparable from me."[6] Beyond the pursuit of novel mathematical knowledge, Lovelace aspired to use this "*Molecular Laboratory*" to produce what she called a "*Calculus* of the *Nervous System.*"[7] By reflecting on the relations between her own body and the intellectual powers it facilitated, she claimed it would be possible to develop an understanding of the physical conditions for mental work in general—though, in fact, she never developed more than a series of preliminary thoughts on the subject.

One could not, and I will not, argue that Lovelace was typical of contemporary women, although the way in which she differed from them makes her of unusual historical interest: in and around her representations of herself were concentrated some of the central cultural issues of the period, and these were explored as she and others tried to interpret her mind, her body, and the relations between them. One should also note that there *was* a group, albeit a small one, which Lovelace might have claimed to represent. One fundamental characteristic that intellectual women shared was their very oddity—the fact that they were not like other women. They constituted a group defined by its individual members' personal peculiarities. But while these individuals did not *stand for* womanhood in general, they could sometimes venture to *speak for* them. Both individually and as a group, intellectual women became the focus of cultural debate. Lovelace in particular regarded herself and was used by those around her as an experiment in how masculinity and femininity could be defined in relation to knowledge making.

This chapter examines how, in the 1830s and early 1840s, Lovelace portrayed and discussed this issue with her family and a prestigious circle of natural philosophers and medical doctors. In these years, her intellectual aspirations developed in parallel with her physical ailments. Lovelace's first serious illness came when she was seven and was followed by other childhood complaints (fig. 6.1). In adolescence there was a three-year spell of paralysis that left her on crutches and intermittently confined her to bed.[8] Later she experienced exhaustingly heavy menstrual periods, excruciating headaches, and a variety of gastrointestinal problems, all of which she mentioned, usually in delicately ambiguous terms, in correspondence. On one occasion, a tutor's instruction to conduct an astronomical observation could not be carried out because Lovelace had lost "all power to

6. Lovelace to Lady Byron, 11 November [1844], cited in Toole, *Ada, the Enchantress of Numbers,* 291.

7. Somerville Papers, Bodleian Library, University of Oxford (henceforth SP), Dep. c. 367, 15 November 1844.

8. Moore, *Ada, Countess of Lovelace,* 28–29, 40.

FIGURE 6.1. Lovelace, early 1820s. Courtesy British Museum.

walk or stand": one of several effects, according to her mother, of "measles, and too rapid growth."[9] In 1832–33 she "used to lie a great deal in a horizontal position" and "was subject to fits of giddiness when she looked down from any height."[10] Concentrated thought could bring on similar physical discomfort. After one tutorial session with Mary Somerville, Lovelace told her that "I was shattered when I left you." "When I am weak," she explained, "I am always so exceedingly terrified at *nobody knows what,* that I can hardly help having an agitated look & manner; & this was the case when I left you."[11] By the mid-1830s, her nervousness, shaking fits, and attacks of terror were being given various and conflicting explanations. One of her most troubling instances of nervousness was a pronounced distaste for her own children's company. This aversion intensified markedly as her passionate attachment to mathematical study grew. By 1840, in what was an extraordinary admission for a young Victorian wife, Lovelace confessed to a prospective governess her "total deficiency in all natural love of children."[12] The 1840s brought a fresh wave of problems, including cold spells, pains in her back, kidney problems, and emotional upsets.[13] Later in the

9. Lovelace Papers, Bodleian Library, University of Oxford (henceforth LP), 71, as cited in Stein, *Ada: A Life,* 29.

10. SP Dep. b. 206, Greig Memoir, as cited in Stein, *Ada: A Life,* 34.

11. SP Dep. b. 367, 20 February [1835].

12. LP 168, fol. 103, 13 December 1840.

13. Stein, *Ada: A Life,* 281–97; Baum, *Calculating Passion,* 91.

decade her health improved intermittently, but in 1851 it declined disastrously and she died the following year.

Throughout the 1830s and 1840s Lovelace's family and friends debated at length how her ailments would affect her ability to realize her ambitions: what sickness prevented her from doing and what it enabled her to do. Examining how they did so, and the circumstances under which she endeavored to occupy the recognized roles of "genius" and "analyst," will reveal some of the profound and paradoxical problems of self-presentation that confronted women intellectuals in the period.[14]

■ *"That Spark of Heaven That Is Not Granted to the Sex"* ■

BY THE MIDDLE DECADES of the nineteenth century, gender had become a particularly significant variable for defining intellectual identities and, for that matter, all social roles. It has been argued that in the early modern period gender, although important, was subservient to gentility in the allocation of credibility.[15] From the late eighteenth century, however, gendered characteristics became increasingly central for distinguishing producers from reproducers, competent from incompetent, domestic from public, and mad from sane.[16] The increasing stress on sexual difference as relevant to the ordering of status and credit coupled with a new stress on sex as generalized throughout an individual's corporeal and mental being had a number of implications.[17] One was that possibilities for intellectual and public activity, which existed for ladies in preceding generations, became more constricted.[18] The decades between 1800 and the mid–nineteenth century saw great debate on the relationship between gendered characteristics and intellectual capacity. Not only was it the case that no stable model of the corporeal dynamics of knowledge could be established, but individuals of all parties and persuasions felt free to mix and match the resources of gender, as and when they needed, to serve a wide range of particular and immediate ends. It is there-

14. While I will use the term *intellectual* to refer to Lovelace, it is important to note that this usage is problematic. There exists no overarching term that refers to individuals who distinguished themselves by original, significant reasoning, as the portrayal of the "deep thinker" varied nationally and regionally in different social contexts. This problem is particularly acute in relation to women thinkers, of course, because (as this chapter emphasizes) the question of how one would define and designate a woman thinker was very much at issue in this period.

15. Shapin, *Social History of Truth,* 86–91; Findlen's study of Laura Bassi ("Science as a Career in Enlightenment Italy") indicates a lack of any extensive discussion by her contemporaries of the significance of her sex for her prominent position.

16. Outram, "Before Objectivity"; Jordanova, *Sexual Visions;* Davidoff and Hall, *Family Fortunes;* Showalter, *Female Malady.*

17. Moscucci, *Science of Woman,* Russett, *Sexual Science.*

18. Myers, *Bluestocking Circle;* Outram, "Before Objectivity."

fore possible to sketch the broad outlines of these representations, but it is necessary at the outset to stress their malleability and the conflicting ways in which they were used.

An especially powerful and long-lived representation portrayed woman as lacking creative powers and therefore as suited only to mechanical forms of intellectual activity. Variations of this representation were devised in a number of different contexts throughout the nineteenth century. Earlier conventions according to which intellectual power could be demonstrated by virtuoso performances of calculation gave way to the notion that calculation could be rendered "mechanical," and that it was therefore properly associated with manual labor.[19] This devalued prodigious calculating feats: such skill became regarded as essentially uncreative, and thus accessible to women.[20] Similarly, with the rise of cheap print in the first half of the nineteenth century, women who wrote as scientific popularizers were regarded as uncreative but reliable reproducers of knowledge. Harriet Martineau, Jane Marcet, and Mary Somerville were presented in this way as laborers—repackaging and disseminating knowledge created by men.[21] Somerville was particularly successful in portraying *herself* as just such a capable but unoriginal intellectual worker. Although renowned for her scientific publications, in particular *The Connexion of the Physical Sciences,* she persistently proclaimed herself "conscious that I had never made a discovery myself, that I had no originality." She could properly lay claim to "perseverance and intelligence," Somerville remarked, but not to "genius, that spark from heaven is not granted to the sex."[22]

Competing with this representation was a claim—again, one with a long history—that women were untrustworthy purveyors of knowledge because their imaginations were *too* powerful. Their bodies were apparently subject to nervous influences that overstimulated the imagination and made it impossible for them to be free agents in the evaluation of evidence. These powers were held to destabilize even unoriginal and repetitive work. Such a resource, however, was also turned to alternative ends and used to show that a powerful imagination could generate original ideas. It was employed to argue that women were particularly suited to the most creative kinds of intellectual work. However, opponents of this view claimed that women's bodies were not strong enough to harness these pow-

19. Daston, "Enlightenment Calculation."
20. Grattan-Guinness, "Work for the Hairdressers." For an American example, see Rossiter, *Women Scientists;* on human "calculators" in the nineteenth century, see Schaffer, "Astronomers Mark Time" and "Babbage's Dancer."
21. David, *Intellectual Women and Victorian Patriarchy;* Winter, "Harriet Martineau"; Patterson, *Mary Somerville;* Proctor, *Light Science for Leisure Hours;* Marcet, *Conversations on Chemistry.*
22. Cited in Baum, *Calculating Passion,* 38.

ers to generate authoritative knowledge.[23] They could not create knowledge, but only fancy. Worse still, women doing intellectual labor were likely to overstrain their delicate bodies; such activity produced madness or other distempers.

The question of the bodily constraints placed on women by their reproductive systems was central to a very wide range of debates over women's education, moral fortitude, and mental health. Women's reproductive parts were seen to be almost always in precarious conditions (while a woman was celibate or during menstruation, pregnancy, and nursing for instance.) The imagination and the female reproductive system had long been seen to be in a pernicious reciprocal interaction, each encouraging the other toward unhealthiness or, through a contest between their energies, limiting the other's ability to function. The nineteenth century saw increasing efforts to analyze reproductive functions and understand the production of nervous disorders, especially in women.[24] The study of these phenomena thus promised a scientific allocation of social identities—including intellectual ones.[25]

By the middle decades of the century, the notion of the conservation of bodily resources or energy was becoming important to discussions about mind-body relations in women, and the idea that only a very strong body—a man's body—could produce sufficient power to support sustained intellectual functioning was growing influential.[26] In particular, by midcentury, a model of "manly" mathematical study obtained, according to which intense concentration would destroy the health of all but those of the most robust physical constitution.[27] It was a common assertion that a woman, by virtue of the fundamental and general fact that her body was smaller than a man's (and had to sustain a reproductive system too), possessed insufficient strength to cope with intellectually stringent demands. This claim was not, of course, limited to mathematical work but was related to a wide range of possible areas of mental activity whose suitability for women (and vice versa) was being debated in this period. It was not only the case that women were thought to be hampered by their generally weaker bodies; they were also hindered in their intellectual pursuits by the specific demands of reproduction. The powers the body could supply to the mind were severely limited by the need to allocate a large proportion of resources to perpetuating the

23. Carter, *Hysteria.*
24. Poovey, *Uneven Developments,* 36–38; Moscucci, *Science of Woman;* Winter, "Harriet Martineau."
25. Schiebinger, *The Mind Has No Sex?;* Showalter, *Female Malady;* Jordanova, *Sexual Visions;* Moscucci, *Science of Woman,* 15.
26. Burstyn, "Education and Sex"; Conway, "Stereotypes of Femininity"; for an American example, see Smith-Rosenberg, "Beauty, the Beast, and the Militant Woman."
27. Warwick, chapter 8 to this volume.

species. In a healthy woman, these limitations were seen to disincline her to intellectual work. Conversely, intellectual activity could compromise a woman's capacity to reproduce. In extreme cases, the mutual strains between body and mind were known to produce insanity, sterility, cancer, or unhealthy offspring.[28]

Both of these representations—women tempest-tossed by their stormy imaginations or trustworthy because they had none—were unstable and contested. Even Somerville's writings could be represented as having profound implications for original scientific work, despite her own protestations. William Whewell famously coined the term "scientist" in a review of her *Connexion of the Physical Sciences.* His well-known criticisms of the tendency of sciences toward overspecialization, and his struggle to characterize its practitioners, took shape in the context of considerations of the fact that, as he put it, there was a "sex in minds" which was directly relevant to these issues. Women could not usually carry out intellectual work, according to Whewell, because in them "the powers of thought are less developed than the instincts of action." But while they might understand less of the natural order, what women did comprehend they understood with a greater "clearness of perception, as far as it goes." [29] When they did theorize, they did so as if from a great height, far removed from the vicissitudes of daily life. Their thoughts (said Whewell, quoting Milton) took shape "In regions mild, of calm and serene air, / Above the smoke and stir of this dim spot. / Which men call earth." [30] Somerville's gender had thus enabled her to envisage her *Connexion* in a way that would have been far more difficult for a man. The work, Whewell proclaimed, stood as a crucial remedy to the current "proclivity" of science for "separation and dismemberment," and even to "disintegration"; without such correctives, he foresaw the possibility of a "great empire falling to pieces." [31] Whewell thus transformed the supposed "generalist" tendencies of women's minds from a thing excluding them from innovative scientific work into something enabling them to contribute to science in a distinct manner. His commentary depended upon the complexity of the notion of women as "diffusers" of knowledge.

The representation of some women as being in thrall to their imaginations also had the potential for positive transformation, for making a picture of the female as prophetess or oracle. Such a transformation, however, was both difficult

28. Cooter, "Dichotomy and Denial"; Scull, *Most Solitary of Afflictions;* Showalter, *Female Malady.*

29. Whewell, "Mrs Somerville," 65.

30. Whewell's quotation came from lines 4–6, spoken by a spirit, of Milton's "Comus" (a masque performed at Ludlow Castle).

31. Whewell, "Mrs Somerville," 59 ff.

and dangerous. It made it possible to describe a feminine role embodying creativity, elevating women to the level of imaginative characters, attributing to them a degree of insight unattainable by less fanciful individuals. Such possibilities were realized in the 1830s to 1850s, especially by women (but not only women) invalids.[32] It is clear from recent studies of such figures as Harriet Martineau, Elizabeth Barrett, and Florence Nightingale, and also from the testimony of those who suffered short-term debilities, like John Stuart Mill and Maria Edgeworth, that invalidism was turned into an experimental resource of particular value in this period.[33] Many invalids, and particularly women, reflexively studied their bodies and minds, to ascertain how their sickly states affected their intellectual and creative powers, and even their public authority.

Some women tempted into such experiments, and many who were not, adopted the role of the priestess, prophetess, or even oracle—figures of otherworldly inspiration, embodying valuable knowledge that they offered to others. Some women found that their encounters with trance phenomena offered them a resource for embodying authority. Extreme examples of this were spiritualist mediums who made a living from trance states during the second half of the century. The paradoxical status of the entranced subject, who drew her authority from the fact that she did not control her own mental state, was crucial to her credibility.[34] One did not have to make a career of entrancement in order to find this a useful state. In describing the beginning of a project, the discovery of a vocation, or a life-changing decision, intellectual women often availed themselves of this role.[35] Harriet Martineau is a case in point. She was one of England's most reliable journalists and popularizers and saw herself as an unambitious "diffusor" of knowledge created by others until she became involved in mesmerism in

32. Winter, "Harriet Martineau"; Owen, *Darkened Room;* Barrow, "Why Were Most Medical Heretics at Their Most Confident?"; Basham, *Trial of Woman;* see also Schaffer, "Self-Evidence."

33. Cooter, "Dichotomy and Denial"; Winter, "Harriet Martineau"; Bailin, *Sickroom in Victorian Fiction.* On Nightingale, see Vicinus and Nergaard, *Ever Yours, Florence Nightingale.* Showalter, "Florence Nightingale's Feminist Complaint"; Pickering, *Creative Malady;* and Smith, *Florence Nightingale,* 90–92, give differing accounts of Nightingale's invalidism; Mill, *Autobiography.*

34. On the significance of the trance in these women's intellectual and professional identities see Owen, "Women and Nineteenth-Century Spiritualism," and idem, *Darkened Room.*

35. For instance, Josephine Butler wrote that her vocation to champion women's rights came to her in a trance; Elizabeth Barrett routinely spoke of herself as carrying out various actions in an entranced state; and Harriet Martineau dictated philosophy while mesmerized. See Butler, *Personal Reminiscences of a Great Crusade,* and Barrett, *Letters of Browning,* 3:60. The best details I have found on Martineau's mesmeric philosophizing are in the manuscript diaries of January–May 1845 of George William Frederic Howard, seventh earl of Carlisle, Castle Howard. For other examples, see Annie Besant's decision to marry and her crisis of faith, in her *Autobiographical Sketches,* and Frances Power Cobbe's conversion to philosophical idealism in the wake of the Irish potato famine, recorded in Cobbe, *The Life of Frances Power Cobbe.* Constance Naden is another example; see Moore, "The Erotics of Evolution."

the mid-1840s. The mesmeric trance, she told correspondents, gave her the ability to resolve profound philosophical issues. During one experiment she "saw the march of the whole human race" and understood the relationship of "all the idolatries of the earth." [36] As one might expect, however, appealing to such resources carried risks. The priestess and prophetess were individuals who consciously wielded authority and imparted knowledge, and, like oracles, were vehicles for the transmission of otherworldly communications. Oracles, however, were direct and exclusive conduits for the expression of supernatural meaning and did not presume to interpret for their audience. The danger of portraying oneself as priestess or prophetess was of openly laying claim to far more authority than was acceptable for a woman; the danger associated with oracular portrayals, on the other hand, was that one had no control over how one's claims (and even oneself) were interpreted. A woman might find herself portrayed as no more than an assemblage of phenomena fit for someone else to explain. So, when Harriet Martineau attempted to draw specific conclusions from her own self-experimentation, one medical commentator scoffed, "It would be somewhat odd for the disease to *give itself* a name." [37]

Given these many, potentially conflicting, representations, it should not be surprising that some women thinkers experienced their bodies as an ongoing experiment in what kinds of physical states would sustain intellectual processes. Individuals content to be portrayed as mere diffusors or replicators of preexisting knowledge did not need to engage in these types of concerns. Somerville's work, for instance, was "disembodied" in the sense that she and others did not reflect on the bodily processes that enabled her work. Whewell did not resort to physiological explanations in discussing how she produced the knowledge that she did. As unusual as her intellectual strengths were, their productions were implicitly accepted as falling within the natural parameters of a woman's mental organization. But women who staked claims to originality were often portrayed, by themselves as well as others, in terms that stressed particular conformations of their bodies. For them, the process of discovering a truth about nature could not be extricated from the process of ascertaining the physiological basis of making that discovery. In consequence, their own physiology, along with its relations to their mental operations, and the implications of such relations for women's mental powers in general, became subjects of investigation. No woman was more ambitious in what she sought to gain from such a course of study than Ada Lovelace.

36. Martineau to Richard Monckton Miles, 2 February 1845, Houghton Papers, Trinity College Cambridge, 16 [61]. For further examples and discussion, see Basham, *Trial of Woman*, 127, 144–45, 181–86.
37. Pathologist, letter.

■ *Mathematical Disciplines and Passions* ■

FROM EARLIEST CHILDHOOD, the issue of Lovelace's potential became a central concern to her elders and, in due course, to herself. Initially, the question asked was, to what degree had she taken the impress of each of her estranged parents. Her mother, Lady Annabella Augusta Noel Byron, was renowned for her piety, her command of those around her, and her unusual gift for mathematics. She wished to encourage the development of whatever her daughter had inherited of these qualities. At the same time, however, she was determined to expunge any legacy left by Lovelace's father. While a talent for poetry might be cultivated, Byron's infamous passions would surely prove even more destructive in a woman than they had in the poet. Thus, while Lovelace was thought to have the seed of genius within her, the obvious archetype for fashioning her as an original thinker was irredeemably tainted by excess.

From her infancy, Lady Byron intended Lovelace's education to be an "operation" on her "brain," designed to cleanse her of Byronic passions.[38] Lady Byron ruminated in her personal journal about how she might inculcate intellectual strength in her daughter, and annihilate every "delusion of the Imagination."[39] It is important to note that this did not merely mean encouraging a resilient independence of mind; in fact, Lady Byron struggled throughout her daughter's short life to maintain parental control. Central to her plans was a rigorous course of mathematics, rather than the sciences of botany or geology which were more conventional areas of study for a young lady, because Lady Byron saw this field as the best means of developing and sustaining mental order and intellectual strength.[40] It was also a continuation and potential fulfillment of her own interests, for Lady Byron reckoned herself an accomplished mathematician. Moreover, this emphasis on mathematics was redoubled after 1833, when the threat Lovelace's passionate nature posed to her moral and intellectual development was forcefully manifested in an attempt to elope with a tutor. After the couple were separated, Lovelace penitently agreed to an intense course of mathematics on the grounds that "nothing but very *close and intense* application to subjects of a scientific nature" would keep her imagination from "running wild."[41] She began her studies with a family friend, a Dr. William King of Brighton (no relation to her future husband). King agreed with Lady Byron that mathematics was the appropriate "mental discipline," because it involved "a natural sequence of ideas, which the mind can work out for itself . . . [having] no connexion with the feel-

38. Baum, *Calculating Passion,* 15; Moore, *Ada, Countess of Lovelace,* 56–57, 61–63.
39. Stein, *Ada: A Life,* 19.
40. Baum, *Calculating Passion,* 9, 11.
41. LP 172, 9 March 1834.

ings of life." Such an activity, he stressed, "cannot possibly lead to objectionable thoughts."[42]

In addition to mathematics, Lovelace's scheme of education focused on the relationship of the mechanical or physical to the living or spiritual. In this subject she soon progressed too far for the meager talents of Dr. King.[43] She therefore exchanged him for the eminent nonconformist savant William Frend (Augustus De Morgan's father-in-law), whose tutelage focused on the natural theological goal of using mathematics to reveal divine Design in nature.[44] She also secured the help of Mary Somerville, who was at that time the only prestigious woman writer on the physical sciences. Despite the emphasis on Design, other experiences presented a distinctively mechanical and materialist portrait of the world. Trips to the Adelaide Gallery and similar venues of public scientific display gave her the opportunity to view the nervous production of electricity by the electric eel and various machines designed to replace artisanal labor.[45] Among these the Jacquard Loom stood out: the automatic drawloom that replaced the labors of tapestry makers and was "skilled" enough to produce finely wrought likenesses of people. The device was to inspire a number of mechanical substitutions for human activities. Perhaps most important here, it supplied the mechanism for Charles Babbage's Analytic Engine.[46] Lovelace's visits to Babbage's salon—with its "thinking machine," as Lady Byron described it, its other automata, and, incidentally, his own Jacquard Loom portrait—presented her with the most ambitious and charismatic representations of the extent to which machines could act for people.[47]

Spurred by such experiences, in the later years of the decade Lovelace solicited the guidance of such individuals as Carpenter, Babbage, and De Morgan, who were becoming preeminent among the savants and doctors who were addressing the question of the degree to which life and the mind could be portrayed or reproduced by mechanism. Carpenter was making a name for himself in his efforts to reform physiology along the lines of the physical sciences, effectively characterizing vital phenomena in terms of mechanical principles.[48] From the early 1830s, Babbage's projects involved demonstrating that everything

42. LP 172, 15 March 1834.
43. Baum, *Calculating Passion*, 28.
44. Stein, *Ada: A Life*, 3.
45. Altick, *Shows of London;* Morus, "Currents from the Underworld."
46. Morus, "Electric Ariel"; Schaffer, "Babbage's Dancer."
47. LP 336, 21 June 1833. Moore, *Ada, Countess of Lovelace,* 41–42, gives an account of this visit and a fuller excerpt from Lady Byron's letter. See also Schaffer, "Babbage's Dancer."
48. Carpenter, *Principles of General and Comparative Physiology.* On Carpenter's physiology see Jacyna, "The Physiology of Mind," and for the controversy over Carpenter's alleged materialism, see Winter, "The Construction of Orthodoxies and Heterodoxies."

previously assumed to belong to qualitatively distinct living, mental, and even spiritual worlds could be mechanically produced. Meanwhile, De Morgan was becoming well known for his work on logic. He embraced, albeit in a qualified manner, the new "analytic" style of mathematics that had been developed earlier in France.[49] In principle, this analysis led to a formally coherent and absolutely certain symbolic mathematical system, but one whose relations to the patterns of the physical world were contested. Many British mathematicians feared that analytic mathematics had the potential to disrupt the links between number and God's natural order which had been integral to the natural-theological tradition. Among those who shared these anxieties was Ada's former tutor William Frend, De Morgan's father-in-law. Such interlocutors as Carpenter, Babbage, and De Morgan were uniquely qualified to guide innovative work on the connections between mind and body. In the early 1840s these figures were being consulted by Lovelace at about the same time. Their advice, as I shall show, was often at odds.

It is important to emphasize that from the beginning of these studies one of Lovelace's most prominent concerns was the constraint her body was deemed to place on her intellectual work. By the mid-1830s, she and her friends subjected her various ailments to a range of conflicting explanations. Some thought they were the result of overwork, others that they might be manifestations of a mental power too great for her body. The power of her own mind was harming her because she was not allowing it sufficient expression (that is, she was not working hard enough). The ailments could be interpreted as inconvenient distractions from work, or a form of mental ill-health that itself produced the drive to carry out intellectual activities inappropriate for and unhealthy in a woman.

It was at this time that Lovelace resolved to use her own life to explicate and test the materialist issues that preoccupied some of her tutors. It was an appropriate time to begin such a program, because her body was undergoing dramatic changes. In the early 1830s, she had been described as "rather stout and inclined to be clumsy, without color and in delicate health."[50] Three years later, she had been transformed from a chubby, sickly girl into a glamorous, delicate young lady (fig. 6.2). In 1835 she married William King, soon to become Lord Lovelace, and rapidly bore him three children: Byron in 1836, Anne Isabella in 1837, and Ralph Gordon in 1839. Yet her transfiguration did not entirely remove her air of eccentricity. Her style of dress deviated sharply from the early Victorian standard of cinched-in waists and leg o' mutton sleeves, and commentators reached for unusual language to describe her. Her manners were described by John Cam Hobhouse, Lord Byron's best friend, as "not those of a woman of the world,"

49. Richards, *Mathematical Visions.*
50. SP Dep. b. 206, Greig Memoir, as cited in Stein, *Ada: A Life*, 34.

Figure 6.2. Lovelace, 1833, dressed for coming out. The chubby, sickly girl was now a glamorous young lady. Sketch by A. E. Chalon. Courtesy British Museum.

indeed, they were rather "fantastic" (fig. 6.3).[51] Henry Acland amused himself in thinking she might be a "Papist" or a "witch."[52] He then told her so.

During these transformative years, Lovelace became not only a bride, but also, as she herself put it, a "Bride of Science."[53] At the very moment when a Victorian bride would be expected to be giving up personal interests for a wholehearted devotion to home, husband, and family, Lovelace's intellectual life began to develop rapidly. True, both her immediate family and her Whig aristocratic social circle were replete with impressive examples of intellectually independent and socially powerful women, including Lady Holland and, of course, Lady Byron herself. They provided links to an older, eighteenth-century bluestocking tradition. But intensifying norms of feminine domesticity were making this a more problematic role than it had been even twenty or thirty years earlier.

Within this context it is striking that soon after her wedding Lovelace wrote to Somerville that matrimony had "by no means lessened my taste" for mathematics.[54] She soon began a correspondence with Babbage, which would lead to her only substantial scientific work. In contrast to Lady Byron's ready confidence in the influence of mathematics as a disciplining activity, the relationship between mathematics and Lovelace's passions proved far from straightforward. The more mathematical work she did, the more she represented it as a "*passion*": a strain, and an exciting one, on the body.[55] The fact that Lovelace launched herself into her most ambitious course of mathematical study so early in her marriage is suggestive, especially since it was accompanied by what she regarded as a relevant enquiry: sustained reflection on the state of her body. Observing the strain of mathematical study acting within her was the basis of Lovelace's interest in the nature and limits of her brain and body. It is unclear whether Lovelace intended this knowledge, once developed, to become as public as (for instance) the "Notes" she published on Babbage's engine, but she certainly meant her conclusions to become known within the intellectual, social, and scientific circles of her large and prestigious acquaintance.

■ *The Necessary Conditions of Mental Power* ■

LOVELACE'S REFLECTIONS gave rise to rival interpretations of her bodily and mental capacities. De Morgan and Carpenter, in offering advice about what kinds of intellectual activity Lovelace could safely undertake, were concerned that she risked completely obliterating herself, body, mind, and soul, by mathematical

51. Quoted in Baum, *Calculating Passion,* 4, xvii.
52. Quoted in ibid., xvii.
53. Ibid., 34; Toole, *Ada, Enchantress of Numbers,* 296.
54. Lovelace to Somerville, SP Dep. c. 367, 1 November 1835.
55. British Library Add. MSS, 37191, fol. 87, November 1839.

FIGURE 6.3. Lovelace, ca. 1839. Engraving based on a sketch by A. E. Chalon. By this time Lovelace was gaining a reputation for her "fantastic" manners. Courtesy the British Library.

study. She corresponded with the former in the early 1840s, and particularly in 1840–42 as she studied calculus and algebra.[56] De Morgan's simultaneous correspondence with Lady Byron expressed both his high estimation of Lovelace's intellectual powers and his worry that encouraging her to develop them might

56. Stein, *Ada: A Life*, 72.

compromise her health. At various points during this time her health did indeed fail. She suffered a variety of physical problems and emotional strains that one of her biographers likens to a "manic-depressive condition."[57] In late 1843, Carpenter, too, corresponded with Lovelace and her mother about possible relations between these problems and her mathematical studies. He suggested that mathematics was eroding her moral sensibilities and her spirituality. Each remedy he proposed involved the renunciation of her dream of becoming an Analyst.

In the early 1840s De Morgan told Lady Byron that he thought Lovelace had the potential to become an "original mathematical investigator, perhaps of first-rate eminence."[58] However, the correspondence between De Morgan and Lovelace makes clear that she found mastering the technical manipulations of calculus and algebra no easy matter and that De Morgan—despite his praise—was well aware of her difficulties. De Morgan's view was that some of the most important work in mathematics involved investigating its philosophical foundations, and it was here that he believed Lovelace's strengths lay, not in technical calculation. She seemed of the same persuasion, saying she wanted to ascertain, for instance, if there were "truths & conclusions which can be derived at by pure analysis, & in no other way" and "how far *abstract analytical expressions* must express & mean *something real*."[59] These were questions fundamental to the status and fate of the new analysis. Lovelace also raised issues touching on De Morgan's specific lines of research. During these years he was developing what he called "double algebra." Lovelace proposed to him that "this extension of Algebra ought to lead to a *further* extension similar in nature to Geometry of *Three Dimensions,* and that perhaps to a further extension into some unknown region, & so on ad-infinitum possibly."[60] This remark described something that De Morgan was at that time struggling (unsuccessfully) to achieve and that William Rowan Hamilton would shortly develop in his system of quaternions.[61] It is worth noting that Hamilton said his quaternions originated in a combination of intellectual pursuits of which Lovelace would have approved: "geometry, algebra, metaphysics and poetry."[62]

De Morgan concluded that Lovelace's "power of thinking" was "utterly out of the common way for any beginner, man or woman." He compared her favorably to Mary Somerville, who was certainly technically competent but, in his

57. Ibid., 289.
58. LP 344, 21 January 1844. Also cited in Baum, *Calculating Passion,* 40.
59. LP 170, 6 February [1841], dated by Stein, *Ada: A Life,* 79.
60. Moore, *Ada, Countess of Lovelace,* 99.
61. Stein, *Ada: A Life,* 79–80; Bloor, "Hamilton and Peacock"; Hankins, "Triplets and Triads."
62. Hankins, "Triplets and Triads," 176.

view, possessed no such understanding of fundamental issues. Lovelace had an ability of a different order altogether, being capable of grasping "strong points and the real difficulties of first principles."[63] However, De Morgan told Lady Byron that he had never communicated this to Lovelace, because he feared it might "promote an application" to her studies. The problem, as he saw it, was that her mind was so strong in relation to her body, that her family should think carefully about whether it was better to "urge or check her obvious determination to try not only to reach but to get beyond, the present bounds of knowledge."[64] So the correspondence between De Morgan and Lady Byron displayed two alternative representations of the corporeal relations of intellectual powers. Having expressed his respect for Lovelace's mathematical potential, De Morgan made it clear that it was *because* she had profound mathematical powers that he was so reluctant to make her aware of them. Her body—and perhaps any woman's body—was not strong enough to cope with the rigors it would have to endure were she to fulfill her potential as an original mathematician. He told Lady Byron that encouraging her daughter would endanger the health of her body and mind. Lovelace, De Morgan claimed, would not be able to emulate Mary Somerville's life of "regulated study, duly mixed with the enjoyment of society, the ordinary cares of life." The limited powers of a Somerville, occupied merely with the "details of mathematical work," could indeed be applied to intellectual labor without placing an undue strain on the body. Lovelace's capabilities were distinct, and, if encouraged, she would "take quite a different course."[65] Her powers were so far beyond what a normal woman's constitution could sustain that they threatened her with destruction:

> All women who have published mathematics hitherto have shown knowledge, & the power of getting it [but few have]. . . wrestled with the difficulties and shown a man's strength in getting over them. The reason is obvious: the very great tension which they require is beyond the strength of a woman's physical power of application. Lady L. has unquestionably as much power as would require all the strength of a man's constitution to bear the fatigue of thought to which it will unquestionably lead her.[66]

Given that De Morgan was the most progressive among mathematicians in his support for women's education, his view demonstrates how deeply held and widespread was the conviction of the necessity of masculine physical strength to

63. LP 344, 21 January 1844. Also cited in Baum, *Calculating Passion*, 40–41.
64. Baum, *Calculating Passion*, 40.
65. LP 344, 21 January 1844.
66. Ibid.

mathematical labor.[67] All women mathematicians had power; only a few were able to deal with the strain this power posed. Even then, their capacity was presumably based in the constitution they were born with, rather than having the potential to be augmented by physical exercise.

De Morgan's warning offers interesting perspectives on the culture of mathematical study then beginning to develop in Cambridge. The intense physical regimes that supported mid-Victorian Wrangler culture would have been impossible for Lovelace, and, of course, the admission of women was decades in the future. Speaking hypothetically, De Morgan told Lady Byron that Lovelace's abilities would have made her among the best of an incoming class of mathematicians at Cambridge. However, her philosophical bent and her relative weakness at calculation would have produced substantial problems.[68] While De Morgan saw Lovelace's mathematical work as a "struggle between the mind and the body,"[69] Lady Byron saw the exercise of the one as necessary to the health of the other. She was thus not in the least worried by De Morgan's warnings. In fact, the opposite problem preoccupied Lady Byron. She was concerned that if her daughter bottled up her intellectual powers, they would damage her body. She pointedly told De Morgan that if her daughter would "but attend to her stomach, her brain would be capable of even more than she has ever imposed on it." The "consciousness of making progress in science seems . . . an essential element in her happiness." Lady Byron told her daughter, "No one who saw you as you are could fancy a *discontinuance* of mathematics necessary."[70]

Lovelace herself skirted the issue of what impact her studies might be having on her health. She cared only that she be able to function intellectually. She told her mother that she would be "perfectly content" to suffer, if pain were "the necessary condition of all that wonderful & available mental power which I see grounds to believe I am acquiring." Pain was bearable, indeed necessary, as a condition of a full intellectual life. She observed, "I conceive that my state of health is in the inevitable *condition* attached to the nature of my mind, & will never be an *impediment* to the development of this."[71] Whatever her physical ailments, Lovelace felt confident that "[m]y intellect will keep me alive, and active." Indeed, were she to curtail her mathematical work, she might begin to "dwell" on herself and her "sensations" in an unhealthy manner.[72]

67. De Morgan, *Three Score Years and Ten;* Richards, "Augustus de Morgan"; Warwick, *Masters of Theory.*
68. Baum, *Calculating Passion,* 20, 42; Warwick, chapter 8 to this volume.
69. Cited in Baum, *Calculating Passion,* 42.
70. LP 57, fol. 3, 17 January [1844].
71. Baum, *Calculating Passion,* 66.
72. Lovelace to Babbage, undated 1843, cited in Toole, *Ada, the Enchantress of Numbers,* 192.

To her family, the most distressing of all her symptoms was a growing intolerance for her own children. In 1843, she and Lord Lovelace began to send their children on extended visits to Lady Byron. During the periods when they were at home it was decided that a tutor would superintend their education. The Lovelaces settled on Carpenter, an up-and-coming (but financially strapped) physician. Carpenter was fast becoming a prominent writer on physiology and comparative anatomy, with a special interest in relations between the intellectual faculties, the will, and the nerves.[73] He was also known for his "progressivist" characterizations of natural law, which barely escaped the charge of materialism.[74]

These interests, and particularly Carpenter's expertise in nervous phenomena, must have been among the reasons Lady Byron solicited his opinions regarding the influence of Lovelace's intellectual work on her attitude toward her children. Between November 1843 and June 1844, Carpenter offered detailed reflections and advice about the course Lovelace's mind must take if her moral and physical condition were to be improved. Her mind, Carpenter concluded, "is so constituted, that the intellectual part of it is more developed than the moral feelings and affections." Expressing her natural talents was her "chief pleasure," and it was therefore not surprising that "she should desire to make the cultivation of these powers the almost exclusive object of her life." But this course would ultimately lead to her own and others' unhappiness. Exercising one mental faculty to the exclusion of others, he warned, was unhealthy. More to the point, "great as is the gratification which the intellectual powers experience in the perception of a new truth, or the glimpse of an elevated generalization, it seems to me less intense—perhaps I might say less natural—than that which results from the culture of the social and benevolent affections." Drawing on a familiar representation of the philosopher's incivility, Carpenter warned that this kind of "exclusive pursuit" of new truths encouraged the mind to "feed too much upon itself." It would become "selfish and egotistical," he declared, "thus lowering the moral tone in some important particulars, whilst higher intellectual views of the Deity and of his operations are being attained." Indulging this tendency therefore had the direct result of "deadening the family affections." He had experienced it himself, "causing my mind to shrink, as it were, within the sphere of its own pursuits,—thus becoming morally contracted, whilst intellectually expanding."[75]

73. See Jacyna, "The Physiology of Mind"; Carpenter, *Principles of General and Comparative Physiology.*
74. Winter, "Construction of Orthodoxies and Heterodoxies."
75. Carpenter to Lady Byron, 14 November 1843, LP 65, fols. 82–87.

Carpenter thus conceived Lovelace's aversion to her children as a problem in the management of mental states. The state of mind in which one appreciated the "pleasures derived from the affections" was one in which "the intellectual kind have no zest whatever but are 'flat stale and unprofitable.'" These states occurred to all individuals with "active intellects"; he predicted that Lovelace's "physical constitution" would make them "recur in her with more than usual frequency and force." She would become increasingly aware of a conflict between one state of mind in which she could be intellectually productive and another in which she could savor her children's love and her reciprocal affection. Yet Carpenter did not advocate that she relinquish her intellectual activities. Rather, he advised that she change their course: "Her pursuit of the exact sciences should not be carried much further; Natural History & Physiology will be the studies best adapted to cherish what is at present most needing development, as bringing her into relation with other sentient beings, and keeping constantly in view the goodness as well as the power & wisdom of the Creator." She should exchange mathematics for what were (conveniently enough) Carpenter's own areas of interest. What Lady Byron thought of this is not clear. Carpenter's tone certainly indicates that he considered he was waging an uphill battle in arguing that mathematics was particularly debilitating.[76] In any case, by early in the summer of 1844, he felt able to assure Lady Byron that her daughter's feelings toward her children were in a "changed state." "I trust that this is an indication that she is gaining a more natural condition," he remarked, "in that and other points."[77]

In this light, Carpenter's relations with Lovelace herself are worth dwelling on. He became rather more than the moral guide that had originally been intended—or perhaps his conception of moral guidance involved more than Lady Byron and Lovelace had anticipated. Carpenter offered himself to her in late 1843 as a confidant, arguing that only if she were completely open with him about her feelings on all subjects could he evaluate the state of her nerves and prescribe an appropriate course of treatment. Lovelace now described to him her various states of depression and exhaustion. In response, he assured her that "I can quite understand your depression, from the action and reaction between your bodily and mental states of discomfort; each aggravating the other." Indeed, all individuals of "active mind" experienced the "desponding tendency" that oppressed her during "intervals of exertion." He could also understand her intolerance of her family from more than a scientific perspective, as he had himself undergone the same despondent states. At such times he had found "the so-

76. Ibid.
77. Carpenter to Lady Byron, undated letter June 1844, LP 169, fols. 100–101.

ciety of my wife, dearly as I love her . . . unbearable to me." He spoke, too, of his recently deceased father, Lant Carpenter, one of the most eminent Unitarian savants of the nineteenth century. Lant Carpenter's death (by drowning, in 1840) occurred during a trip to the Continent recommended by physicians as a restorative, after a period of overwork that had led to a breakdown of his health. In him, Carpenter vouchsafed, "these states were of no unfrequent occurrence, and always resulted from overexertion of mind, or some great demands upon his very strong sympathies." The remedy for Lovelace's problems was thus to maintain a "watchfulness over yourself," aided by confiding in Carpenter as the helper best equipped both to trace her mental problems to their source and to identify their possibly destructive consequences.[78]

Such a proposal was hardly without risk. Carpenter's suggestion tested the bounds of the deferential relationship appropriate between a struggling young doctor and an aristocratic lady, and for that matter between a tutor and the mother of his charges. By making this intimacy between himself and Lovelace a prerequisite for her improvement, Carpenter placed Lovelace's mental and physical constitution in direct conflict with her moral propriety. He needed the freedom of a mentor in communicating with her, but, even as he worked to establish the intimacy that he claimed was necessary for her mental health, he destroyed what intimacy already existed.

Initially, Carpenter expressed confidence that a deeper intimacy was just what Lovelace needed and that indeed she desired it. She had indicated in correspondence, he asserted at one point, that she felt "a growing disposition to give me your entire confidence, and as strong a friendly regard as it is in your nature to cherish for any one." This was "the best proof to me that I have those qualities which will enable me to exercise a beneficial influence over you, and to make your life happier to yourself, by directing your intellectual powers towards the steady prosecution of noble ends, and by cherishing (but not forcing) your latent feelings of affection towards those who have a claim upon them." He coaxed her to admit that she wished she felt the affectionate feelings toward her family that she presently lacked. Carpenter told her that the "tears of pleasure" which he himself often shed "at the perception of a new and elevated Truth" were never as "sweet, as those which have been drawn from me by the affection of a devoted wife, and of an innocent engaging child." And "having experienced also the good effects of self-regulation, I can recommend it to you with more weight, as well as direct you better how to exercise it, than most others could do." Significantly, he maintained that this same insight permitted him simultaneously to reassure her

78. Carpenter to Lovelace, 27 November 1843, LP 169, fols. 19–24.

that their relations would never breach the bounds of propriety since "the nature and strength of my regard depend upon its reciprocation."[79] Anyone who could "cherish a hopeless affection" had "a constitution very different from mine. Now I know you to be constitutionally incapable of forming any attachment to me, which should be unbecoming our mutual relations; and thereby you have a guarantee (even if there were no other) that I would never go too far."[80]

Intimacy and empathy were Carpenter's necessary conditions for Lovelace's self-knowledge and cure. He would soon discover that they were incompatible with the social and professional relationship he formally enjoyed with Lovelace and her family. He had based his claim to prescribe such intimacy on two grounds—his personal experience of the same state and his scientific expertise in the diagnosis and treatment of such states in general. But as the privacy of the message indicates, the closeness he wished to encourage was greater even than that obtaining between Lovelace and her husband, and these qualifications were little match for the force of convention. The comparison becomes even more striking if one juxtaposes Carpenter's counsel to Lovelace to advice she received from a friend of her mother just before the marriage. This friend had advised her to be "always completely open and unreserved"—with her husband. Yet even with him, she should "preserve in thought[,] word and deed that delicacy which is one of the first and most attractive charms of the female character." Above all she must remember that a "refined and feminine mind" would never risk loss of reputation "by uttering or permitting that which a delicate young woman must shrink from."[81] One suspects that this injunction included exactly the kind of communication that Carpenter hoped to elicit. The comparison between Carpenter's instructions and what was a conventional enough piece of advice to a young bride about conduct toward her husband (and no one else) suggests that he was venturing a dangerously indelicate form of openness. His recipe for the evaluation and management of her mental states could easily be read as improper amorous advance.

Lovelace did so interpret his advice. Over the next few weeks Carpenter complained that she was not replying to his letters.[82] When, finally, she did reply, she made it clear that she felt Carpenter to have overreached the bounds of propriety. In his defense he declared he was convinced that some such tutelage was in order from a moral guide such as himself, since her "peculiarity of physical

79. Carpenter to Lovelace, 6 December 1843, LP 169, fols. 47–60.
80. LP 47–60, 6 December 1843; box 65 of Dep. Lovelace Byron, fols. 82–215.
81. Moore, *Ada, Countess of Lovelace*, 69.
82. Carpenter to Lovelace, 13 December 1843, LP 169, fols. 69–72, and 15 December 1843, LP 169, fols. 73–77.

constitution" had had "more influence, than I think you are yourself aware of, upon your general principles and habits of thought and action." He also insisted that he had always framed such advice in the form of suggestions, rather than as "direct admonition or attempted guidance." He had not been aware of seeking more knowledge or intimacy than she herself wished to give. She had only *imagined* that his "feelings of attachment to [her] had more transgressed those bounds, which propriety and guidance interpose." He retreated to the assertion that his very suggestions of intimacy had been for practical purposes: "My hope of utility to you must be confined, as at first, to your intellect."[83]

A flurry of letters to and from Lord Lovelace and Mrs. Carpenter attempted to smooth over the incident—including one from Mrs. Carpenter to Ada Lovelace assuring her that her husband could never possibly think an improper thought about another woman.[84] Over the next several months the relationship between Carpenter and Lovelace's family became scrupulously formal. Carpenter's correspondence with Lovelace never again lapsed from punctilious reserve, confining itself to requests about the dedication of books such as his *Popular Cyclopedia* and enclosures on human physiology and sciences of mind.[85] Social convention proved stronger than medical conviction.

One might straightforwardly represent this incident as arising out of Lovelace's flirtatiousness and Carpenter's unscrupulousness (or naïveté). Yet there may also have been another factor in play. Arguably, the need to study and manage the relations between Lovelace's body and her intellect produced unstable situations of authority for everyone who got close to her. This was a particular example of a predicament posed by a class of relationships in which mental and physical health were to be produced by mental or physical intimacy—a class which was becoming larger over the course of the nineteenth century. Experimental sciences of mind like mesmerism and hypnotism were very prominent in Britain from the 1830s onward. Their phenomena were produced by, and were even thought by some to *consist of,* a physical or psychological intimacy. During this period moral therapies were also replacing physical ones in the treatment of the mad and of those suffering other disorders of the mind and nerves. As alienists sought to restore the senses of their charges via "moral treatment," moral relationships between doctor and patient (rather than physical restraints)

83. Carpenter to Lovelace, 24 December 1843, LP 169, fols. 78–84, and 28 December 1843, LP 169, fols. 85.

84. Louisa Carpenter to Lovelace, LP 169, fols. 93–94; Carpenter to Lovelace, discussing a letter he has received from Lord Lovelace in this matter, January 1844, LP 169, fols. 116–21.

85. Carpenter to Lovelace, 9 January 1844, LP 169, fols. 91–92; 4 April 1844, 169, fols. 137–38; 11 April 1844, 169, fols. 138, 144–47; 23 October 1844, 169, fols. 212–13.

were expected to produce the desired effects. But this left practitioners open to suspicions that the *wrong sort* of intimacy was being encouraged between healer and patient—hence the not-uncommon accusations of seduction, rape, and otherwise inappropriate behavior which haunted mesmerists, and to a lesser extent, alienists.[86] In this case Carpenter's professional reputation was the victim. He was perhaps bound to encounter problems as soon as he proposed using a personal relationship to probe the dynamics of Lovelace's nerves, morals, and intellect. His ambiguous position of unofficial doctor to her mind-body relations was further compromised by his definitively subservient position within the family as the children's resident tutor.

Taken together, the concerns of De Morgan and Carpenter show how complete were the perceived obstacles to Lovelace's intellectual fulfillment, even in the eyes of her scientific friends. The regulation of her life was predicated on concerns for the physical, mental, and moral consequences of her mathematical interests: De Morgan worried that her work would wreck her body, Carpenter that it would destroy her moral being. Both considered it essential for her to relinquish mathematics in deference to the duties of Victorian womanhood. In particular, both urged her to abandon the intensity of her quest for original knowledge. Faced with such analyses, Lovelace saw her challenge as developing new knowledge and producing an account of her own physical and moral powers that would elude such proscriptions and legitimate her own status as knowledge producer. She found the inspiration for such an effort in her collaboration with Charles Babbage on another profoundly problematic Victorian knowledge producer, the Analytic Engine.

■ *Foreseeing* ■

IN THE 1840s Lovelace lived out the definitive issues of philosophic materialism: whether the phenomena of the mind were mechanically produced and whether they could be mechanically mimicked. She wanted to know what it was about her that could or could not be mechanically represented. The collaboration Lovelace maintained with Babbage during the early and mid-1840s provided the perfect forum for exploring this issue. Babbage was studying the extent to which the operations of machines could replicate human actions. His proposed Analytic Engine would be capable of carrying out human intellectual operations, since it possessed what Babbage claimed were the two defining features of intelligence: memory and "foreseeing"—the ability to extrapolate future developments from given information.

86. Showalter, *Female Malady*, 77; [Wakley], "Animal Magnetism," 450, wrote of mesmerists' "libidinous propensities."

His use of the word *foreseeing* was striking. In this period, the most vivid and controversial form of foreseeing was the pronouncements of subjects—usually girls or women—in the altered states produced by mesmerism or hypnotism. The mesmeric subject was taken by believers as an oracular figure: a supposedly passive vehicle for the expression of natural or supernatural truths, including knowledge of the future. Debates over the status of trances and of the knowledge voiced in them—and the arguments surrounding these subjects were prolonged and embittered—raised just the issue that Babbage was so concerned to address. Lovelace likewise focused on this issue, and in so doing completed her transformation into a "prophetess." For Babbage and Lovelace were addressing the corporeality of reasoning by working on two aspects of the same question: the degree to which the human could be rendered machinelike and the extent to which machines could be made to carry out human functions.[87]

Lovelace's initial proposal to Babbage did not reach beyond the parameters of scientific work as exemplified by individuals like Somerville. By January 1841 she had a scrapbook of mathematics to show Babbage and soon persuaded him to let her make her brain be "subservient" to his "purposes and plans."[88] Immediately, she began to draw up a set of "Notes" on his Analytic Engine, which would explain it to an educated audience. Unusual terms entered her exchanges with Babbage as they both sought to find an appropriate way of defining her intellectual role. She referred to herself as a magical or ethereal being, and Babbage followed suit. Their letters spoke of her as a "fairy," a "High Priestess," an "Enchantress of Number."[89] Such language represented both a continuity with older representations of the philosopher as ethereal being and also evoked the volatile status of the mesmeric subject—a perhaps-mechanical, perhaps-magical testifier with privileged access to the truths of nature who did not enjoy the authoritative social status to validate these claims. This metaphorical language, then, offered possible representations of intellectual capacity that allowed Lovelace to retain her femininity.

Lovelace's ambitions quickly expanded: she laid claim to more analytical powers than she had hitherto asserted and to more authority with respect to their exercise. As she revised her "Notes" on Babbage's engine in late 1842 she announced to him that she would soon be "really something of an Analyst."[90] She now claimed that her "brain" was something "more than mere mortal; as time

87. See Peter Dear's discussion of the Cartesian problematic in chapter 2 of this volume.

88. 12 January 1841, in Toole, *Ada, the Enchantress of Numbers,* 140.

89. Lovelace to Babbage, 26 July 1843, cited in Toole, *Ada, the Enchantress of Numbers,* 210–11; Baum, *Calculating Passion,* 78, 87–88; 10 September 1843 to Babbage, cited in Toole, *Ada, the Enchantress of Numbers,* 263–64; 9 September 1843, cited in Toole, *Ada, the Enchantress of Numbers,* 236.

90. Baum, *Calculating Passion,* 62.

will show; (if only my breathing & some other etceteras do not make too rapid a progress towards instead of from mortality)." It would allow her to develop a systematic treatment of the nature, organization, and history of natural knowledge and "to define & to classify all that is to be legitimately included under the term discovery." She also renounced her previously deferential relations with Babbage, deciding instead that his projects could succeed only if he accepted the guidance of a strong patron (herself) who could keep him focused on one piece of work at a time. He refused to accept this arrangement, and while their friendship continued to flourish, their scientific collaboration came to an end.[91]

Meanwhile, Lovelace had her own analytic engine to understand. Her body offered her a rich field for direct observation, and one to which she had unique access. Her family physician, Charles Locock, had long since given up his attempts at diagnosis, pronouncing her ailments beyond the reach of current medical knowledge.[92] In the early 1840s, Lovelace experimented with such psychologically potent stimuli as opium and mesmerism. Few details survive of her day-to-day physical condition, but it is clear that her menstrual cycle and her digestive organs were giving her trouble. In August 1843 she discussed physiological changes in a passage which may have referred to the possibility of early menopause:

> There seems to be more ground for believing in the change of constitution; & altogether things appear more hopeful. There has however been a very *frightful crisis;* & the last occurrence of the period very nearly brought me into danger; not at the time, but in its *subsequent* effects for weeks. It seems that the period paralyses the digestive functions almost completely. I hardly think it can occur again, tho' I conclude I must not *build* on this hope; since it often takes years to stop completely.[93]

The headaches and nervous strains that had plagued her for years were at least as bad as ever, as this letter of early 1844 (to Woronzow Greig) reveals:

> I had been very queer of late in some of the sensations I have had about my head & yesterday morning the whole throat & face suddenly swelled (in one instant) to an enormous size, & I felt threatened with instant annihilation. . . . This seizure lasted 3 minutes, & then subsided; but left behind it (of course) most strange feelings in the brain & eyes. I fancy it must be of the cataleptic nature.[94]

91. Ibid., 87–88; Lovelace to Babbage, 14 August 1843, cited in Toole, *Ada, the Enchantress of Numbers,* 227–32. See also Lady Byron, cited in ibid., 218–84.

92. 25 July 1843, cited in Toole, *Ada, the Enchantress of Numbers,* 209–10; Moore, *Ada, Countess of Lovelace,* 185 ff. Locock was a physician and accoucheur to royalty. See Maulitz, "Metropolitan Medicine."

93. Quoted in Toole, *Ada, Enchantress of Numbers,* 224.

94. Lovelace to Greig, 10 February 1844, cited in Moore, *Ada, Countess of Lovelace,* 215.

Lovelace decided that if she could not get rid of such symptoms, they could become the occasion, or even the focal point, of a course of research into nervous physiology. She now read current and standard physiological works and also more specialized publications that addressed her particular interest in the relations between physical forces and the mind.[95] She also decided to master the "Practical" skills necessary to carry out "experimental tests" of the "brain, blood, & nerves, of animals." These studies suggested that her initial mesmeric experiments of 1841 might have been the cause of her current complaints and that opium was, of the two, by far the better therapy. Opium, she proclaimed, could render her "philosophical, & so takes off all fretting eagerness & anxieties"; it harmonized her "whole constitution," such that each "function" acted in "a just proportion; (with judgment, discretion, moderation)."[96] In November 1844 she began to carry out investigations of the relationship between electricity and life at the home of Andrew Crosse:

> I am anxious to consult you about the most convenient and manageable and portable forms for obtaining constantly acting batteries; not great intensity, but continual and uninterrupted action. Some of my own views make it necessary for me to use electricity as my prime-minister, in order to test certain points experimentally as to the nature and putting together (con-sti-tu-tion) of the molecules of matter.[97]

Crosse was known primarily for his claims in the late 1830s to have produced life—in the form of a rare genus of insects—by running electricity through chemical solutions for long periods.[98] Lovelace clearly had similar ambitions.

The summit of her studies was self-experimentation. Lovelace was particu-

---

95. In the early 1840s Lovelace read works such as Lamarck's *Philosophie Zoologique* (Toole, *Ada, the Enchantress of Numbers,* 156) and Robert Christison's 1829 *Treatise on Poisons* (for its discussions of the effects of opium, see Stein, *Ada: A Life,* 132). In the mid-1840s she moved on to works primarily devoted to the relationship between physical forces and nervous phenomena, such as Carlo Matteuchi's 1844 *Traité des Phénomènes Electro-Physiologiques* (Stein, *Ada: A Life,* 152), and the work of the Austrian chemist Carl von Reichenbach. Reichenbach claimed to have found evidence of a new imponderable fluid when he noticed that individuals of unusual nervous sensitivity could perceive lights issuing from the poles of magnets. Lovelace, who was trawling for German physiological works in 1844 and 1845, probably read Reichenbach's claims in their original and lengthy German (published in early 1845), but certainly read and discussed William Gregory's *Abstract of "Researches on Magnetism"* [1846]. She was convinced of the reality of the phenomena Reichenbach and Gregory recorded (LP 175, fols. 211, 223, as quoted in Stein, *Ada: A Life,* 150–52); it is likely that, like many onlookers, she regarded Faraday's announcement of the polarization of light in September 1845 as confirmation of Reichenbach's earlier claims. Lovelace wrote to Babbage in November asking urgently for further information about Faraday's experiments (Toole, *Ada, the Enchantress of Numbers,* 318).

96. Autumn 1844, cited in Toole, *Ada, the Enchantress of Numbers,* 287–89.

97. Quoted in Stein, *Ada: A Life,* 143.

98. Secord, "Extraordinary Experiment."

larly suited to develop an understanding of the physiology of knowledge because, she claimed, she had "a frame so susceptible that it is an experimental laboratory always about me, & inseparable from me. I walk about, not in a snail-shell, but in a Molecular Laboratory."[99] The study of her own body in this "laboratory" would, she anticipated, transform the processes that had initially seemed to place limitations on her capacity for knowledge into a source of overarching knowledge. She was attuned to the physiological developments that were effecting this transformation. Her mathematics had given her "a strength of head & attention which I never expected." The very fact that she was sufficiently self-conscious to register these phenomena was itself testimony to the "strength of head" she had achieved. She was now asserting that her own intellectual labor was a physiological process that could monitor and experimentally test itself.

Her woman's frailties, ironically, allowed her to go where Babbage could not in his studies of mechanism and intelligence. He could not treat his own body as a laboratory in the way that Lovelace could. She, on the other hand, could carry out this laboratory work because her mind uniquely united "habits of *matter of fact* reasoning & observation, with the *highest imagination*." The life sciences required the stimulus of a "Newton for the Molecular Universe," she averred, if they were to fulfill their potential to illuminate the workings of body and mind; and Ada Lovelace was superbly equipped to become that Newton.[100]

■ *"Getting Cerebral Phenomena"* ■

I HAVE ARGUED that Lovelace's research was partly related to the difficulty she had in finding a ready-made intellectual role, yet there *was* an available cultural identity that could accommodate many of the characteristics Lovelace wished to claim for herself. Her work produced an "immense development of imagination," she noted, "so much so, that I feel no doubt if I continue my studies, I shall in due time be a Poet." Because the Romantic poet simultaneously possessed insight into nature and humanity, as well as self-knowledge, that identity was compatible with her wish not only to produce mathematics but to understand the epistemological and corporeal dynamics that this entailed. It was also a role that could define her as a genuine maker, rather than a mere purveyor, of knowledge. Moreover, it offered her a means of constructing a positive form of "Byronism," allowing her, in principle, to incorporate her father's gifts while surpassing his achievements and avoiding his moral failings.

In the 1840s her father's legacy obsessed Lovelace more than ever. She regularly reflected on the ways in which his gifts and weaknesses (but particularly the

99. Lovelace to Lady Byron, 11 November [1844], cited in Toole, *Ada, the Enchantress of Numbers,* 291.

100. Toole, *Ada, the Enchantress of Numbers,* 293.

former) were manifest in her. In principle, Byron raised major problems for Lovelace's ambitions to genius. His tainted reputation, and the horrific picture of his mind and life that Lady Byron systematically encouraged, made his archetypal genius something to be warded off rather than embraced. Lovelace sought a way of positively characterizing this inheritance. She told Andrew Crosse that a dominant will was one of her most distinctive and prized family traits: "You know I believe no creature ever could WILL things like a Byron. And perhaps that is at the bottom of the genius-like tendencies in my family. We can throw our whole life and existence for the time being into whatever we will to do and accomplish."[101]

Her Byronic will, she anticipated, would give her the strength to realize the potential that had been corrupted in her father. Lovelace hoped to redeem Byron's achievements by producing a more overarching, less partial, less qualified, and certainly less tainted version of them. She had told her mother in 1841 that she aspired to "do greater things than ever he did." There was "less flash about me & much more depth." If she had inherited any of Byron's "genius," she insisted, she would "use it to bring out great truths & principles. I think he has bequeathed this task to me!"[102] The mathematical "poetry" she would produce to this end would, then, be of a "unique" kind. It must be more "philosophical" than existing mathematics, "& higher in it's [sic] nature than aught the world has perhaps yet seen." Her ambition extended from this to encompassing within an overarching system not just mathematics, nor even just natural science, but every field of study—all the more striking, then, that Lovelace never advanced even to the point of setting out clearly how such a project could take shape, much less carrying out the work that might be involved in bringing it to fulfillment.[103]

Lovelace did claim at one point to have a very specific project in mind. She had concluded that "cerebral matter need be more unmanageable to the mathematicians than sidereal & planetary matter & movements."[104] She set down on paper a number of scattered reflections on the nature of perception and imagination, for instance, a materialist speculation included in an 1844 letter fragment: "There may possibly be simply a different law for the *propagation* of impressions thro their substance. The molecules may *move* differently amongst each other. The Creator may have ordained that difference of sensation shall accompany each different law of *Molecular* movement of this description."[105]

Different "impressions," then, produce mechanically different motions of

101. Lovelace to Greig, 15 November 1844, SP Dep. c. 367. See also Baum, *Calculating Passion*, 57–59.

102. To her mother, 3 March [1841], cited in Toole, *Ada, the Enchantress of Numbers*, 156.

103. Baum, *Calculating Passion*, 58.

104. Lovelace to Greig, 15 November 1844, SP Dep. c. 367, cited in Toole, *Ada, Enchantress of Numbers*, 293.

105. Ibid.

bodily chemicals and tissues that affect the nervous system differently, so as to produce varieties of sensation. One interesting feature of these reflections was that they were developed as part of a process of *self*-experimentation. She claimed to have "hopes, & very distinct ones too, of one day getting *cerebral* phenomena such that I can put them into mathematical equations; in short a law, or *laws,* for the mutual actions of the molecules of brain; (equivalent to the *law of gravitation* for the *planetary* & *sidereal* world). I am proceeding on a track quite peculiar & my own, I believe." [106]

In Lovelace's proposal, the "getting" of "cerebral phenomena" referred simultaneously to the production of phenomena in her own brain and to the analytical manipulations to which she would subject them in the process of characterizing them mathematically. Moreover, the choice of term is also an example of her recurrent choice of reproductive language (the association with *beget*). The actions of producing and analyzing cerebral phenomena were not separable, since what she wished to study consisted of those processes by which she understood her thought and the object of her thought. It was this reflexive process which would allow her to "bequeath to the generations a Calculus of the Nervous System." [107]

Accordingly, her project involved a portrayal of her own role in her researches that was as reminiscent of the oracular mesmeric subject as it was of the "analyst" she had mentioned in her earlier correspondence with Babbage:

> I am simply the *instrument* for the divine purposes to act *on* & *thro';* happening to be appropriate for the object. Like the Prophets of old, I shall but *speak the voice* I am inspired with.
>
> I may be the Deborah, the Elijah of *Science.*—
>
> The only merit that can ever be due to *me,* is that of putting myself, & maintaining myself, in such a *state* (physically & mentally) that God & His agents, can use me as their *vocal* organ for the ears of mortals. [108]

Having embraced the role of oracle, however, she explicitly renounced control over her own mental processes. This made it difficult for her to claim the role of a natural philosopher and also invited imputations that she was mad or suffering some less extreme form of mental debilitation that would completely undermine her credibility.

Lovelace's biographers have represented these reflections as the meaningless ravings of someone in the throes of a nervous breakdown. That she was subject to "manias and whims," and that these were partly the result of her mathemati-

106. Ibid.
107. Ibid.
108. LP 42, fols. 152, 11 November 1844, cited in Toole, *Ada, the Enchantress of Numbers,* 291.

cal work, was something that Lovelace herself once claimed, but this was only one (and not the last) of several representations she made of her own mental states, most of which stressed the positive and powerful capabilities of her unusual mind.[109] Yet despite the biographers, Lovelace's ramblings can be seen as laying open to question what it was for the nerves and the mind to be in a state that either compromised or augmented the reasoning powers. She could not succeed in making her brain and body into a laboratory precisely because this involved renouncing control over the very processes that produced knowledge. Her attempt and self-acknowledged failure were coextensive with what was designated as her nervous breakdown. This is not to say that that the effort itself in some straightforward, clinical sense "really" destroyed her health, but that one of the many appropriate and reasonable ways of depicting what was going on in these passages was that Lovelace was failing to come up with a viable mode of being self-evidential. Indeed, Lovelace made no further progress in developing her physiology of knowledge. She began an affair with John Crosse (Andrew Crosse's son), one of those many intellectual contacts whose knowledge of German metaphysics and physiology she had hoped would serve as a guide to the literature most relevant to her project. In his company she used her mathematical talents for gambling, and he used the debts she incurred and their relationship as the basis for continuing blackmail.[110]

■ Conclusion ■

IN MAY 1851, extreme vaginal hemorrhaging led her to consult a series of doctors. There was a typical lack of agreement about how to interpret her body. A Dr. Lee diagnosed uterine cancer, and a Dr. Malcolm (family doctor of Lord Zetland) an intermittent matutinal fever; the eminent Drs. James Clark and James South were each consulted, and each requested another opinion. Her family doctor and accoucheur, Locock, took a more optimistic view: the bleeding was merely a local problem (Lovelace referred to it as "the local state").

The treatments appropriate to these conditions varied accordingly. Uterine cancer was usually fatal, and the new treatment of hysterectomy was rarely performed and highly dangerous. Dr. Malcolm dosed her with quinine and nitrate of silver and advised a healthy restful outdoor life for next two years. Above all, he reckoned that writing was harmful. Lovelace's mother thought her problems were caused by overexcitement; Lovelace herself ascribed them to the "miserable East Winds" and the "high pressure of the present age & epoch & state of society." She told the publisher John Murray that she was suffering the knock-on effects

109. 21 December 1844, cited in Toole, *Ada, the Enchantress of Numbers,* 309–10.
110. Baum, *Calculating Passion,* 92.

of "spasms of the heart" and that she had had this problem for twenty years.[111] Locock prescribed leeching and the consumption of wine and other restoratives that were intended to make up for her loss of blood.

Lovelace followed Malcolm's advice for some time, but her allegiance was to Locock despite the many occasions on which he had been proved wrong or failed to make a diagnosis. She had full confidence in his prescriptions even as she owned that "the treatment appears to do harm sometimes." The leeches were intended to effect "local depletion" of the "sore" inside her. The stimulants he advised included

> Bark & Port Wine 4 times a day always, & unless a very high diet is kept up, the SORES become threatening, & indeed I break out all over. And yet, local bleeding is necessary. One thing: there has been no PERIOD now for several months. . . . Many symptoms look like the departure of the periods altogether;—particularly the hemorrhages I had for a year, previous to the present total stoppage.[112]

The "sloughing" produced a hemorrhage that "would have alarmed me much; but for his absolute confidence and assurance that it was all as it should be & as he wished."[113] In July, Locock carried out an internal examination (an unusual step to take, and one that was generally carried out only in the most serious of cases) and saw an "extensive & deep-seated sore"; but, he assured Lord Lovelace, "it is a healthy sore."[114]

Lovelace's other major treatments consisted of such consciousness-altering agents as opium, chloroform, cannabis, and mesmerism. These certainly helped with the pain, but Lovelace also regarded them as giving her a mental fortitude that replenished powers depleted by the loss of blood. Referring to the care of a Dr. Cape, an advocate of mesmerism, she told her mother that one of her main problems had been "a continuous current drawn off from the Brain. I often felt great CONFUSION, & great difficulty in concentrating my ideas. . . . I was (partially) dead; but I have come back to life again."[115]

The only common ground among Lovelace's medical advisers and her mother was a sense that her condition was worsening. Lovelace herself, despite increasingly hollow words of optimism, worried that she would not have sufficient time to develop the great system of knowledge to which she aspired. And she had other anxieties. Her gambling debts were great, and Crosse's blackmail

111. Stein, *Ada: A Life,* 218.

112. Lovelace to Lady Byron, LP 44, fol. 69, cited in Stein, *Ada: A Life,* 225.

113. Lovelace to Lady Byron, 19 June 1851. LP 43, fol. 195. For a fuller discussion of the circumstances, see Stein, *Ada: A Life,* 218.

114. Locock to Lord Lovelace, 24 July 1851, LP 166, fol. 169.

115. LP 43, fols. 219 [16 August 1851], as cited in Stein, *Ada: A Life,* 221.

FIGURE 6.4. Lovelace on her deathbed. Sketch by Lady Byron, 1852. Courtesy Laurence Pollinger Ltd. and the Bodleian Library.

further devastated her finances. Her husband had become aware of these debts over the previous year, but Lovelace was desperate that her mother not learn of them. Her mother did finally learn of her debts, and this led to feuding between Lady Byron and Lord Lovelace throughout Lovelace's last days. Lady Byron's relations with Babbage were similarly fraught, and Lovelace's last months were plagued with quarrels over her among her closest friends. At the same time her claims to knowledge grew pathetically grandiose. She told her mother that

> I think . . . when you see certain productions, you will not even despair of my being *in time* an *Autocrat,* in my own way, before whose *marshalled regiments* some of the iron rulers of the earth may even have to give way. But of *what materials my regiments* are to consist, I do not at present divulge.
>
> I have however the hope that they will be most *harmoniously* disciplined troops;—consisting of vast *numbers,* & *marching* in irresistible power to the sound of Music. Is not this very mysterious? Certainly *my* troops must consist of *numbers* or they can have no existence at all, & would cease to be the particular sort of troops in question.[116]

In reality she no longer had any way to marshall such troops. Knowledge, being corporeal, dies with the body. The end came in November 1852 (fig. 6.4). Her obituary in *The Times* referred to the great "patience and fortitude" with

116. Lovelace to Lady Byron, 29 October 1851, LP 43, fol. 246, cited in Baum, *Calculating Passion,* 93–94.

which she had borne her last illness and informed readers that she left behind a husband and three children.[117] No mention was made of her intellectual work.

117. 27 November 27 1852, cited in Baum, *Calculating Passion*, 94.

■ **ACKNOWLEDGMENTS** ■

Thanks to the following for helpful discussions, comments on drafts, or other help: Janet Howell, Adrian Johns, Ludmilla Jordanova, Mac Pigman, Simon Schaffer, James Secord, Ingeborg Sepp, Helen Small, John Sutherland, Andrew Warwick, and Cindy Weinstein. I would also like to thank the Division of Humanities and Social Sciences at California Institute of Technology for generous institutional support, and Somerville College, University of Oxford, the British Library, Master and Fellows of Trinity College Cambridge, Laurence Pollinger Ltd., and the Byron family for permission to quote from archival material.

■ **REFERENCES** ■

Abir-Am, Pnina, and Dorinda Outram, eds. *Uneasy Careers and Intimate Lives: Women in Science 1789–1979*. New Brunswick, N.J.: Rutgers University Press, 1989.

Altick, R. D. *The Shows of London*. Cambridge: Harvard University Press, 1978.

Bailin, Miriam. *The Sickroom in Victorian Fiction: The Art of Being Ill*. Cambridge: Cambridge University Press, 1994.

Barrett, Elizabeth. *Letters of Elizabeth Barrett Browning to Mary Russell Mitford 1836–1854*, ed. M. B. Raymond and M. R. Sullivan. 3 vols. Waco, Texas: Armstrong Browning Library; Wellesley, Mass.: Wellesley College, 1983.

Barrow, Logie. "Why Were Most Medical Heretics at Their Most Confident around the 1840s? (The Other Side of Mid-Victorian Medicine)." In *British Medicine in an Age of Reform*, ed. Roger French and Andrew Wear, 165–85. Cambridge: Cambridge University Press, 1991.

Basham, Diana. *The Trial of Woman: Feminism and the Occult Sciences in Victorian Literature and Society*. New York: New York University Press, 1992.

Baum, Joan. *The Calculating Passion of Ada Byron*. Hamden, Conn.: Archon Books, 1986.

Besant, Annie. *Autobiographical Sketches*. London: Unwin, 1893.

Bloor, David. "Hamilton and Peacock on the Essence of Algebra." In *Social History of Nineteenth-Century Mathematics*, ed. Herbert Mehrtens, Hank Bos, and Ivo Schneider, 202–32. Boston: Birkhäuser, 1981.

Burstyn, Joan. "Education and Sex: The Medical Case against Higher Education for Women in England, 1870–1900." *Proceedings of the American Philosophical Society* 117 (1973): 89–97.

Butler, Josephine. *Personal Reminiscences of a Great Crusade*. New ed. London: H. Marshall, 1898.

Carpenter, William Benjamin. *Nature and Man: Essays Scientific and Philosophical*. Ed. J. Estlin Carpenter. London: Kegan, Paul, Trench and Co., 1888.

———. *Principles of General and Comparative Physiology*. London: John Churchill, 1839.

———. *Principles of Mental Physiology*. London: H. S. King and Co., 1875.

Carter, Robert Brudenell. *On the Pathology and Treatment of Hysteria*. London: John Churchill, 1853.

Cobbe, Frances Power. *Life of Frances Power Cobbe.* Boston: Houghton, Mifflin, 1895.

Conway, Jill. "Stereotypes of Femininity in a Theory of Sexual Evolution." *Victorian Studies* 14 (1970): 47–62.

Cooter, Roger. "Dichotomy and Denial: Mesmerism, Medicine and Harriet Martineau." In *Science and Sensibility: Gender and Scientific Enquiry 1780–1945,* ed. Marina Benjamin, 144–73. Oxford: Blackwell, 1991.

Daston, Lorraine. "Enlightenment Calculations." *Critical Inquiry* 21 (1994): 182–202.

David, Deirdre. *Intellectual Women and Victorian Patriarchy: Harriet Martineau, Elizabeth Barrett Browning, George Eliot.* Ithaca: Cornell University Press, 1987.

Davidoff, Leonore, and Catherine Hall. *Family Fortunes: Men and Women of the English Middle Class, 1780–1850.* Chicago: University of Chicago Press, 1987.

De Morgan, Sophia. *Three Score Years and Ten: Reminiscences of the Late Augustus De Morgan, and Others.* Ed. Mary A. De Morgan. London: Richard Bentley, 1895.

Findlen, Paula. "Science as a Career in Enlightenment Italy: The Strategies of Laura Bassi." *Isis* 84 (1993): 441–69.

Gould, Paula. "'A Thing Inexpedient and Immodest': Women and the Culture of University Physics in Late Nineteenth Century Cambridge." Unpublished undergraduate dissertation, Whipple Museum Library, Cambridge, 1993.

Grattan-Guinness, Ivor. "Work for the Hairdressers: The Production of de Prony's Logarithmic and Trigonometric Tables." *Annals of the History of Computing* 12 (1990): 177–88.

Hankins, Thomas L. "Triplets and Triads: Sir William Rowan Hamilton on the Metaphysics of Mathematics." *Isis* 68 (1977): 175–93.

Harrison, J. F. C. "Early Victorian Radicals and the Medical Fringe." In *Medical Fringe and Medical Orthodoxy,* ed. W. F. Bynum and Roy Porter, 198–215. London: Croom Helm, 1987.

Jacyna, L. S. "The Physiology of Mind, the Unity of Nature, and the Moral Order in Late Victorian Thought." *British Journal for the History of Science* 14 (1981): 109–32.

———. "Principles of General Physiology: The Comparative Dimension to British Neuroscience in the 1830s and 1840s." *Studies in the History of Biology* 17 (1984): 13–48.

Jordanova, Ludmilla. *Sexual Visions: Images of Gender in Science and Medicine between the Eighteenth and Twentieth Centuries.* London: Harvester Wheatsheaf, 1989.

Lightman, Bernard. "'The Voices of Nature': Popularizing Victorian Science." In *Contexts of Victorian Science,* ed. Bernard Lightman. Chicago: University of Chicago Press, forthcoming.

Marcet, Jane Haldimand. *Conversations on Chemistry: in which the Elements of that Science are Familiarly Explained & Illustrated by Experiments.* London: Longman, Hurst, Rees and Orm, 1806.

Marchand, Leslie, ed. *"Famous in My Time": Byron's Letters and Journals.* Cambridge: Harvard University Press, 1973.

Martineau, Harriet. *Life in the Sick-Room: Essays by an Invalid.* London: E. Moxon, 1844.

Maulitz, Russell C. "Metropolitan Medicine and the Man-Midwife: The Early Life and Letters of Charles Locock." *Medical History* 26 (1982): 25–46.

Menabrea, L. F. "Sketch of the Analytical Engine Invented by Charles Babbage. With Notes upon the Memoir by the Translator, Ada Augusta, Countess of Lovelace." *Bibliothèque Universelle de Genève,* 82. Reprinted in *Charles Babbage and his Calculating Engines: Selected Writings by Charles Babbage and Others,* ed. Philip Morrison and Emily Morrison, 225–97. New York: Dover Books, 1961.

Moore, Doris Langley-Levy. *Ada, Countess of Lovelace: Byron's Legitimate Daughter.* London: John Murray, 1977.

Moore, James R. "The Erotics of Evolution: Constance Naden and Hylo-Idealism." In *One Culture: Essays on Science and Literature,* ed. George Levine, 225–57. Madison: University of Wisconsin Press, 1987.

Morus, Iwan. "Currents from the Underworld: Electricity and the Technology of Display in Early Victorian England." *Isis* 84 (1993): 50–69.

————. "The Electric Ariel: Technology and Commercial Culture in Early Victorian England." *Victorian Studies* 39 (1996): 359–78.

Moscucci, Ornella. *The Science of Woman: Gynaecology and Gender in England, 1800–1929.* Cambridge: Cambridge University Press, 1990.

Myers, Sylvia Harcstark. *The Bluestocking Circle: Women, Friendship, and the Life of the Mind in Eighteenth-Century England.* Oxford: Clarendon Press, 1990.

Oppenheim, Janet. *"Shattered Nerves": Doctors, Patients, and Depression in Victorian England.* New York: Oxford University Press, 1991.

Outram, Dorinda. "Before Objectivity: Wives, Patronage, and Cultural Reproduction in Early Nineteenth-Century French Science." In Abir-Am and Outram, 19–30.

Owen, Alex. *The Darkened Room: Women, Power and Spiritualism in Late Nineteenth-Century Britain.* London: Virago, 1989.

————. "Women and Nineteenth-Century Spiritualism: Strategies in the Subversion of Femininity." In *Disciplines of Faith: Studies in Religion, Politics and Patriarchy,* ed. J. Obelkevich, L. Roper, and R. Samuel, 130–53. London: Routledge and Kegan Paul, 1987.

Pathologist. Letter to the editor. *Lancet* ii (1844): 304.

Patterson, Elizabeth Chambers. *Mary Somerville and the Cultivation of Science, 1815–1840.* The Hague: Nijhoff, 1983.

Pickering, Andrew, and Adam Stephanides. "Constructing Quaternions: On the Analysis of Conceptual Practice." In *Science as Practice and Culture,* ed. Andrew Pickering, 139–67. Chicago: University of Chicago Press, 1992.

Pickering, George. *Creative Malady: Illness in the Lives and Minds of Charles Darwin, Florence Nightingale, Mary Baker Eddy, Sigmund Freud, Marcel Proust, Elizabeth Barrett Browning.* London: George Allen and Unwin, 1974.

Poovey, Mary. *Uneven Developments: The Ideological Work of Gender in Mid–Victorian England.* Chicago: University of Chicago Press, 1988.

Proctor, Richard Anthony. *Light Science for Leisure Hours, Second Series: Familiar Essays on Scientific Subjects, Natural Phenomena, & c. with a Sketch of the Life of Mary Somerville.* London: Longmans, Green, 1873.

Richards, Joan L. "Augustus De Morgan, the History of Mathematics, and the Foundations of Algebra." *Isis* 78 (1987): 7–30.

————. *Mathematical Visions: The Pursuit of Geometry in Victorian England.* Boston: Academic Press, 1988.

Rosenberg, Charles, and Carroll Smith-Rosenberg. "The Female Animal: Medical and Biological Views of Women." In *No Other Gods: on Science and American Social Thought,* ed. Charles Rosenberg, 54–70. Baltimore: Johns Hopkins University Press, 1976.

Rossiter, Margaret W. *Women Scientists in America: Struggles and Strategies to 1940.* Baltimore: Johns Hopkins University Press, 1982.

Russett, Cynthia Eagle. *Sexual Science: The Victorian Construction of Womanhood.* Cambridge: Harvard University Press, 1989.

Schaffer, Simon. "Astronomers Mark Time: Discipline and the Personal Equation." *Science in Context* 2 (1988): 115–45.

————. "Babbage's Dancer and the Impresarios of Mechanism." In *Cultural Babbage: Technology, Time and Invention,* ed. Francis Spufford and Jenny Uglow, 53–80. London: Faber and Faber, 1996.

————. "Self-Evidence." *Critical Inquiry* 18 (1992): 327–62.

Schiebinger, Londa. *The Mind Has No Sex? Women in the Origins of Modern Science.* Cambridge: Harvard University Press, 1989.

————. *Nature's Body: Gender in the Making of Modern Science.* Boston: Beacon Press, 1993.

Scull, Andrew. *The Most Solitary of Afflictions: Madness and Society in Britain 1700–1900.* New Haven: Yale University Press, 1993.

Secord, James A. "Extraordinary Experiment: Electricity and the Creation of Life in Early Victorian England." In *The Uses of Experiment: Studies in the Natural Sciences,* ed. David

Gooding, Trevor Pinch, and Simon Schaffer, 337–83. Cambridge: Cambridge University Press, 1989.

Shapin, Steven. "'A Scholar and a Gentleman': The Problematic Identity of the Scientific Practitioner in Early Modern England." *History of Science* 29 (1991): 279–327.

———. *A Social History of Truth: Civility and Science in Seventeenth-Century England.* Chicago: University of Chicago Press, 1994.

Shapin, Steven, and Barry Barnes. "Head and Hand: Rhetorical Resources in British Pedagogical Writing." *Oxford Review of Education* 2 (1976): 231–54.

Shapin, Steven, and Simon Schaffer. *Leviathan and the Air-Pump: Hobbes, Boyle, and the Experimental Life.* Princeton: Princeton University Press, 1985.

Showalter, Elaine. *The Female Malady: Madness, Women and English Culture 1830–1980.* London: Virago, 1987.

———. "Florence Nightingale's Feminist Complaint: Women, Religion and Suggestions for Thought." *Signs* 6 (1981): 395–412.

Smith, F. B. *Florence Nightingale: Reputation and Power.* New York: St. Martin's Press, 1982.

Smith-Rosenberg, Carroll. "Beauty, the Beast and the Militant Woman." *American Quarterly* 23 (1971): 562–84.

———. "The Hysterical Woman." *Social Research* 39 (1972): 652–78.

Somerville, Mary. *The Connexion of the Physical Sciences.* London: John Murray, 1834.

Stein, Dorothy K. *Ada: A Life and a Legacy.* Cambridge: MIT Press, 1985.

Toole, Betty A. *Ada, the Enchantress of Numbers: A Selection from the Letters of Byron's Daughter and Her Description of the First Computer.* Mill Valley, Calif.: Strawberry Press, 1992.

Vicinus, Martha, and Bea Nergaard. *Ever Yours, Florence Nightingale: Selected Letters.* Cambridge: Harvard University Press, 1990.

[Wakley, Thomas]. "Animal Magnetism." *Lancet* i (1838–39): 450–51.

Warwick, Andrew. *Masters of Theory.* Cambridge: Cambridge University Press, forthcoming.

Whewell, William. "Mrs Somerville on the Connexion of the Sciences." *Quarterly Review* 51 (1834): 54–68.

Winter, Alison. "The Construction of Orthodoxies and Heterodoxies in the Early Victorian Life Sciences." In *Contexts of Victorian Science,* ed. Bernard Lightman. Chicago: University of Chicago Press, forthcoming.

———. "Harriet Martineau and the Reform of the Invalid in Victorian England." *Historical Journal* 38 (1995): 597–616.

SEVEN

I COULD HAVE RETCHED ALL NIGHT

Charles Darwin and His Body

JANET BROWNE

CELEBRATED VICTORIAN THINKERS usually knew how to arrange their ill health. A day in the life of one prominent man went something like this:

> His custom was to work in his official room from 9 to about 2.30, though in summer he was frequently at work before breakfast. He then took a brisk walk, and dined at about 3.30. This early hour had been prescribed and insisted upon by his physician, Dr Haviland of Cambridge, in whom he had great confidence. He ate heartily, though simply and moderately, and slept for about an hour after dinner. He then had tea, and from about 7 to 10 he worked in the same room with his family. . . . He would then play a game or two at cards, read a few pages of a classical or historical book, and retire at 11. . . . He was very hospitable, and delighted to receive his friends in a simple and natural way at his house. . . . But he avoided dinner parties as much as possible—they interfered too much with his work—and with the exception of scientific and official dinners he seldom dined away from home. His tastes were entirely domestic, and he was very happy in his family. With his natural love of work, and with the incessant calls upon him, he would soon have broken down had it not been for his system of regular relaxation.[1]

We could be forgiven for thinking that this was a pretty accurate description of Charles Darwin grinding through his days at Down House in Kent. On the contrary, however, it was George Biddell Airy, the astronomer royal: the man who ran professional astronomy in the British empire, president of the Royal Society in 1871, Plumian Professor at Cambridge, director of the university observatory, and author of eleven books; a man who was often extremely ill but whose illnesses barely figure in our collective historical memories.

Anyone can find similar passages in the period's voluminous sets of lives and letters. A great number of Victorian scientists were unwell, some of them worse than Darwin, some less so. All of them, like Darwin, had to negotiate ways

---

1. Airy, *Autobiography,* 8–9.

to work while suffering from ill health. He was not alone. Yet even during his own lifetime Darwin's illnesses became something special, something unusual. How many contemporaries worried about Airy's physical troubles, for example? Or Thomas Henry Huxley's? Or Adam Sedgwick's? Darwin was not so much an invalid as a *famous* invalid. More than this, his fame became closely intertwined with what people thought about his invalidism and his intellect.

These intertwinings are still manifest. Few modern readers need reminding of the extensive literature published during recent decades on Darwin's illnesses, ranging from the flurry of interest in the 1950s in possible biological conditions such as Chagas' disease to the psychological and nervous conundrums of the late 1970s and 1980s. All of these come together as a decided genre in medical and historical writing.[2]

Nevertheless, much of the work on Darwin's medical state tends to limit itself, for one reason or another, to identifying or discussing the conditions from which he may have been suffering. Even the well-known notable exceptions, such as Ralph Colp, George Pickering, and Adrian Desmond and James Moore, focus mainly on the complex interrelations between ill health and Darwin's extraordinarily fertile inner life. Yet the whole question surely cries out for some broader attention to his ailing body as a cultural phenomenon along the lines set out in the other essays in this volume. How did ill health, celebrity status, and brains interlock in the nineteenth century, for example, and how did Darwin's very public life of the shawl mesh with Victorian cultural commitments of wider relevance? Some of the simplest inquiries along these wider lines can be revealing. When did the ordinary man or woman in the street, for instance, realize that Darwin was an ill man? Because their purpose mainly lies elsewhere, neither Pickering's *Creative Malady* nor Colp's *To Be an Invalid* can tell us. However, the first public announcement of his unhealthy condition, appropriately enough, seems to be in Darwin's *On the Origin of Species* (1859). In the introduction, after speaking of his return from the *Beagle* voyage and of the years spent puzzling over species, the mystery of mysteries, Darwin said his work was nearly finished.

2. Any survey would include the psychological and psychosomatic interpretations as presented by Hubble, "Darwin and Psychotherapy"; idem, "Life of the Shawl"; Keith, *Darwin Revalued;* Pickering, *Creative Malady;* Kempf, "Charles Darwin"; and Colp's fine study, *To Be an Invalid,* which additionally surveys the preexisting literature. In *Charles Darwin,* John Bowlby opts for hyperventilation. Johnston, "Ill-Health," discusses neurasthenia. Darwin's relations with his father are examined by Good, "Life of the Shawl," and Greenacre, *Quest for the Father.* Chagas' disease is discussed by Adler, "Darwin's Illness," and Bernstein, "Darwin's Illness," and disputed by Woodruff, "Darwin's Health," and Keynes, *Beagle Diary,* 263, 315. Multiple allergy is proposed by Smith, "Darwin's Ill Health" and "Darwin's Health Problems"; hypoglycemia by Roberts, "Reflections." Darwin would make a good subject for a historical analysis along the lines of Bynum and Neve, "Hamlet on the Couch."

"But as it will take me two or three more years to complete it, and as my health is far from strong, I have been urged to publish this abstract."[3] Deeply symbolic in his choice of moment, Darwin made his bad health a primary reason for publishing the *Origin*, putting his precarious physical state well before any remarks about Alfred Russel Wallace's having arrived at "almost exactly the same general conclusions." Illness therefore became an integral part of the radical text that followed, a persuasive device of the first order. Such an announcement in the *Origin*—on the first page of the *Origin* no less—can take on real meaning in the light of the cultural uses of medicine.

This larger question about Darwin's health and his public renown surely hinges on the way his illnesses were simultaneously experienced, presented, and interpreted—on the way notable men and women apparently integrated their faltering states into more comprehensive campaigns for engaging the attention of the Victorian community. Darwin was certainly, on many occasions, horribly ill. But he was also adept at deploying nearly everything that came to hand for promoting evolutionary theory. This chapter therefore suggests a few avenues that might be explored by considering Darwin's sick body as one further professional resource in a rich repertoire of resources: not so much in the individual sense, where continued unwellness undoubtedly etched his character and notion of self-identity and contributed significantly to his dogged determination to publish, although these were profoundly significant in Darwin's case; and not really in the context of his immediate family environment, which positively reveled in ill health; but much more in relation to the demanding and multifaceted public eye.[4] In other words, how did Darwin's afflictions enter into the construction of

3. Darwin, *Origin of Species*, 1. This sentence remained unchanged through all subsequent editions; see Peckham, *Origin: A Variorum Text*, 71. Ill health is not mentioned again in any of his publications until a footnote to the introduction to his *Variation under Domestication*, 1:2: "the great delay in publishing this work has been caused by continued ill health." Otherwise, Darwin rarely discussed his condition except in private correspondence, e.g., Burkhardt and Smith, *Correspondence*, vols. 2–9, passim. It is clear that he did not deliberately employ illness as a means of engaging public sympathy for his ideas, although the friends to whom he wrote undoubtedly made up a large proportion of the influential, elite scientific community. The image of him as an invalid seems to have been constructed more by the interweaving of private information, his own statements, his acquaintances' statements, and public interest. Reviewers certainly referred to his health in writing about the *Origin*: for example, W. B. Carpenter, "Darwin on the Origin of Species," in Hull, *Darwin and His Critics*, 92 ("as soon as his imperfect health should permit"); Heinrich Bronn, review of the *Origin of Species*, in ibid., 124 ("The author's poor health"); and an anonymous reviewer in the *Athenaeum*, 19 November 1859, 659 ("sicklied o'er with the pale cast of thought"). Such ill health was codified, as it were, for the public in Darwin's autobiography, in which he talked frankly about continued sickness, first published in F. Darwin, *Life and Letters*, 1:26–107. This was accompanied by Francis Darwin's reminiscences, which also include references to his father's ill health, ibid., 108–60, esp. 159–60.

4. Described, for example, in Morris, *Culture of Pain*, and Taylor, *Sources of the Self*. For the Darwin family's ill health, see F. Darwin, *Life and Letters*, 1:159–60, and Raverat, *Period Piece*.

social relations with his scientific colleagues, with his doctors, and with his readers? How did he feel about it all? Oddly enough, as Dorinda Outram points out in a recent essay, the one thing that is frequently ignored in accounts of bodies is the owner's subjective experience of just such an item.[5] Darwin seems to have put his body under the Victorian spotlight just as concretely as he presented his mind through the *Origin of Species*. In the process, it seems that this "public" body gradually came to evoke the disembodied quality of thought. Darwin—and Darwin's body—consequently offer a good point of entry into some of the complex forces at play when we try to talk about the presentation of intellectual authority, authority that invariably reaches far beyond the usual framework of printed books.

■ *The Illnesses* ■

IN COMMON WITH most sick people, Darwin was noticeably conscious of his outer frame and never shy of describing its miseries to intimates. Though his symptoms came and went over the years and varied in intensity, they remained more or less within the same parameters. Putting it succinctly, these were predominantly gastrointestinal. "For 25 years," he wrote, "extreme spasmodic daily & nightly flatulence; occasional vomiting, on two occasions prolonged during months. . . . All fatigue, especially rocking, brings on the head symptoms . . . cannot walk above ½ mile—always tired—conversation or excitement tires me most."[6]

The truth was that, as he said to Joseph Hooker and Thomas Henry Huxley, he suffered from incessant retching or vomiting, usually brought on by fatigue; and from painful bouts of wind that churned around after meals and obliged him to sit quietly in a private room until his body behaved more politely. Reading between the lines, his guts were noisy and smelly. "I feel nearly sure that the air is generated somewhere lower down than stomach," he told one doctor plaintively in 1865, "and as soon as it regurgitates into the stomach the discomfort comes on."[7] He was equally forthright with his cousin William Darwin Fox: "all excitement & fatigue brings on such dreadful flatulence that in fact I can go nowhere." When he did go somewhere, he needed privacy after meals, "for, as you know, my odious stomach requires that."[8]

5. Outram, "Body and Paradox." See also her *Body and the French Revolution*.

6. Medical notes supplied by Darwin to Dr. John Chapman, 16 May 1865, University of Virginia Library, transcribed in Colp, *To Be an Invalid*, 83–84. By "rocking" Darwin means the motion of horse-drawn carriages and railway trains: a nineteenth-century version of travel sickness.

7. Darwin to Chapman, 7 June 1865, University of Virginia Library.

8. Darwin to W. D. Fox, 24 October 1852, transcribed in Burkhardt and Smith, *Correspondence*, 5:100; Darwin to J. D. Hooker, 17 June 1847, ibid., 4:51.

He also had trouble with his bowels, frequently suffering from constipation and vulnerable to the obsession with regularity that stalked most Victorians. He developed crops of boils in what he called "perfectly devilish attacks" on his backside, making it impossible to sit upright, and occasional eczema. There were headaches and giddiness. He probably had piles as well.[9]

Not surprisingly, when these debilitating signs of weakness arrived in a batch, Darwin felt terribly dejected, almost as if his physical shell was taking over. Despite all the care and attention he lavished on it, the body rebelled: illness was an alien presence that robbed him of his power over himself. "I shd suppose few human beings had vomited so often during the last 5 months," he gasped early in 1864.[10] "It is astonishing the degree to which I keep up some strength.... I have had a bad spell vomiting every day for eleven days and some days many times after every meal."[11] His body was not particularly pleasant for him to be with. Considerately, he tried to keep it out of other people's way as well. But he was often preoccupied with it to the exclusion of almost everything else in ordinary life. He was sure that such prolonged misery indicated a physiological disorder.

It fell to Joseph Hooker, his botanical friend at Kew Gardens who was previously trained as a physician, to ask the obvious question. "Do you actually throw up, or is it retching?" It was both, Darwin replied, but food hardly ever came up. "You ask after my sickness—it rarely comes on till 2 or 3 hours after eating, so that I seldom throw up food, only acid & morbid secretion." "What I vomit [is] intensely acid, slimy (sometimes bitter), corrodes teeth." Doctors puzzled, he added defiantly.[12]

Hooker did not venture a diagnosis. Nor did Huxley, or any other of Darwin's closest medically trained friends, although they offered constant sympathy and practical advice. They kept themselves out of what might be a difficult situation. Instead, they suggested the names of leading doctors he might wish to consult. Acting on this advice over the years, Darwin sought out a number of London physicians, most of whom he knew by reputation or through his scientific work; the two exceptions were his father, Robert Darwin of Shrewsbury, when he was alive, and Henry Holland, a second cousin on the Wedgwood side. With the usual prerogative of the wealthy classes, he tended to choose doctors with a reputation for having studied some topics in greater detail than usual. In

9. Darwin's symptoms, and the various remedies and treatments he tried over the years, are fully described by Colp, *To Be an Invalid,* esp. 109–44.

10. Darwin to Hooker, 27 January 1864, Cambridge University Library, Darwin Archive (hereafter DA) 115:217. For illness as an alien presence, see particularly Leder, *Absent Body,* and Shilling, *Body and Social Theory.*

11. Darwin to Hooker, 5 December 1863, DA 115:213.

12. Hooker to Darwin, 5 February 1864, DA 101:180; Darwin to Hooker, DA 115:219.

fact, the variety of Darwin's physicians through the 1860s and 1870s is an interesting theme in itself, one that the continuing publication of his correspondence will reveal in detail.[13] Darwin also had more doctors than might be expected, another perquisite of the rich. Like many patients searching for an acceptable diagnosis, he moved constantly among different medical men, trying their remedies and their diets, their purges, mineral acids, and magnesium salts for a couple of months before giving up and turning in despair to another expert and another treatment. In a medical world barely beginning to fragment into specialties, this movement was perhaps inevitable. It was certainly a regular feature of Darwin's life and also of others'.

Such constant medical attention was addictive. In his time, Darwin sought out physicians with an interest in stomachs, skin, urine, blood, nerves, and gout, and on one occasion the entrepreneurial publisher of the *Westminster Review,* John Chapman, who qualified at Saint Andrews medical school when his publishing business tottered and claimed to have found a cure for seasickness and nausea in icebags applied to the spine.[14] Darwin was very taken with Chapman and his therapy and was sorry when a month's application of icebags along his spine (three times a day for an hour and a half at a time) made no difference to his retching. "We liked Dr. Chapman so very much we were quite sorry the ice failed for his sake as well as ours," wrote Emma Darwin.[15] Apart from all the other things, Darwin's body was starting to represent an expensive medical investment.[16]

These doctors mostly agreed that Darwin suffered from an intestinal or stomach disorder of a chronic recurrent nature, probably involving the nerves supplying the gut. Darwin's rapid trajectory through them reveals something of his own belief that his nervous system, his brain, and his stomach were uniformly implicated—if one doctor neglected to include all the elements he felt were failing he soon moved elsewhere for another opinion. From very early on, he believed that too much work brought on bouts of vomiting: that "the noddle and the stomach are antagonistic powers." As a Darwin aunt said, "his health is

13. Burkhardt and Smith, *Correspondence,* vols. 1–9, which cover the years 1821 to 1861. The entire correspondence is listed in summary form in Burkhardt and Smith, *Calendar.* See also F. Darwin, *Life and Letters,* and Darwin and Seward, *More Letters.*

14. Nausea (including seasickness), according to Chapman's theories, was caused by a rush of blood to the spinal cord. See Chapman, *Sea Sickness* and *Neuralgia,* for what he calls neuro-dynamic medicine, and Haight, *George Eliot and John Chapman,* 113–16.

15. Emma Darwin to Hooker, 18 July 1865, DA 115:272v.

16. From September 1862 to September 1863, for example, his expenditure on medical treatment was £129 11s. 6d. Compare this with around £50 for "Science," £10 for "Books," and £158 for "Manservants." The following year was less expensive at £56 4s. 10d. Darwin's Classified Account Books, Down House Archives, Kent.

always affected by his nerves."[17] Some physicians, like William Brinton, the eminent physiologist, called his condition dyspepsia and prescribed magnesium to counteract excess acid secretions.[18] Others, like Henry Bence Jones, thought the problem more to do with the physiology of digestion indicating an imbalance of acids and alkalies in the blood—what Bence Jones and Henry Holland called "suppressed gout," a state defined by them as too much uric acid remaining in the blood.[19] So Darwin had his urine tested, followed Bence Jones's special diet, and dosed himself with colchicum, a dangerously corrosive specific for gout.[20] Dr. Chapman treated the base of his spinal cord with ice. Dr. Gully at the Malvern water cure treated the top of it with cold water.[21] Dr. Lane at Moor Park and then Dr. Smith at Ilkley sat him in freezing hip baths.[22] Dr. William Jenner prescribed podophyllum and other drastic purgatives. Dr. George Busk thought the problem was primarily mechanical—the stomach did not push on its contents as rapidly as it ought to.[23] On the other hand, Dr. Engleheart, the physician in Down village, told Darwin to look to his drains. Almost despairing of a cure, he eventually sent some vomit on a slide to John Goodsir for him to search for pathogenic vegetable spores and was disappointed to hear there were none.[24] It seems probable that toward the end he was suffering just as much from overmedication and a surfeit of conflicting advice as from his own special combination of physical disorders and medical neuroses.

Darwin's doctors also acknowledged the importance of his nervous system and tactfully dealt with his delicate mental constitution as well as they could. In the post-*Origin* years, they knew they were dealing with a famous thinking man, however modest and unassuming in personal demeanor. It was important for

17. Litchfield, *Emma Darwin*, 2:142.
18. Brinton was prominent in the field of stomach disorders. Darwin recorded paying ten guineas for a consultation on 22 November 1862 (Account Book, Down House Archives). See also Darwin to Hooker, 10 November 1863, DA 115:208.
19. Jones, *On Gravel*, discusses the role of diet in diminishing nonnitrogenous principles in the blood. See also his *Animal Chemistry*. Holland, *Medical Notes*, 239–69, discusses gout more generally. "Latent" gout was a common diagnosis in midcentury and was understood as a metabolic disorder characterized by raised uric acid in the blood with few obvious clinical features; see Colp, *To Be an Invalid*, 109–12, 236, and Porter and Rousseau, *Gout*.
20. Holland, *Medical Notes*, 258–69, and Scudamore, *Colchicum Autumnale*. On colchicum's injurious effects, see Rennie, *Observations on Gout*. William Jenner prescribed "enormous quantities of chalk, magnesia & carb. of ammonia"; Darwin to Hooker, 13 April 1864, DA 115:229.
21. Gully, *The Water Cure*, and Browne, "Spas and Sensibilities."
22. Colp, *To Be an Invalid*, and Burkhardt and Smith, *Correspondence*, vols. 6 and 7, passim. For general accounts of hydrotherapy, see Metcalfe, *Hydropathy*, Turner, *Taking the Cure*, and Rees, "Water as a Commodity."
23. Busk to Darwin, October 1863, DA 170.
24. Goodsir to Darwin, 21 August 1863, 26 August 1863, and 28 August 1863, DA 165.

both physician and patient to reach a diagnosis they felt mutually comfortable with; and that therapy should err on the side of professional caution. No one wanted to go down in history as the man who killed the Newton of nineteenth-century biology. It was not so many years earlier that Sir Richard Croft, accoucheur to Princess Charlotte, had committed suicide after the princess's unfortunate death in childbirth in 1818.[25] Bence Jones, for example, recommended the traditional upper-class remedy of a change of scene. Failing that, he said, Darwin should "get a pony and be shaken once daily to make the chemistry go on better."[26] This Darwin did, and enjoyed the exercise until a riding accident required that he should call it a day. Perhaps some yachting, Bence Jones blithely proposed a year or two later. Mental distraction, he thought, was a crucial part of the answer. Andrew Clark reiterated the same instructions in a different form. "Do not notice your own sensations . . . struggle to avoid self-scrutiny & self-consciousness."[27] Above all, he said, try to relax and take time off from the punishing self-imposed schedule of scientific work. But Darwin dismissed these suggestions. His work, he claimed, was the only thing that took his mind off his sickness; and in later life, a game of billiards.

However prominent they were, these physicians evidently found it difficult to be blunt with him; in Darwin's case the customary negotiation between doctor and patient had to accommodate his intellectual status as much as anything else.[28] And naturally enough, the doctors wanted to keep Darwin's custom. Andrew Clark, for instance, in his eagerness to become a great man's physician, nearly overstepped the mark early in their medical relationship. In 1873 he felt it necessary to write an abject letter to Darwin apologizing for "pushing too close." Nevertheless, he said, he hoped one day to be of service.[29] During the late 1870s he achieved that aim in becoming Darwin's primary physician, warmly praised by Francis Darwin in his edition of Darwin's *Life and Letters,* and eventually attending Darwin on his deathbed. It must have been daunting, furthermore, for these men to find themselves simultaneously sucked into Darwin's scientific projects. "Can you persuade the resident doctor in some hospital," Darwin asked James Paget when consulting him on his own behalf, "to observe a person retch-

25. Price, *Critical Inquiry.*
26. Henry Bence Jones to Darwin, 10 February 1866, DA 168. See also Jones to Emma Darwin, 1 October 1867, and Jones to Darwin, 2 August 1870, DA 168.
27. Clark to Darwin, 8 July 1876, DA 210:21.
28. Detailed analyses of the interrelations between doctors and patients are given by Peterson, *The Medical Profession,* Porter, *Patients and Practitioners,* Digby, *Making a Medical Living,* and Oppenheim, *"Shattered Nerves."*
29. Clark to Darwin, 3 September 1873, DA 161. In the same letter he reports that Darwin's urine was loaded with uric acid but showed no albumin—a favorable diagnosis.

ing violently, but throwing nothing from the stomach." Darwin wanted to know whether tears came to their eyes, a trait he was investigating for the *Expression* book. "From my own personal experience I do think that this is the case."[30] In his usual methodical way, he made his illnesses, and his intellectual project, inseparable from the more general problem of being treated by medical experts.

Still, as Pickering, Colp, and others have noted, Darwin's illnesses usually took the external form that would be most useful.[31] A weak stomach was a very good reason for avoiding dinner parties, much better than an attack of rheumatism or unmentionable boils. Repeated vomiting after a train journey was well suited to avoiding trips to London. A night of retching after ten minutes' talk was a valid obstacle to prolonged social activity: half an hour with Ernst Haeckel, or even close friends like Hooker, Huxley, or Lyell, or any selection of agreeable Down House neighbors, could literally make Darwin sick. Many of Darwin's disabilities were, in this sense, socially relevant ones. Much of their circumstantial value lay in their diversity, applicability, and lack of diagnosis. Such illnesses, as novelists like Jane Austen and Elizabeth Gaskell readily recognized, act as a mode of social circulation as well as instruments of domestic tyranny. Perpetually ailing and complaining, *Emma*'s Mr. Woodhouse was the focus of some of Austen's most pointed observations.

Darwin took advantage of his versatile failings. In the most general manner, of course, his avowed need to stay quiet allowed him to keep apart from the controversies surging around the *Origin*. Ill health permitted him to discourage all but the most wanted visitors to Down House and to choose whom he saw when he went into London. It allowed him to fall asleep during piano recitals and novel readings. It excused him from boring evenings at scientific societies. It sanctioned his retreat after dinner (with the ladies) instead of sitting up with cigars and wine in masculine company. In these subtle ways, he let ill health carry the brunt of displaying a preoccupation with other, more intellectual concerns: a nineteenth-century counterpart to ascetic philosophers described here by Shapin (chapter 1) who gave themselves up wholly to the search for truth. Like them, Darwin's belly was at the opposite pole to mentality. The poet William Allingham got it just right when, after meeting Darwin in 1868, he said that: "he has his meals at his own times, sees people or not as he chooses, has invalid's privileges in full, a great

30. Darwin to Paget, 4 June 1870, Wellcome Institute for the History of Medicine, London, Western MS 5703, item 38. For Darwin's theories of expression, see his *Expression of the Emotions*.

31. Pickering, *Creative Malady*, 71, 77–80, Colp, *To Be an Invalid*, 122–26, 141–44. Useful accounts in this area are given by Berrios, "Obsessional Disorders," Bynum, "Rationales for Therapy," Inglis, *Diseases of Civilization*, Lopez Pinero, *Neurosis*, and Sicherman, "Uses of Diagnosis." See also Wiltshire, *Jane Austen and the Body*, and Bailin, *The Sickroom in Victorian Fiction*.

help to a studious man."[32] Other passing acquaintances were less impressed. "Why drat the man," said old Mrs. Grote at Chevening Court, "he's not as bad as I am."[33]

Ill health also helped turn Darwin's absences into a statement. It provided a reason for not going into London to receive the Royal Society's Copley Medal in 1864, the award that generated intense debate in the council on whether the *Origin* should be acknowledged or not.[34] If it was acknowledged, as Darwin's friends in the council demanded, Huxley believed there would be a public reprimand from the president about the book's dangerous opinions and promised Darwin that he would provide a spirited defense. Too ill to go, Darwin sent Hugh Falconer (his proposer) and his brother Erasmus Alvey Darwin along instead. They all knew why. "What a pity you can't be there," Erasmus wrote sardonically. "And yet if you were it could not be done so well."[35]

Darwin employed the same panic-stricken tactic when his old friend and professor, John Stevens Henslow, was at death's door.

> I write now only to say that if Henslow . . . would really like to see me I would of course start at once. The thought had [at] once occurred to me to offer, & the sole reason why I did not was that the journey, with the agitation, would cause me probably to arrive utterly prostrated. I shd be certain to have severe vomiting afterwards, but that would not much signify, but I doubt whether I could stand the agitation at the time. I never felt my weakness a greater evil. I have just had a specimen, for I spoke a few minutes at Linnean Society on Thursday & though extra well, it brought on 23 hours vomiting. I suppose there is some Inn at which I could stay, for I shd not like to be in house (even if you could hold me) as my retching is apt to be extremely loud.[36]

Not many people could persist in asking Darwin to visit after receiving a letter like that. By staying away, the implication goes, he was helping Henslow

32. Allingham, *Diary* (7 February 1868), 185.

33. George Darwin's reminiscences, DA 112 (ser. 2): 23.

34. Bartholomew, "Copley Medal," and MacLeod, "Of Medals and Men." See also F. Darwin, *Life and Letters*, 3:27–28; and Erasmus Alvey Darwin to Darwin, 9 November 1863, DA 105 (ser. 2): 13.

35. E. A. Darwin to Darwin, undated 1864, DA 105 (ser. 2): 33. The proposal and citation were prepared by Falconer, who wished to mention Darwin's illness: "Dr Falconer . . . wants dates of your voyage. Also, what I should think would not be judicious to bring forward, when your sickness came on & how long it lasted & whether in consequence of it FitzRoy persuaded you to give up the voyage." E. A. Darwin to Darwin, 27 June 1864, DA 105 (ser. 2): 28. Darwin's health was not mentioned in the published citation for the award.

36. Darwin to Hooker, 23 April 1861, DA 115:98.

far more than if he arrived. In both instances, if only for a critical moment, his body's significance lay in its absence. There are many other examples in his correspondence.

It appears too easy then to write off all these complaints as mere hypochondria, although a strong dash of it definitely ran in the family. Darwin was as much aware of the family foible as anyone and perfectly capable of joking about it with his wife and relatives. His sister Caroline, he would say, was intensely irritating when she became heroic about her illnesses: he and Emma much preferred people speaking up.[37] Equally, it was Erasmus Alvey Darwin, Darwin's older brother, who was considered the irretrievable family hypochondriac, not Darwin. This Erasmus Darwin was a witty, cynical man, enjoying what he called his misanthropy in Marlborough Street. "I have been lying on the sofa in a state of utter torpor," he wrote once. "I mean to go out today to see if I am well or not . . . If the present beautiful weather continues I shall be compelled to go and be happy in the country but at present I prefer being miserable in London."[38]

Darwin's bulletins about his own symptoms need to be fitted into this gently self-mocking and intelligent family pattern. "Charles came up yesterday," said Erasmus, "and went out like a dissipated man to a tea party." We need to remember that people can have a sense of humor about illness, which they sometimes direct at themselves. When Darwin signs himself at the end of a letter as "your insane and perverse friend," he does not mean it literally.

It appears almost too easy, as well, to assume that all these illnesses were somehow a consequence of Darwin's deep-seated anxieties about evolutionary theory and its religious consequences. We seem to expect individuals with radical new ideas to be tormented by them to the exclusion of other anxieties— other anxieties we retrospectively perceive as having lesser importance. We are probably right to expect a great deal of that mental conflict to emerge as illness. But for the sake of argument, Darwin's nights of retching might just as well be related to financial preoccupations as to any known personal crisis about the metaphysical implications of his theories. Although he was rich, he invariably worried about where his next penny was coming from. The ups and downs of his investments in railway stock, the tortuous ramifications of family trust funds, and profit-sharing arrangements with the publisher John Murray, for example, often coincided with his best-documented bouts of sickness. One very bad attack came a year or two after establishing his eldest son, William, as a partner in

37. Burkhardt and Smith, *Correspondence,* 2:314.
38. E. A. Darwin to Frances Wedgwood, Wedgwood/Mosely Collection, Keele University Library.

a privately owned bank in Southampton, which had required Darwin, as his father, to promise £10,000 as security against a run on deposits. In 1863, when Darwin became ill, an act of Parliament was passed allowing the establishment of joint-stock banking concerns that spread the risk and promised only limited liability for its members, a new state of affairs which jeopardized William's old-style partner-led syndicate. At the same time, railway amalgamations were creating vast, unregulated monopolies by swallowing up many of the smaller companies in which Darwin had a stake; and interest rates on cash capital that year were particularly unstable.[39] Darwin was a cautious investor, keen to avoid risks. The prospect of gambling his substantial fortune, and the future inheritances of his children, on the vagaries of the City of London and nebulous entities like public confidence was, to him, a very serious question. Only a few years before, his own London bank, the Union Bank, temporarily collapsed after an immense fraud carried out by a clever clerk.[40]

Alternatively, or simultaneously, his illnesses may well have acted as a mediator in married life—something we are closely attuned to in accounts of literary couples like Robert and Elizabeth Browning, Thomas and Jane Carlyle, and Charles and Catherine Dickens but seemingly ignore when dealing with an evolutionist.[41] He might have been dismayed about getting old, or going bald, or about the lack of future occupations for his unhealthy sons and daughters. There was the Civil War in North America and the implications of continued slavery to worry about. If Darwin had been a famous *female* invalid, moreover, like Harriet Martineau, Florence Nightingale, or Ada Lovelace as described here by Alison Winter (chapter 6), our interpretation of the disorders would also be very different.[42] If he had been plain old Professor Airy, his illnesses would not be

39. English law confined the number of partners in a bank to six until the Joint Stock Banks and Companies Act of 1863 allowed some redistribution of the risk in the wake of the crashes of 1857–58. See Anderson and Cottrell, *Money and Banking*. For railways, see Barker and Savage, *An Economic History of Transport*, 87.

40. These financial details are drawn from the *Annual Register*, 1863.

41. See, for example, Markus, *Dared and Done*, and Checkland, *The Gladstones*. General analyses are in Wohl, *The Victorian Family*; Graham, *Women, Health and the Family*; Peterson, *Family Love*; Davidoff and Hall, *Family Fortunes*. In some ways Darwin and his wife reversed the customary roles of the "weak" woman and "strong" man while still expressing much of their relationship through the preoccupations of invalid and nurse. For comments on the ways in which ideas of male health were based on self-control and of female ill health on an inability to control the body, see Shuttleworth, "Female Circulation."

42. For example, Cooter's study of Martineau, "Dichotomy and Denial," Pickering on Nightingale, *Creative Malady*, 122–77, and Micale, "Hysteria Male/Hysteria Female." See also Edel, *Diary of Alice James*, and Trombley, *All That Summer She Was Mad*. Interpretations of women's diseases are discussed in Bailin, *The Sickroom in Victorian Fiction*, 17–19, Digby, "Women's Biological Straitjacket," Wood, "Fashionable Diseases," Ehrenreich and English, *Complaints and Disorders*, and Figlio,

interesting in the wider sense at all. And have we really eliminated the possibility of a long-lasting subclinical problem or constellation of problems? It is not doing full justice to the richness and complexity of Darwin's situation to opt for one or another "cause" without careful analysis of what we really wish to claim about medical embodiment. As Rob Iliffe stresses in chapter 4, in the case of Isaac Newton, richly diverse sets of discourses can envelop one and the same individual.

■ *The Body* ■

NONETHELESS, DARWIN was acutely aware of his body and all its failings. It was the primary focus of his attention. It was the focus of his friends' and doctors' attention. It dominated the domestic arrangements of his wife and family. It was something he told the readers of the *Origin* about. And he found it a useful device for avoiding tiresome social obligations and unpleasant scientific controversy. Darwin's ill health was doing a lot of "work" in the modern sense. Yet at the same time it remains difficult to say precisely how this ill health materially contributed to his special genius. Personally, he recognized that too much study made him ill. But equally, he claimed that he was capable of dissociating himself from his physical disorders only through abstract thought.[43] In terms of the simple dualism that attracted and helped many Victorian invalids through their daily lives, Darwin, so to speak, invariably rose above the malfunctions of the flesh to engage in what his contemporaries celebrated as an extraordinarily active life of the mind. Where some intellectuals displayed their intellectuality by neglecting their bodies, as Shapin (chapter 1) and Iliffe (chapter 4) demonstrate, and the physicians described here by Christopher Lawrence (chapter 5) carefully evoked the physical image of a learned gentleman, Darwin's "disembodiment" emerged out of a determined, and in time heroic, conquering of his inadequate frame. As all the essays in this volume variously show, this too is what we would call "work."[44]

Even so, few historians ask how far that idea of personal dissociation, the triumph of mind over matter, filtered into the public realm and became part of what people thought about Darwin. What were Victorian men and women of-

"Chlorosis and Chronic Disease." See also Bynum, "The Nervous Patient"; Micale, *Approaching Hysteria;* and Oppenheim, *"Shattered Nerves."* Authoritative studies of bodies and gender can be found in Ehrenreich and English, *For Her Own Good;* Schiebinger, *Nature's Body,* Vicinus, *Suffer and Be Still;* Gallagher and Laqueur, *The Making of the Modern Body;* Laqueur, *Making Sex;* and Jordanova, *Sexual Visions.*

43. Leder, *Absent Body.*

44. See particularly Haley, *Healthy Body;* Goffman, *Presentation of Self;* Gilman, *Disease and Representation;* and idem, *Health and Illness.* General studies of the field are Turner, *Body and Society;* and Porter, "History of the Body." See also Outram, "Body and Paradox."

fered in the way of visual information about his mind and body? Did they see intellect or illness?

Judging from the mass-reproduced photographs available in archive collections, they principally saw a well-to-do gentleman in dark, sober suits, a gentleman with little regard for fashionable taste (figs. 7.1–7.3). He always wears warm clothes, sometimes a cape or an overcoat on top, a waistcoat underneath, perhaps a scarf draped over the shoulders, a felt hat. He is mostly sitting down. There are no symbolic props to supply clues about the figure's special calling: no books open on the knee, no microscope, no dogs or plants beside his chair, no spectacles on the nose. His accoutrements, or his lack of accoutrements, tentatively suggest a philosopher—a careworn and modest philosopher at that.[45] Yet he could as easily be a member of any one of a number of solidly prosperous Victorian professions: a university don or schoolteacher, a member of Parliament, a banker, a lord, or a country gentleman.[46]

Such photographs are not often reproduced in twentieth-century histories and biographies—they are not sufficiently striking to appeal to modern tastes or perhaps do not fully resonate with what is now expected of a Darwin illustration. But they were how Darwin presented himself to his public. Figures 7.1 through 7.3, for example, were taken expressly to cater to the surge of contemporary interest in portrait photography and served a specific public purpose. Darwin, as much as anyone, was gripped by the craze for exchanging cartes de visite with his correspondents. By 1865 or so photographic technology in Britain, France, and America had diversified sufficiently to allow the mass production of studio portraits on small cards, often incorporating a facsimile of the sitter's signature. Darwin posed several times for cards like these and made use of them as a kind of autograph to send through the post.

What is perhaps less well known is just how much commercial activity surrounded the carte de visite business.[47] Professional photographers naturally supplied the sitter with a few packets of cards for private use. But they also sold

45. Clarke, *The Portrait in Photography*; Linkman, *The Victorians*; Piper, *Personality and the Portrait*; Fyfe and Law, *Picturing Power*; and Tagg, *The Burden of Representation*. Other forms of visual presentation are discussed in Adler and Pointon, *The Body Imaged*; Stafford, *Body Criticism*; Porter, "Bodily Functions"; Lynch and Woolgar, *Representation in Scientific Practice*; Goffman, *Presentation of Self*; Gilman, *Seeing the Insane*; and Cowling, *The Artist as Anthropologist*. Symbolism in early portraiture is analyzed in Gent and Llewellyn, *Renaissance Bodies*. Photographic imagery is thoroughly discussed in Fox and Lawrence, *Photographing Medicine*; Edwards, *Anthropology and Photography*; and Weaver, *British Photography*.

46. Cunnington and Cunnington, *Handbook of English Costume*. The clothes of less prosperous groups are illustrated in Cunnington and Lucas, *Occupational Costume*. See also Hollander, *Seeing through Clothes*; and Harvey, *Men in Black*.

47. The rise of commercial portrait photography is discussed by Bolas, *The Photographic Studio*; Prescott, "Fame and Photography"; and Darrah, *Cartes de Visite*. See also Lee, "Victorian Stu-

ELLIOTT & FRY   Copyright   55, BAKER S[
PORTMAN SQ[

FIGURE 7.1. Charles Robert Darwin. Carte de visite by Elliott and Fry, ca. 1878. These cartes de visite became hugely popular after 1865 and brought likenesses of scientific authors before a wide general public. They were sold in photographic shops as well as distributed privately by the sitter. Note the copyright label. Courtesy Wellcome Institute Library, London. V0026271B00.

FIGURE 7.2. Charles Darwin. Carte de visite by Barraud, possibly December 1881. The photograph's condition is poor. Many agencies built up a lasting professional reputation by taking portrait photographs of famous Victorians. Courtesy Wellcome Institute Library, London. V0028472B00.

FIGURE 7.3. Commercial photograph of Darwin, copyright to the London Stereoscopic and Photographic Company. Courtesy Wellcome Institute Library, London. V0026271B00.

them for profit in their shops. The *Photographic Journal* for 1862 recorded that one London studio was selling £50 worth of portrait cards daily and that more than fifty thousand items passed through the hands of another dealer in a single

dio"; Pritchard, "Commercial Photographers"; and Welford, "Cost of Photography." In 1856 Maull and Polyblank charged 5s. for an albumin print, 8 × 6 inches. Three cartes de visite cost 2s. 6d. from Ernest Edwards in the mid-1860s.

month.[48] "The public are little aware," said a surprised author in *Once a Week*, "of the enormous sale of the *cartes de visite* of celebrated persons. An order will be given by a wholesale house for 10,000 of one individual—thus £400 will be put into the lucky photographer's pocket who happens to possess the negative."[49] Not surprisingly, a photograph of Darwin after the *Origin of Species* was published was a sound commercial proposition. In October 1862, the photographer Polyblank (of Maull and Polyblank) wrote to Darwin, via Erasmus, asking for general permission to reproduce and sell the one taken by the firm some years earlier. Erasmus reported to Darwin that "Polyblank says that for some he has a general order to sell & for others he requires special permission so I shd. think you might as well give a general order as it is a good photograph."[50]

Furthermore, the shop windows where these pictures were displayed, said one literary magazine, were better than the National Portrait Gallery (itself only opened in 1859)—more egalitarian, for one thing, where an engineer could be seen beside the queen of Naples, or Mrs. Fry cheek by jowl with Lord Brougham. Or Huxley and Samuel Wilberforce, if we did but know it. Coming to the same conclusion from a different angle, the *London Review* ran an article entitled "The New Picture Gallery" criticizing the lack of artistic merit in such cartes de visite.

Darwin resisted the craze for several years before capitulating to having his own cartes de visite made. Before then he sent out copies of a photograph taken by his son William, a keen amateur photographer whose hobby was financed by Darwin. But as the swapping and requests for pictures accumulated, he had his cartes made and updated them every few years thereafter. He also started his own album in 1864 for mounting the photographs sent in return by scientific friends.[51] In the process, Darwin became aware of his own value as a public image, though unassuming and diffident about it on most occasions.

Even the earliest nonphotographic portrait of him was relatively widely distributed. The well-known study by Thomas Maguire, a lithographic print taken from life in 1849, was from the outset a commercial project (fig. 7.4).[52] This was

48. "Miscellanea," *The Photographic Journal*, 15 March 1862, 21.

49. Wynter, "Cartes de Visite," 376. See also "Commercial Photography," *British Journal of Photography*, 2 January 1867, 47. According to anecdote, five thousand portraits of John Wilkes Booth were sold after the assassination of Abraham Lincoln.

50. E. A. Darwin to Darwin, undated, October 1862, DA 105 (ser. 2): 9. It is not clear to which photograph he refers although judging from the wide circulation of a number of subsequent reissues, it was probably the one with checked trousers (fig. 7.10). See note 56.

51. The whereabouts of this album is unknown.

52. T. H. Maguire, the Irish painter and lithographer, was appointed lithographer to the queen in 1851 and subsequently took portraits of Prince Albert and the royal children. Throughout his career he exhibited genre scenes at the Royal Academy, eventually issuing *The Art of Figure Drawing* in 1869.

FIGURE 7.4. Lithograph of Darwin, 1849, by Thomas Maguire for the British Association series, published 1851. Prints were produced in sufficient numbers to make the project a commercial one. Darwin's dress and pose emulate the masculine sobriety of his friends Henslow and Lyell in the same series (figs. 7.7 and 7.8). Courtesy Wellcome Institute Library, London. V0001461B00.

one of a series of fifty or so portraits of scientists taken in honor of a British Association meeting at Ipswich in 1851. The series was conceived by George Ransome, the head of the agricultural machinery company and local secretary of the association meeting, as a paying venture to mark the Ipswich occasion and the opening of the Ipswich Museum.[53] Some sitters were drawn twice; and multiple prints and sets were offered for sale both at that time and later. Prints were still circulating in the commercial sense as late as 1864 when Darwin was offered two "most beautiful" proof impressions of his own portrait for 7s. 6d.[54]

In the Maguire portrait Darwin appears prosperous and confident, not at all ill in his outward appearance. However, more modish philosophical gentlemen of the period looked quite different, usually sporting a fashionably "lank" hairstyle, a shortened form of frock coat, and a stock fastened with a tiepin (fig. 7.5).[55] In the same Maguire series, Edward Forbes, a romantic philosophical naturalist, adopted the latter style as a badge of his poetic predisposition, possibly also as a sign of his French and German scientific affiliations (fig. 7.6). In choosing the clothes and masculine pose that he did, Darwin patently aligned himself with the sturdy, no-nonsense faction of nineteenth-century scientific life represented by men such as Henslow, Sedgwick, and Lyell (figs. 7.7 and 7.8).

The first commercial photograph of Darwin was taken in the studio of Henry Maull in 1855 or so—the first readily reproduced photographic image of him, so to speak (fig. 7.9). This was semipublished in the sense that Maull released it as part of a set of photographs under the title of *Literary and Scientific Portrait Club* (published in parts from around 1854). Darwin thought it made him look "atrociously wicked,"[56] and it is not known if he ever ordered any duplicates for private use: there are none in the Darwin archive at Cambridge,

53. There is no record of a direct payment to Maguire in Darwin's Account Book of 1849–51, although he recorded payment, on subscription, to the portraits of the bishop of Norwich and George Ransome in the same series. A sum of £1 6d. was paid to Lovell Reeve, the publishers of the series, on 17 November 1849.

54. E. A. Darwin to Darwin, 9 April 1864, DA 105 (ser. 2): 25.

55. Cunnington and Cunnington, *Handbook of English Costume*; and Lurie, *The Language of Clothes.* Shortland, "Bonneted Mechanic," addresses the theme of dress in greater detail.

56. This set of photographs was issued on subscription, loose in a folder. A copy is preserved at the National Portrait Gallery, London. The exact date of the first photograph (fig. 7.9) is difficult to ascertain. Freeman, *Companion,* 97, gives a tentative date of ca. 1854. However, a sitting to Maull in 1855 is apparently confirmed in Burkhardt and Smith, *Correspondence,* 5:339, with a payment to Maull and Polyblank of £3 1s. 6d. in Darwin's Account Book, 31 December 1855, Down House Archives. The set is discussed in Prescott, "Fame and Photography." Some later engravings of this photograph incorporate the date 1854: this date cannot be independently substantiated. Francis Darwin thought the second photograph (fig. 7.10) was also taken in 1854 and reproduced it as such as the frontispiece to *Life and Letters,* vol. 1. For the dates of Maull's various partnerships, see Pritchard, *Directory of London Photographers.*

Figure 7.5. Costume plate, 1840. Note the longer hair, modish frock coat, and soft cravat, all indicative of poetic, intellectual style. From Cunnington and Cunnington, *Handbook of English Costume in the Nineteenth Century,* published by Faber and Faber Ltd. Copyright © The Estate of Cecil Willett Cunnington and Phillis Cunnington, 1959.

FIGURE 7.6. Edward Forbes. Lithograph by Thomas Maguire for the British Association series, 1849. Forbes's dress, which includes an academic gown, and his pose, including a plant, suggests his allegience lay with the poets, as in fig. 7.5. Courtesy Wellcome Institute Library, London. V0001959B00.

FIGURE 7.7. John Stevens Henslow. Lithograph by Thomas Maguire for the British Association series, 1851. Henslow looks every inch the sturdy university professor. Courtesy Wellcome Institute Library, London. V0002695B00.

for example, and very few sets of the original publication have survived. The second photograph, taken a short while later by the same firm, was generally the one he preferred and this, once taken, became something to send out to friends (fig. 7.10). The general effect of the second photograph is strikingly polychromatic. Darwin wears a necktie, waistcoat, and trousers in the noisy checks which

FIGURE 7.8. Charles Lyell. Lithograph by Thomas Maguire for the British Association series, 1851. In these Maguire portraits, Henslow, Darwin, and Lyell reveal something of their scientific calling by the glasses, either in hand or slung on a ribbon around their necks. Courtesy Wellcome Institute Library, London. V0003723B00.

were greatly favored toward 1860, usually dubbed "Great Exhibition" checks. Once again he did not bother with scientific props. Other men, perhaps less certain of their status, or with more avowedly polemic things to express, were less retiring. Richard Owen, the comparative anatomist, posed with a bone, Michael Faraday with his scientific apparatus and bench, Alexander von Humboldt in his

FIGURE 7.9. Charles Darwin. Photograph ca. 1855 by Maull and Polyblank. This was photographed for a series, and distributed by subscription in parts, with a brief accompanying letterpress. Darwin felt he looked "atrociously wicked." There are no complete sets left in existence. From Maull and Polyblank, *Literary and Scientific Portrait Club*. Private collection.

FIGURE 7.10. Charles Darwin. Photograph ca. 1857 by Maull and Polyblank. Darwin much preferred this photograph and purchased it in several sizes and formats over a number of years. Maull also made it into a commercial carte de visite after the *Origin of Species* was published. Courtesy Wellcome Institute Library, London.

FIGURE 7.11. Richard Owen. Photograph ca. 1856 by Maull and Polyblank for the same series as fig 7.9. Conscious of his position in natural history circles, Owen invariably wore his Royal College of Surgeons gown in portraits. In this photograph he is also accompanied by an obvious tool of his trade. Courtesy Wellcome Institute Library, London. V0026949B00.

FIGURE 7.12. James Scott Bowerbank. Photograph ca. 1855 by Maull and Polyblank for the same series as figs. 7.9 and 7.11. Like Owen, but unlike Darwin, Bowerbank has gone to considerable trouble to convey his scientific interest in sponges and microscopy. Courtesy Wellcome Institute Library, London. V0027568B00.

study with traveling boxes and a large map of the world, Alfred Russel Wallace with a globe, and Ernst Haeckel amid a plentiful supply of natural history apparatus (figs. 7.11–7.14).

After the *Origin,* and with the major technical advances of the 1860s, commercial reproductions of photographs of Darwin proliferated. "Such heaps of

FIGURE 7.13. Ernst Haeckel in Italy, 1860. This depiction of Haeckel as a young natural-history collector is very different in tone from Darwin's presentation as a naturalist. From Ernst Haeckel, *Italienfahrt. Briefe an die Braut 1859–1860* (Leipzig, 1921). Courtesy Wellcome Institute Library, London. L0025212B00.

FIGURE 7.14. Alexander von Humboldt in his study. This is one of a pair of colored lithographs, published 1852. They capitalized on a widespread interest in seeing great men in their working environs. Humboldt's study displays the icons of a great traveler and philosophical writer. Author's collection.

people want to know what you are like," said Hooker. Yet he thought in general "the photographs are not pleasing."[57]

In 1866, for example, Darwin was asked if he would sit for Ernest Edwards for a series of photographs and biographical memoirs edited by Lovell Reeve (a series soon taken over by Edward Walford, the genealogist from Balliol), called *Portraits of Men of Eminence* (1863–67), and he said he would be proud to do so. This book of portrait photographs was one of a number of lavishly produced albums issued around this time, mostly reissues and compilations of texts and previous studio studies by Walford and others.[58] He sat again for another photograph (or possibly it was taken at the previous sitting) to be included

57. Hooker to Darwin, 24 January 1864, DA 101:176.
58. A copy of this Edwards photograph is in the National Portrait Gallery, London, neg. 28523. Prescott describes the publication of several such series, e.g., Mason and Co., *The Bench and the Bar*, and *The Church of England Portrait Gallery*, from 1858 to 1861; and J. E. Mayall, *Royal Album*, 1860. Darwin's picture is in Reeve and Walford, *Portraits of Men of Eminence*, vol. 5, opposite p. 49. A fee of £1 for Ernest Edwards, the photographer for this volume, is recorded in Darwin's Account Book, 2 March 1866, and another sum of £3 8s. 6d. on 5 September.

FIGURE 7.15. Darwin, photographed by Ernest Edwards for Edward Walford, *Representative Men in Literature, Science and Art.* Sitting for Edwards, later one of America's most famous photographers, Darwin shows the ravages of a long illness. The handle of his walking stick can be glimpsed bottom left. He grew the beard in 1862. Courtesy Wellcome Institute Library, London. L0024920B00.

in a selection reprinted by Walford in 1868, called *Representative Men in Literature, Science and Art,* each print available separately priced at 1 s. 6d. (fig. 7.15). This one is interestingly different from the pre-*Origin* photographs. Darwin looked less confident, less well dressed, more anxious, more like an invalid, especially when the handle of the walking stick is glimpsed on the left. Eventually Darwin became quietly aware of the fact that photographers hoped to make money out of his face. In 1869, when George Charles Wallich proposed that he should be included in a new edition of his *Eminent Men of the Day,* he politely refused.[59]

Julia Margaret Cameron nevertheless managed to make capital out of him. Her famous photograph of Darwin (fig. 7.16), and its less famous mate (reproduced as the frontispiece to Richard Freeman's *Companion*), were taken by Cameron in July 1868, when Darwin was on holiday with his family and brother Erasmus on the Isle of Wight.[60] The Darwins occupied a villa rented from the Camerons in Freshwater Bay, the fashionable artistic center that Julia Cameron and Alfred Tennyson had between them created. The holiday was hardly a secluded rest. During those six weeks Darwin was visited by Tennyson, Longfellow, and Thomas Appleton, as well as socializing with the Camerons themselves. Mrs. Cameron often took the opportunity of photographing any celebrated visitors. "She thinks it is a great honour to be done by her," said one guest. "Sitting to her was a serious affair, not to entered lightly upon . . . she expected much from her sitters."[61] Mrs. Cameron also photographed Erasmus Darwin and Horace Darwin, Darwin's youngest son.

One of these photographs was the one Darwin liked better than any other portrait of him and he wrote a sentence to that effect (fig. 7.16). Eventually,

59. Wallich, *Eminent Men,* issued in 1870. Wallich knew Darwin through his natural-history researches; see Darwin to Wallich, 18 April 1869, American Philosophical Society Library. Later, Darwin asked Wallich for photographs that he could use in his *Expression of the Emotions;* Darwin to Wallich, 24 February 1872, Cleveland Health Sciences Library, Ohio; and Darwin to Wallich, 20 March 1872, Northumberland Record Office. These arrangements suggest a certain amount of collaboration between sitters and photographers/publishers. Celebrity photography clearly generated its own rules.

60. F. Darwin, *Life and Letters,* 3:92, 102; and Litchfield, *Emma Darwin,* 2:220–22.

61. Hopkinson, *Julia Margaret Cameron;* and Hinton, *Immortal Faces,* 33. See also Cameron, *Alfred, Lord Tennyson and His Friends;* Weaver, *Julia Margaret Cameron;* and Woolf and Fry, *Victorian Photographs.* For Cameron's photographing of Erasmus Alvey Darwin and Horace Darwin, see Litchfield, *Emma Darwin,* 2:220. Cameron was not the only art photographer to photograph Darwin. In 1871 Darwin approached Oscar Rejlander for help with his studies on the expression of the emotions and established a close working relationship with him. See Browne, "Darwin and Expression of the Emotions," and Jones, *Father of Art Photography.* Rejlander made a fine portrait study of Darwin in or around 1871, ultimately engraved and reproduced in *Nature,* 4 June 1874. Rejlander taught Cameron and C. L. Dodgson something of their technique.

Figure 7.16. Darwin, photographed by Julia Margaret Cameron, 1868, during a holiday at Freshwater, Isle of Wight. Copies of this fine portrait photograph were available for sale through Colnaghi's London gallery, with its stamp (not visible here) at bottom left. Mrs. Cameron's inscription, bottom right, suggests that this copy was given away privately. Darwin's comments at the bottom read, "I like this Photograph very much better than any other which have been taken of me." The words appear to be mechanically reproduced. Reproduced by permission of the President and Council of the Royal Society, London.

either he or Cameron included the remark as a mechanically reproduced inscription at the bottom.[62] But Cameron was notorious for forcing her sitters into some kind of public endorsement along these lines, which she then used to promote sales of authenticated prints through Colnaghi's London gallery. The Cameron photograph was consequently just as much in the public arena as the cartes de visite Darwin employed—although more expensive and more artistic. It should further be noted that Darwin paid £4 7s. for this photograph, and other sums later on for various reproductions.[63] During her time in England, Cameron's photographic sales were virtually her only means of supporting herself and her husband.

They were both excellent portrait studies. Julia Cameron subscribed to the "men of genius" school of thought, and her studies of other Victorian thinkers like Tennyson and Herschel showed a deep appreciation of masculine intellect. Her portraits of women and children made the reverse point rather more vividly, in that these rarely carried any proper names and the sitters were usually dressed or posed for some allegorical purpose—"Alethea," perhaps, or "The gardener's daughter." These sentimental *tableaux vivants* were often criticized and ridiculed in her own day. Furthermore, she tended to use a light color wash and give full rein to her special trick of fuzziness: "very daring in style," said the *Photographic News* reviewing her first exhibition in 1864; "out of focus," complained the *British Journal of Photography*.[64] These images are a marked contrast to her other, far more rugged and individually named projections of notable men.

Yet despite the air of biographical transparency, Cameron's male sitters were just as carefully posed as her female subjects—she made John Herschel wash and fluff up his hair for his sitting, and draped Robert Browning in a velvet cloak.[65] Darwin's costume was evidently his own and mostly conveys his careful precautions against the cold, especially on holiday on the British south coast in July. His dress was that of a respectable middle-aged gentleman: a man whose clothes signaled that, beyond going to a good tailor, he was relatively uninterested in clothes. On the whole, however, they are but a minor part of the composition. Because of her careful use of ceiling light, one could almost say that there is nothing in the photographs except Darwin's massive forehead, top-lit to emphasize the great dome of his skull, the brow creased in thought, and his luxurious

62. The Royal Society copy, on the original Colnaghi mount, with blind stamp, is the only copy I have seen with this mechanically reproduced text at the bottom.

63. F. Darwin, *Life and Letters*, 3:102; and Classified Account Books, 19 August 1868, Down House Archives.

64. *Photographic News* 8 (3 June 1864): 266; *British Journal of Photography* 11 (1864): 261.

65. Hopkinson, *Cameron*, 68.

beard. As in classical paintings, the effect was of a softened, extremely wise, subject. More than anyone else Cameron created the visual image of Darwin as a great abstract mind.

Darwin's beard is one of the most interesting aspects of this kind of public representation. He grew it in the late summer of 1862 with the expressed intention of soothing his eczema: "Mamma [Emma] says I am to wear a beard," he told William in July. Constant shaving irritated his face. "Charles, Emma and Lenny slept here on Monday," Erasmus Darwin wrote to Fanny Wedgwood some months later: "Emma in a splendid wig, Lenny bald & Charles in a fine grey beard."[66]

Its impact, however, was much more subtle and pervasive than a mere family event. When Darwin distributed a photograph of himself with this new asset (taken by his son William), Hooker replied immediately. "Glorified friend! Your photograph tells me where Herbert got his Moses for the fresco in the House of Lords—horns & halo & all. . . . Do pray send me one for Thwaites, who will be enchanted with it. Oliver is calling out too for one."[67] Funnily enough, said Darwin, his sons declared it made him look like Moses too. The botanist Asa Gray agreed. "Your photograph with the venerable beard gives the look of your having suffered, and perhaps, from the beard, of having grown older. I hope there is still much work in you—but take it quietly and gently!"[68]

It was a philosopher's beard, as Cameron, Hooker, and Gray plainly saw, with strongly religious overtones. Darwin was delighted with the idea: "Do I not look reverent?" he teased relatives. For himself, he hardly gave the possible motives for growing it or the symbolism of such a patristic outgrowth a second thought.[69] Yet at some fairly obvious level, it must have served as an external disguise on a par with his notorious personal shyness—a form of evading difficult confrontations by hiding behind a smokescreen of hair, not just a literal disguise, but a metaphysical one as well. The beard helped keep many of his thoughts pri-

66. Darwin to William Darwin, 4 July 1862, DA, 210.6; and E. A. Darwin to F. M. Wedgwood, 1 October 1862, Wedgwood/Mosely Collection, Keele University Library. Although Darwin may well have shaved intermittently, he had a full beard when his friends saw him again in 1864. He earlier sported a large black beard on the *Beagle* while surveying in Tierra del Fuego; see Browne, *Charles Darwin,* 217, 246. With regard to the wig, Emma and Leonard Darwin suffered from scarlet fever that summer and had been shaved during the skin-peeling period.

67. Hooker to Darwin, 11 June 1864, DA 101:225. The attribution to William is in a letter from Darwin to Asa Gray, 28 May 1864, Harvard University, Gray Herbarium, 79.

68. Asa Gray to Darwin, 11 July 1864, DA 165.

69. As they became more common after the Crimean War, beards were often discussed in Victorian literature, e.g., Hannay, "The Beard." Something of the history of beards is given by Asser, *Historic Hairdressing;* Cooper, *Hair;* Corson, *Fashions in Hair,* esp. 398–461; and Reynolds, *Beards.*

vate in the same way as his autobiographical writings avoided any penetrating self-analysis.[70] Such a beard allowed him, if he wished, to become a sage or a prophet. Or a sphinx.

This beard came to be featured more and more in the photographs and their subsequent reproductions in magazines through the 1870s and early 1880s (figs. 7.1–7.3, 7.17). As it got larger, and more patriarchal generally, it began to codify many of the things Victorians were told or thought about Darwin: it represented precisely the paradoxical fact that he was simultaneously a gentleman and a revolutionary, with distinguished philosophical antecedents in Plato and Socrates.[71] Such a beard made it clear that he was no fresh-faced radical, no dangerously groomed Frenchman.[72] It was reassuring in its religious demeanor, its benevolence, its suggestion of a wise father and patient friend. Such a beard hinted at hermits and holy men, even the apostles. Popular depictions of ancient Greek philosophers invariably emphasized the same features of beard and expansive forehead (for example, fig. 7.18). Nor is it too far fetched to allude to Father Christmas, at the start of his mythical existence in Victoria and Albert's England. It conveyed sagacity. It was, moreover, a dramatically masculine beard, a very visual symbol of the real seat of Victorian power, and one of the most obvious outward results of what Darwin went on to describe as sexual selection among humans.[73] It was a gift to the cartoonists when they got to work on his theories of monkey ancestry. Above all, it signified intensely deep qualities of mind—those qualities of the Victorian masculine intellect that a set of whiskers could never hope to represent in a similar context.

By the 1880s, and the last two years of Darwin's life, virtually all that the

70. See Neve, *Charles Darwin's Autobiography,* introduction; and Colp, "'I Was Born a Naturalist'"; and idem, "Notes on Darwin's *Autobiography.*" Berg, *Unconscious Significance of Hair,* summarizes the psychoanalytic view.

71. Constable, "Beards in History," surveys beards and hair from antiquity to the Middle Ages in the West. See also Reynolds, *Beards,* 48–49. Pliny speaks of the respect and fear inspired by the beard of Euphrates, a Syrian philosopher. The Roman satirists more usually ridiculed the relationship between beards and wisdom, e.g., Herod Atticus, "I see the beard and the cloak, but I do not see the philosopher." Williams, *The Hairy Anchorite,* discusses the religious symbolism.

72. Reynolds, *Beards,* 267–68, on the seditious mustache; and Pointon, "Dirty Beau." In 1854, Hannay, "The Beard," 49, stated that the beard was at that time the symbol of "revolution, democracy and dissatisfaction with existing institutions . . . only a few travellers, artists, men of letters and philosophers wear it." Francis II of Naples forbade beards because of their association with Garibaldi.

73. Darwin, *Descent of Man,* 2:317–23, 372, 379–80. See also Mangen and Walvin, *Manliness and Morality;* Roberts, "Paterfamilias"; and Shortland, "Bonneted Mechanic." Masculinity in art is discussed by Kestner, *Mythology and Misogyny;* and idem, *Masculinities in Victorian Painting;* and in science by Richards, "Darwin and the Descent of Woman"; and idem, "Huxley and Women's Place in Science."

FIGURE 7.17. Wood engraving of Darwin from a photograph, *Illustrated London News,* 1871, p. 244. As reproduction techniques improved, pictures of Darwin and other famous individuals regularly appeared in the popular press. Courtesy Wellcome Institute Library, London. V0001462B00.

SOCRATES

FIGURE 7.18. Socrates, "the bearded master." Note the stylistic device of portraying ancient philosophers with a capacious forehead, bald pate, and beard: features that are also reflected in Cameron's study of Darwin, fig. 7.16. The vignette at lower right shows Socrates drinking the hemlock. From Thomas Stanley, *The History of Philosophy*, 3d ed. (London, 1701), 74. Courtesy Wellcome Institute Library, London.

FIGURE 7.19. One of the last photographs of Darwin, taken by Elliott and Fry at Down House, 1881. This portrait shows Darwin at his most enigmatic. Private collection.

FIGURE 7.20. Darwin's study, after an etching by Axel Haig, 1882. The invalid couch and crowded tables clearly convey the two main features of Darwin's life. From A. C. Seward, *Darwin and Modern Science* (Cambridge, 1909). Courtesy Wellcome Institute Library, London. L0025093B00.

public saw in published photographs and photogravures were his beard, his hat, and his eyes (fig. 7.19). Darwin, as a physical presence, had almost disappeared. All that was left was the intense impression of mind. The final photograph of his life, probably taken by Clarence E. Fry, the senior photographic partner of the firm of Elliot and Fry,[74] who must have visited Down House in 1880 or so and taken at least three different portrait shots of Darwin on the veranda, speaks powerfully of wisdom and frailty combined, a last evocative statement in the gradual, progressive sequence of Darwin's disengagement from his malfunctioning body.

The process of focusing in, as it were, on mind, or what Christopher Shilling calls an absent presence,[75] reached its apogee with representations of Darwin's study (fig. 7.20). The place where his knowledge was created seems to have become as interesting to Victorians as the mental attributes and personality of the man himself; and magazine articles about his life and times often featured

74. Hillier, *Victorian Studio Photographs*.
75. Shilling, *Body and Social Theory*. The idea of "absent presence" relates in some degree to Foucault, *The History of Sexuality*, introduction; and Ostrander, "Foucault's Disappearing Body." The theme of death in visual culture is discussed in Llewellyn, *The Art of Death*. See also Elias, *The Loneliness of the Dying*.

pictures of his house, garden, greenhouse, and study, usually devoid of human figures. While making allowance for stylistic conventions in conveying domestic interiors, these empty places or spaces for generating knowledge suggest that Darwin's intellect was, by then, seen by the public as almost entirely disembodied. The room did not have Darwin in it. Instead, it is filled with signs of his mind at work—the plants, the papers, the books, the prints of scientific friends and family on the walls; and with signs of his unhealthy body—the fire, the shawl on the chair, the chaise longue. It would be quite a different kind of picture if Darwin were present, as in the engravings published in the *Illustrated London News* of Lubbock or Hooker bristling with activity at their desks. This picture depends on his absence.[76]

Through a long and arduously medicalized life, Darwin had created an idea of himself in which, at the end, he could be recognized—and venerated—by an empty room.

76. One comparable scene would be Freud's study, of which photographs were issued in 1938, reproduced in Engleman, *Berggasse 19*. Discussions of the creation of intellectual spaces can be found in Smith, *Making Space;* Livingstone, "Spaces of Knowledge"; Shapin, "The Mind Is Its Own Place"; and Ophir and Shapin, "Place of Knowledge." The history of studies as places for generating knowledge is much neglected in the literature, although it is addressed briefly in Ophir and Shapin; Girouard, *Life in the English Country House;* and Thornton, *Authentic Decor.* Thomson, "Some Reminiscences," reproduces interesting photographs. Such private masculine spaces (for smoking, business papers, writing, and reading) were seemingly a Victorian development that went hand in hand with the diversification of room use and the division of labor in larger country houses, and are not as clearly related to the traditional use of academic spaces like libraries, monastic cells, and college rooms as we might perhaps expect. For Darwin's use of his study as a place of experiment, see Chadarevian, "Laboratory Science." See also Marsh, *Writers and Their Houses.*

■ ACKNOWLEDGMENTS ■

I gratefully acknowledge permission to reproduce material from the Syndics of Cambridge University Press, the publishers of the *Correspondence of Charles Darwin,* and from the Syndics of Cambridge University Library, where the vast majority of Darwin's papers and letters are held. The Darwin Correspondence Project has also kindly allowed me to use material that will be published in future volumes of the *Correspondence.* I am extremely grateful to the Wellcome Institute Library, London, for permission to cite manuscripts and to reproduce illustrative material in its collection; and to the Royal Society of London and to John Johnson Ltd.

■ REFERENCES ■

Adler, Kathleen, and Marcia Pointon, eds. *The Body Imaged: The Human Form and Visual Culture since the Renaissance.* Cambridge: Cambridge University Press, 1993.
Adler, Saul. "Darwin's Illness." *Nature* 184 (1959): 1102–3.

Airy, George B. *Autobiography of Sir George Biddell Airy.* Ed. Wilfred Airy. Cambridge: Cambridge University Press, 1896.

Allingham, William. *A Diary.* Ed. Helen Allingham and D. Radford. London: Macmillan, 1907.

Anderson, Bruce L., and Philip L. Cottrell. *Money and Banking in England: The Development of the Banking System, 1694–1914.* Newton Abbot: David and Charles, 1974.

Asser, Joyce. *Historic Hairdressing.* London: Sir Isaac Pitman, 1966.

Bailin, Miriam. *The Sickroom in Victorian Fiction: The Art of Being Ill.* Cambridge: Cambridge University Press, 1994.

Barker, Theodore C., and Christopher I. Savage. *An Economic History of Transport in Britain.* London: Hutchinson, 1974.

Bartholomew, Michael. "The Award of the Copley Medal to Charles Darwin." *Notes and Records of the Royal Society of London* 30 (1976): 209–18.

Benjamin, Marina, ed. *Science and Sensibility: Gender and Scientific Enquiry, 1780–1945.* Oxford: Blackwell, 1991.

Berg, Charles. *The Unconscious Significance of Hair.* London: George Allen and Unwin, 1951.

Bernstein, Ralph B. "Darwin's Illness: Chagas Disease Resurgens." *Journal of the Royal Society of Medicine* 77 (1984): 608–9.

Berrios, Germen E. "Obsessional Disorders during the Nineteenth Century: Terminological and Classificatory Issues." In Bynum, Porter, and Shepherd, 1:166–87.

Bolas, Thomas. *The Photographic Studio.* London: Marion and Co., 1895.

Bowlby, John. *Charles Darwin: A Biography.* London: Hutchinson, 1990.

Browne, Janet. *Charles Darwin: A Biography.* Vol. 1, *Voyaging.* London: Jonathan Cape; New York: Alfred Knopf, 1995.

———. "Darwin and the Expression of the Emotions." In *The Darwinian Heritage,* ed. David Kohn, 307–26. Princeton: Princeton University Press in association with Nova Pacifica, 1985.

———. "Spas and Sensibilities: Darwin at Malvern." In *The Medical History of Waters and Spas,* ed. William F. Bynum and Roy Porter. *Medical History,* supplement 10 (1990): 102–13.

Burkhardt, Frederick H., and Sydney Smith, eds. *Calendar of the Correspondence of Charles Darwin.* Rev. ed. Cambridge: Cambridge University Press, 1994.

———, eds. *The Correspondence of Charles Darwin.* 9 vols. to date. Cambridge: Cambridge University Press, 1983–.

Bynum, William F. "The Nervous Patient in Eighteenth and Nineteenth Century Britain: The Psychiatric Origins of British Neurology." In Bynum, Porter, and Shepherd, 1:88–102.

———. "Rationales for Therapy in British Psychiatry, 1785–1830." *Medical History* 18 (1974): 317–34.

Bynum, William F., and Michael Neve. "Hamlet on the Couch." In Bynum, Porter, and Shepherd, 1:289–304.

Bynum, William F., Roy Porter, and Michael Shepherd, eds. *The Anatomy of Madness: Essays in the History of Psychiatry.* 3 vols. London: Tavistock, 1985.

Cameron, Henry H. H. *Alfred, Lord Tennyson and His Friends.* With 25 portraits by Julia Cameron. London: T. Fisher Unwin, 1893.

Chadarevian, Soraya de. "Laboratory Science versus Country House Experiments: The Controversy between Julius Sachs and Charles Darwin." *British Journal for the History of Science* 29 (1996): 17–41.

Chapman, John. *Neuralgia and Kindred Diseases of the Nervous System.* London: Churchill, 1873.

———. *Sea Sickness: Its Nature and Treatment.* London: Trubner, 1864.

Checkland, Sydney G. *The Gladstones: A Family Biography, 1764–1851.* Cambridge: Cambridge University Press, 1971.

Clarke, Graham, ed. *The Portrait in Photography.* London: Reaktion Books, 1994.

Colp, Ralph. "'I Was Born a Naturalist': Charles Darwin's 1838 Notes about Himself." *Journal of the History of Medicine* 35 (1980): 8–39.

———. "Notes on Charles Darwin's *Autobiography.*" *Journal of the History of Biology* 18 (1985): 357–401.

———. *To Be an Invalid: The Illness of Charles Darwin.* Chicago: University of Chicago Press, 1977.

Constable, Giles. "Beards in History." In *Apologia duae,* ed. Robert B. C. Huygens, 47–56. Corpus Christianorum. Continuatio Medievalis 62. Turnhout, Belgium: Brepols, 1985.

Cooper, Wendy. *Hair: Sex, Society, Symbolism.* London: Aldus Books; New York: Stein and Day, 1971.

Cooter, Roger. "Dichotomy and Denial: Mesmerism, Medicine and Harriet Martineau." In Benjamin, 144–73.

Corson, Richard. *Fashions in Hair: The First Five Thousand Years.* London: Peter Owen, 1965.

Cowling, Margaret. *The Artist as Anthropologist: The Representation of Type and Character in Victorian Art.* Cambridge: Cambridge University Press, 1989.

Cunnington, Cecil W., and Phillis Cunnington. *Handbook of English Costume in the Nineteenth Century.* 3d ed. London: Faber and Faber, 1970.

Cunnington, Phillis, and Catherine Lucas. *Occupational Costume in England.* London: Black, 1967.

Dale, Philip M. *Medical Biographies: The Ailments of Thirty-Three Famous Persons.* Norman: University of Oklahoma Press, 1952.

Darrah, William C. *Cartes de Visite in Nineteenth Century Photography.* Gettysburg, Pa.: W. C. Darrah, 1981.

Darwin, Charles. *The Descent of Man and Selection in Relation to Sex.* 2 vols. 1871. Facsimile, with an introduction by John T. Bonner and R. M. May, Princeton: Princeton University Press, 1981.

———. *The Expression of the Emotions in Man and Animals.* London: John Murray, 1872. Reprinted with an introduction by Konrad Lorenz, Chicago: University of Chicago Press, 1965.

———. *On the Origin of Species by Means of Natural Selection: A Facsimile of the First Edition.* 1859. Reprinted with an introduction by Ernst Mayr, Cambridge: Harvard University Press, 1964.

———. *The Variation of Animals and Plants under Domestication.* 2 vols. London: John Murray, 1868.

Darwin, Francis, ed. *The Life and Letters of Charles Darwin, Including an Autobiographical Chapter.* 3 vols. London: John Murray, 1887.

Darwin, Francis, and Albert C. Seward, eds. *More Letters of Charles Darwin.* 2 vols. London: John Murray, 1903.

Davidoff, Leonore, and Catherine Hall. *Family Fortunes: Men and Women of the English Middle Class 1780–1850.* Chicago: University of Chicago Press, 1987.

Desmond, Adrian, and James Moore. *Darwin.* London: Michael Joseph, 1992.

Digby, Anne. *Making a Medical Living: Doctors and Patients in the English Market for Medicine, 1720–1911.* Cambridge: Cambridge University Press, 1994.

———. "Women's Biological Straitjacket." In *Sexuality and Subordination: Interdisciplinary Studies of Gender in the Nineteenth Century,* ed. Susan Mendus and Jane Rendell, 192–220. London: Routledge, 1989.

Edel, Leon, ed. *The Diary of Alice James.* Harmondsworth: Penguin, 1982.

Edwards, Elizabeth, ed. *Anthropology and Photography, 1860–1920.* New Haven: Yale University Press, 1992.

Ehrenreich, Barbara, and Deidre English. *Complaints and Disorders: The Sexual Politics of Sickness.* London: Writers and Readers Publishing Cooperative, 1973.

———. *For Her Own Good: 150 Years of the Experts' Advice to Women.* London: Pluto, 1979.

Elias, Norbert. *The Loneliness of the Dying.* Oxford: Blackwell, 1985.

Engleman, Edmund. *Berggasse 19: Sigmund Freud's Home and Offices, Vienna 1938.* New York: Basic Books, 1976.

Figlio, Karl. "Chlorosis and Chronic Disease in Nineteenth-Century Britain: The Social Constitution of Somatic Illness in a Capitalist Society." *Social History* 3 (1978): 167–97.

Foucault, Michel. *The History of Sexuality.* Vol. 1, *Introduction.* Harmondsworth: Penguin, 1981.

Fox, Daniel, and Christopher Lawrence. *Photographing Medicine: Images and Power in Britain and America since 1840.* New York: Greenwood Press, 1988.

Freeman, Richard. *Charles Darwin: A Companion.* Folkestone: Dawson, 1978.

Fyfe, Gordon, and John Law, eds. *Picturing Power: Visual Depiction and Social Relations.* Sociological Review Monograhs, no. 35. London: Routledge, 1988.

Gallagher, Catherine, and Thomas Laqueur, eds. *The Making of the Modern Body: Sexuality and Society in the Nineteenth Century.* Berkeley and Los Angeles: University of California Press, 1987.

Gent, Lucy, and Nigel Llewellyn, eds. *Renaissance Bodies: The Human Figure in English Culture c. 1540–1660.* London: Reaktion Books, 1990.

Gilman, Sander. *Disease and Representation: Images of Illness from Madness to AIDS.* Ithaca: Cornell University Press, 1988.

———. *Health and Illness: Images of Difference.* London: Reaktion Books, 1995.

———. *Seeing the Insane.* New York: Brunner, Mazel, 1982.

Girouard, Mark. *Life in the English Country House.* Harmondsworth: Penguin, 1980.

Goffman, Erving. *The Presentation of Self in Everyday Life.* 2d ed. London: Allen Lane, 1969.

Good, Rankine. "The Life of the Shawl." *Lancet* i (1954): 106–7.

Graham, Hilary. *Women, Health and the Family.* Brighton: Harvester Press, 1984.

Greenacre, Phyllis. *The Quest for the Father: A Study of the Darwin-Butler Controversy as a Contribution to the Understanding of the Creative Individual.* New York: International Universities Press, 1963.

Gully, James M. *The Water Cure in Chronic Disease.* London: J. Churchill, 1842.

Haight, Gordon. *George Elliot and John Chapman: With Chapman's Diaries.* New Haven: Yale University Press, 1940.

Haley, Bruce. *The Healthy Body and Victorian Culture.* Cambridge: Harvard University Press, 1978.

Hannay, James. "The Beard." *Westminster Review* 62 (1854): 48–67.

Harvey, John. *Men in Black.* London: Reaktion Books, 1995.

Hillier, Bevis. *Victorian Studio Photographs from the Collections of Studio Bassano and Elliott and Fry.* Boston: D. R. Godine, 1976.

Hinton, Brian. *Immortal Faces: Julia Margaret Cameron on the Isle of Wight.* Newport, Hants: Isle of Wight County Press, 1992.

Holland, Henry. *Medical Notes and Reflections.* 3d ed. London: Longman, 1855.

Hollander, Anne. *Seeing through Clothes.* Berkeley and Los Angeles: University of California Press, 1975.

Hopkinson, Amanda. *Julia Margaret Cameron.* London: Virago, 1986.

Hubble, Douglas. "Charles Darwin and Psychotherapy." *Lancet* i (1943): 129–33.

———. "The Life of the Shawl." *Lancet* ii (1953): 1351–54.

Hull, David L., ed. *Darwin and His Critics: The Reception of Darwin's Theory of Evolution by the Scientific Community.* Cambridge: Harvard University Press, 1973.

Inglis, Brian. *The Diseases of Civilization.* London: Hodder and Stoughton, 1981.

Johnston, W. W. "The Ill-Health of Charles Darwin: Its Nature and Its Relation to His Work." *American Anthropologist,* n.s., 3 (1901): 139–58.

Jones, Edgar Yoxall. *Father of Art Photography: O. G. Rejlander, 1813–1875.* Newton Abbot: David and Charles, 1973.

Jones, Henry Bence. *On Animal Chemistry in its Application to Stomach and Renal Diseases.* London: John Churchill, 1850.

———. *On Gravel, Calculus and Gout.* London: Taylor and Walton, 1842.

Jordanova, Ludmilla. *Sexual Visions: Images of Gender in Science and Medicine between the Eighteenth and Twentieth Centuries.* Madison: University of Wisconsin Press, 1989.

Keith, Arthur. *Darwin Revalued.* London: Watts, 1955.

Kempf, Edward. "Charles Darwin: The Affective Sources of His Inspiration and Anxiety Neurosis." *Psychoanalytic Review* 5 (1918): 151–92.

Kestner, Joseph. *Masculinities in Victorian Painting.* London: Scolar Press, 1995.

———. *Mythology and Misogyny: The Social Discourse of Nineteenth-Century British Classical-Subject Painting.* Madison: University of Wisconsin Press, 1989.

Keynes, Richard D., ed. *Charles Darwin's Beagle Diary.* Cambridge: Cambridge University Press, 1988.

Laqueur, Thomas. *Making Sex: Body and Gender from the Greeks to Freud.* Cambridge: Harvard University Press, 1990.

Leder, Drew. *The Absent Body.* Chicago: University of Chicago Press, 1990.

———, ed. *The Body in Medical Thought and Practice.* Dordrecht: Kluwer Academic Press, 1992.

Lee, David. "Victorian Studio." *British Journal of Photography* 133 (1986): 152–65, 188–99.

Linkman, Audrey E. *The Victorians: Photographic Portraits.* London: Tauris Parke Books, 1990.

Litchfield, Henrietta E., ed. *Emma Darwin, Wife of Charles Darwin: A Century of Family Letters.* 2 vols. Cambridge: privately printed at Cambridge University Press, 1904.

Livingstone, David. "The Spaces of Knowledge: Contributions towards a Historical Geography of Science." *Environment and Planning D: Society and Space* 13 (1995): 5–34.

Llewellyn, Nigel. *The Art of Death: Visual Culture in the English Death Ritual, c. 1500–c. 1800.* London: Reaktion Books, 1994.

Lopez Pinero, Jose M. *Historical Origins of the Concept of Neurosis.* Trans. D. Berrios. Cambridge: Cambridge University Press, 1983.

Lurie, Alison. *The Language of Clothes.* New York: Random House, 1981.

Lynch, Michael, and Steve Woolgar, eds. *Representation in Scientific Practice.* Cambridge: MIT Press, 1990.

MacLeod, Roy. "Of Medals and Men: A Reward System in Victorian Science, 1826–1914." *Notes and Records of the Royal Society of London* 26 (1971): 81–105.

Mangan, James, and James Walvin, eds. *Manliness and Morality: Middle-Class Masculinity in Britain and America 1800–1940.* Manchester: Manchester University Press, 1987.

Markus, Julia. *Dared and Done: The Marriage of Elizabeth Barrett and Robert Browning.* London: Bloomsbury, 1995.

Marsh, Kate, ed. *Writers and Their Houses.* London: Hamish Hamilton, 1993.

Maull, Henry, and Polyblank, [1855?]. *Literary and Scientific Portrait Club.* London: Maull and Polyblank.

Metcalfe, Richard. *The Rise and Progress of Hydropathy in England and Scotland.* London: Simpkin Marshall, Hamilton, Kent and Co., 1906.

Micale, Mark. *Approaching Hysteria: Disease and Its Interpretations.* Princeton: Princeton University Press, 1995.

———. "Hysteria Male/Hysteria Female: Reflections on Comparative Gender Construction in Nineteenth-Century France and Britain." In Benjamin, 200–239.

Morris, David B. *The Culture of Pain.* Berkeley and Los Angeles: University of California Press, 1991.

Neve, Michael, ed. *Charles Darwin's Autobiography: With an Introduction.* Harmondsworth: Penguin, forthcoming.

Ophir, Adi, and Steven Shapin. "The Place of Knowledge: A Methodological Survey." *Science in Context* 4 (1991): 3–21.

Oppenheim, Janet. *"Shattered Nerves": Doctors, Patients, and Depression in Victorian England.* New York: Oxford University Press, 1991.

Ostrander, G. "Foucault's Disappearing Body." In *Body Invaders: Sexuality and the Postmodern Condition,* ed. Arthur Kroker and Marilouise Kroker, 169–82. Basingstoke: Macmillan Education, 1988.

Outram, Dorinda. "Body and Paradox." *Isis* 84 (1993): 347–52.

———. *The Body and the French Revolution: Sex, Class and Political Culture.* New Haven: Yale University Press, 1989.

Peckham, Morse, ed. *The Origin of Species by Charles Darwin: A Variorum Text.* Philadelphia: University of Pennsylvania Press, 1959.

Peterson, M. Jeanne. *Family Love and Work in the Lives of Victorian Gentlewomen.* Bloomington: Indiana University Press, 1989.

———. *The Medical Profession in Mid-Victorian London.* Berkeley and Los Angeles: University of California Press, 1978.

Pickering, George W. *Creative Malady: Illness in the Lives and Minds of Charles Darwin, Florence Nightingale, Mary Baker Eddy, Sigmund Freud, Marcel Proust, Elizabeth Barrett Browning.* London: George Allen and Unwin, 1974.

Piper, David. *Personality and the Portrait.* London: BBC Publications, 1973.

Pointon, Marcia. "The Case of the Dirty Beau: Symmetry, Disorder and the Politics of Masculinity." In Adler and Pointon, 175–89.

Porter, Roy. "Bodily Functions." *Tate: The Art Magazine* 3 (1994): 42–47.

———. "History of the Body." In *New Perspectives on Historical Writing,* ed. Peter Burke, 206–32. Cambridge: Polity Press, 1991.

———, ed. *Patients and Practitioners.* Cambridge: Cambridge University Press, 1985.

Porter, Roy, and George Rousseau. *Gout: The Patrician Malady.* Princeton: Princeton University Press, 1997.

Prescott, Gertrude. "Fame and Photography: Portrait Publications in Great Britain, 1856–1900." Ph.D. thesis, University of Texas, Austin, 1985.

Price, Rees. *A Critical Inquiry into the Nature and Treatment of the Case of H.R.H. the Princess Charlotte.* London: Chapple, 1817.

Pritchard, Michael. "Commercial Photographers in Nineteenth Century Britain." *History of Photography* 11 (1988): 213–15.

———. *A Directory of London Photographers, 1841–1908.* Rev. ed. Watford: PhotoResearch, 1994.

Raverat, Gwen. *Period Piece: A Cambridge Childhood.* London: Faber and Faber, 1952.

Rees, Kelvin. "Water as a Commodity: Hydropathy at Matlock." In *Studies in the History of Alternative Medicine,* ed. Roger Cooter, 28–45. London: Macmillan, 1988.

Reeve, Lovell Augustus, and Edward Walford. *Portraits of Men of Eminence in Literature, Science and Art.* With Biographical Memoirs by Edward Walford. 6 vols. London: Lovell Reeve, 1863–67.

Rennie, Alexander. *Observations on Gout . . . with Practical Remarks on the Injurious Effects of Colchicum.* London: Underwood, 1825.

Reynolds, Reginald. *Beards: An Omnium Gatherum.* London: George Allen and Unwin, 1950.

Richards, Evelleen. "Darwin and the Descent of Women." In *The Wider Domain of Evolutionary Thought,* ed. David Oldroyd and Ian Langham, 57–111. Dordrecht: D. Reidel, 1983.

———. "Huxley and Women's Place in Science: The 'Woman Question' and the Control of Victorian Anthropology." In *History, Humanity and Evolution: Essays for John C. Greene,* ed. James Moore, 253–84. Cambridge: Cambridge University Press, 1989.

Roberts, David. "The Paterfamilias of the Victorian Governing Classes." In Wohl, 59–81.

Roberts, Hyman J. "Reflections on Darwin's Illness." *Journal of Chronic Diseases* 19 (1966): 723–25.

Schiebinger, Londa. *Nature's Body: Gender in the Making of Modern Science.* Boston: Beacon Press, 1993.

Scudamore, Charles. *Observations on the Use of the Colchicum Autumnale in the Treatment of Gout.* London: Longman, 1825.

Shapin, Steven. "'The Mind Is Its Own Place': Science and Solitude in Seventeenth-Century England." *Science in Context* 4 (1991): 191–218.

Shilling, Christopher. *The Body and Social Theory.* London: Sage, 1993.

Shortland, Michael. "Bonneted Mechanic and Narrative Hero: The Self-Modelling of Hugh Miller." In *Hugh Miller and the Controversies of Victorian Science,* ed. Michael Shortland, 14–75. Oxford: Clarendon Press, 1996.

Shuttleworth, Sally. "Female Circulation: Medical Discourse and Popular Advertising in the Mid–Victorian Era." In *Body/Politics: Women and the Discourses of Scientific Knowledge,* ed. Sally Shuttleworth, Mary Jacobus, and Evelyn Fox Keller, 47–68. London: Routledge, 1990.

Sicherman, Barbara. "The Uses of Diagnosis: Doctors, Patients, and Neurasthenia." *Journal of the History of Medicine* 32 (1977): 33–54.

Smith, Crosbie, ed. *Making Space for Science.* London: Macmillan, forthcoming.

Smith, Fabienne. "Charles Darwin's Health Problems: The Allergy Hypothesis." *Journal of the History of Biology* 25 (1992): 285–306.

———. "Charles Darwin's Ill Health." *Journal of the History of Biology* 23 (1990): 443–59.

Stafford, Barbara. *Body Criticism: Imaging the Unseen in Enlightenment Art and Medicine.* Cambridge: MIT Press, 1991.

Tagg, John. *The Burden of Representation: Essays on Photographies and Histories.* Amherst: University of Massachusetts Press; London: Macmillan, 1988.

Taylor, Charles. *Sources of the Self: The Making of Modern Identity.* Cambridge: Harvard University Press, 1989.

Thomson, Joseph John. "Some Reminiscences of Scientific Workers of the Past Generation and Their Surroundings." *Proceedings of the Physical Society* 48 (1936): 217–46.

Thornton, Peter. *Authentic Decor: The Domestic Interior, 1620–1920.* London: Weidenfeld and Nicolson, 1984.

Trombley, Stephen. *"All That Summer She Was Mad": Virginia Woolf and Her Doctors.* London: Junction Books, 1981.

Turner, Bryan S. *The Body and Society: Explorations in Social Theory.* Oxford: Blackwell, 1984.

Turner, Ernest S. *Taking the Cure.* London: Michael Joseph, 1967.

Vicinus, Martha, ed. *Suffer and Be Still: Women in the Victorian Age.* Bloomington: Indiana University Press, 1972.

Walford, Edward. *Photographic Portraits of Living Celebrities; Executed by Maull and Polyblank, With Biographical Notices.* London: Maull and Polyblank, 1856–59.

———. *Representative Men in Literature, Science, and Art. The Photographic Portraits from Life by Ernest Edwards.* London: A. W. Bennett, 1868.

Wallich, George C. *Eminent Men of the Day. Photographs by G. C. Wallich.* London: Van Voorst, 1870.

Weaver, Michael. *British Photography in the Nineteenth Century: The Fine Art Tradition.* Cambridge: Cambridge University Press, 1989.

———. *Julia Margaret Cameron, 1815–1879.* London: Herbert Press, 1984.

Welford, Samuel. "The Cost of Photography in the Period 1850–1897." *PhotoHistorian* 92 (1991): 15–17.

Williams, Charles A. *Oriental Affinities of the Legend of the Hairy Anchorite.* Urbana: University of Illinois Press, 1925.

Wiltshire, John. *Jane Austen and the Body: "The Picture of Health."* Cambridge: Cambridge University Press, 1992.

Winter, Alison. "Harriet Martineau and the Reform of the Invalid in Victorian England." *Historical Journal* 38 (1995): 597–616.

Wohl, Anthony, ed. *The Victorian Family: Structure and Stresses.* London: Croom Helm, 1978.

Wood, Ann Douglas. "The Fashionable Diseases: Women's Complaints and Their Treatment in Nineteenth Century America." In *Clio's Consciousness Raised: New Perspectives on the History of Women,* ed. Mary S. Hartmann and Lois Banner, 1–22. New York: Harper Colophon Books, 1974.

Woodruff, Alan W. "Darwin's Health in Relation to His Voyage to South America." *British Medical Journal* i (1965): 745–50.

Woolf, Virginia, and Roger Fry. *Victorian Photographs of Famous Men and Fair Women.* London: Harcourt Brace, 1926. Reprint, London: Hogarth Press, 1973.

Wynter, Andrew. "Cartes de Visite." *Journal of the Photographic Society of London* 7 (1862): 375–77; also in *Once a Week* 6 (1862): 1134–37.

Zanker, Paul. *The Mask of Socrates: The Image of the Intellectual in Antiquity.* Berkeley and Los Angeles: University of California Press, 1995.

EIGHT

EXERCISING THE STUDENT BODY
Mathematics and Athleticism in Victorian Cambridge

A N D R E W   W A R W I C K

A German student is not a pleasant companion, he is unclean, eats "Blut-und-Leber Wurst" and is usually not "begeistert." On the other hand the English Undergrad is, as a rule clean, lives healthily and learns in a healthy fashion. The aesthetic ideal, mind and body combined, is on our side.
—Karl Pearson, 1880.[1]

■ Introduction ■

SHORTLY AFTER BEING placed third wrangler in the Cambridge Mathematical Tripos of 1879, the twenty-two-year-old Karl Pearson (future biometrician and statistician) set off to the recently united Germany to further his education.[2] Convinced of the superiority of German professors, German universities, and the sheer scale of German learning, he studied physics and metaphysics in Heidelberg and law and Darwinism in Berlin. But as he wrote to his friend and mentor Oscar Browning back in Cambridge, his intellectual quest seemed to have ended in disappointment. Pearson found German intellectual life uncultured and overconcerned with "facts." In Berlin he found the art, the architecture, and the people "barbaric"; the streets of Berlin flowed not with "Geist," he lamented, but with "sewage."[3] Pearson's comments should not, of course, be taken at face value. After eight months of itinerant study in a new land and a new language, he was tired, homesick, and increasingly anxious about his professional future. His youthful hopes of finding intellectual enlightenment beyond what he had seen as the competitive mathematical myopia and outdated religiosity of Anglican Cambridge had faded.[4] The collapse of his belief in the intellectual superiority of the new German Reich had reduced him to racist insults.

1. Karl Pearson to Oscar Browning, 16 January 1880, Oscar Browning Collection, King's College Cambridge (hereafter cited as OB-KCC).
2. Pearson, *Karl Pearson*. The term "wrangler" is explained below.
3. Pearson to Browning, 16 January 1880, OB-KCC.
4. Pearson to his mother, 22 July 1880, Karl Pearson Collection, University College, London (hereafter cited as KP-UCL).

In another sense, however, Pearson had been enlightened. He confided to Browning that his studies in German universities had taught him the merits of the very Cambridge system he had so recently abandoned. Pearson now reckoned Oxbridge professors every bit as cultured and learned as their German counterparts. He had, moreover, acquired a new respect for Cambridge mathematical discipline. Pearson's remarkable technical proficiency in mathematics—drilled into him by his Cambridge coach, Edward Routh—had given him a satisfying edge over his German peers when it came to tackling problems in physics.[5] In Berlin, he often found himself idly "scribbling" such problems when he ought to have been studying Roman law, longing once more to be "working with symbols and not words."[6] He even conceded to Browning that mathematical physics seemed somehow "true" in a sense that eluded other disciplines, as if in physics one were "struggling with nature herself."[7] Pearson's German adventure had given him a new perspective on his undergraduate experience in Cambridge. As a student he had accepted, albeit reluctantly, the tough regime of private coaching and competitive examination in mathematics demanded of the most ambitious undergraduates.[8] He knew from the start that, after ten terms of residence, he would be assessed solely by the number of marks he could accrue in nine days of grueling examination in the University Senate House. What had been less visible to him as a student was the Cambridge ideal of combining bodily and intellectual endeavor. As an undergraduate, Pearson had carefully balanced hard mathematical study against such physical activities as walking, skating, and hockey; only in Germany had he recognized this practice as peculiar to Cambridge life.[9]

Pearson's insistence on the importance of an "aesthetic ideal" of "mind and body combined" in a university that prized mathematical studies above all others might seem peculiar to the modern reader. Mathematics and mathematical physics are, after all, generally regarded as archetypical of disembodied disciplines whose efficacy ought not depend on the physical fitness or state of health of the practitioner. Furthermore, the culture of athleticism that became so characteristic of British educational establishments, including Cambridge, during the mid-Victorian era, has generally been *contrasted* with scholarly culture, as if these two cultures straightforwardly reflected a long-standing division between serious

5. Pearson, *Karl Pearson,* strongly supports the Mathematical Tripos.

6. Pearson to Browning, 16 January 1880, OB-KCC.

7. Ibid.

8. Pearson's undergraduate correspondence reveals his discomfort under the competitive training system; see, for example, Pearson to his mother, 15 December 1878 (KP-UCL).

9. On Pearson's undergraduate sporting activities see, for example, Pearson to his brother Arthur, 22 December 1878 (KP-UCL).

scholars committed to academic excellence and the sons of the aristocracy who sought a good time and a pass degree.[10] Yet, as the epigraph reveals, Pearson had come to recognize the *combination* of fully developed body and fully developed mind as constitutive of the robust Cambridge intellect. In this chapter I explore the relationship between training in the classroom and on the playing field in order to highlight the new economy of the student mind and body that became definitive of mathematical study in early Victorian Cambridge. In the following sections I begin with an overview of the emergence of progressive study and written examination in mathematics as the primary means of assessing the undergraduate elite. I then show that these changes in undergraduate teaching, described by contemporaries as an effective *industrialization* of the learning process, were not accomplished without cost. Using contemporary student diaries and correspondence, we shall see that undergraduates privately railed, though never publicly rebelled, against a system of competitive study that routinely pushed them to the point of emotional and intellectual breakdown. It was, I argue, in the context of this highly demanding system, especially after 1815, that undergraduates first began to employ regular physical exercise both to regulate the working day and in the belief that it preserved a robust constitution. We shall also see that, during the 1830s, physical and intellectual endeavor became even more closely linked as new team sports added a complementary, competitive dimension to physical exercise. By the 1840s, the combined pursuits of mathematics and sport had become constitutive of the "liberal education" through which good undergraduate character was formed. Emphasis on both bodily and mental development in a university that prized manly individuals of rational intellect also enabled the respective ideals of physical and intellectual endeavor to pervade one another: success in competitive sport became a hallmark of the rational body while the hard study of mathematics became a manly pursuit. The research style of Cambridge mathematicians was also shaped, in part, by these same ideals of competition, fair play, and manliness. I conclude by considering the extent to which these ideals remained prevalent in late nineteenth- and twentieth-century mathematical science.

■ *Mathematics, Competitive Study, and a Liberal Education* ■

THE DECADES AROUND the turn of the nineteenth century witnessed major changes in undergraduate studies at both Oxford and Cambridge. During the eighteenth century, the traditional role of these wealthy institutions as Anglican

---

10. On sport in the Victorian education system see Haley, *The Healthy Body;* Mangan, *Athleticism in the Victorian and Edwardian Public School;* and esp. Curthoys and Jones, "Oxford Athleticism."

seminaries and educators of the landed gentry was increasingly called into question.[11] As democratic rule and industrialization undermined the power of both the aristocracy and the Anglican church, the ancient universities sought a new and more visibly meritocratic role in British life.[12] One important way in which Cambridge responded to this challenge was by developing a new system of assessing and ranking academic ability. During the second half of the eighteenth century the traditional method of assessing academic performance, the Latin disputation, was gradually replaced by written examinations in the vernacular. This change produced a new intellectual meritocracy in which academic success was measured not by a public display of eloquent and quick-witted wrangling, but by the accurate reproduction and skilled application of technical knowledge on paper. From the mid–eighteenth century, the results of the annual "Senate House" (or "Tripos") examination were also published, the names of the students being strictly ranked according to the examiner's assessment of their relative abilities.[13] The successful honors students were divided into three classes, "wranglers" (first class), "senior optime" (second class), and "junior optime" (third class). The remaining students, the majority, were deemed unworthy of honors and known as "hoi polloi" (the "many") or "poll men." [14]

The ranking of students by written examination was accompanied by other closely related changes in undergraduate teaching.[15] First, the Tripos examination became increasingly competitive. Toward the end of the eighteenth century, college fellowships began to be awarded in recognition of high placing in the Tripos. Bright and ambitious students from the middle classes saw the examination as a reliable route to a college fellowship or other secure professional future and competed fiercely for a place among the top wranglers.[16] Second, the system of ranking by marks gave increasing prominence to mathematics in undergraduate studies. Mathematics had always been part of a Cambridge education, but the discipline was especially well suited to a system that sought to discriminate

11. On the changing educational context at Cambridge see Becher, "William Whewell"; Gascoigne, "Mathematics and Meritocracy"; Rothblatt, "The Student Sub-culture"; idem, "Failure in Early Nineteenth-Century Oxford and Cambridge"; Williams, "Passing on the Torch."

12. A wide range of factors of possible relevance to the introduction of written examinations in Oxford and Cambridge is discussed in Rothblatt, "The Student Sub-culture."

13. On the term "Tripos" see Ball, *A History of the Study of Mathematics,* 217–19.

14. On the origin of the terms "wrangler," senior and junior "optime," and "hoi polloi" see Wordsworth, *Scholae Academicae.*

15. For a good general discussion of the changes in Cambridge undergraduate social life that accompanied the introduction of written examinations see Rothblatt, "The Student Sub-culture."

16. Gascoigne, "Mathematics and Meritocracy." By the early Victorian period the middle classes were enormously overrepresented among wranglers; see Becher, "Social Origins and Post-graduate Careers of Cambridge Intellectual Elite."

among the performances of large numbers of well-prepared students. Skilled examiners could manufacture mathematical problems in ascending levels of difficulty until they had produced a single continuous ranking of all candidates. Unlike answers to problems in theology or moral philosophy, those in Euclidian geometry and Newtonian mechanics could be assessed according to an agreed marking scheme, thereby circumventing accusations of subjectivity or partiality on the part of the examiners. Third, the emergence of a competitive written examination in mathematics gradually changed the way undergraduates were taught. In the late eighteenth century, most students in the university were taught either in large elementary classes by a college lecturer or in small groups by a private tutor (usually a young college fellow seeking to supplement his meager stipend). As the standard of mathematics required to guarantee honors success got higher, however, college lectures became irrelevant to ambitious students and, during the second and third decades of the nineteenth century, responsibility for the preparation of the elite wranglers passed almost entirely to a small group of professional private tutors.

Finally, the arrival of French analytical mathematics at Cambridge during the second decade of the nineteenth century, and the rapid increase in student numbers after 1815, accelerated the trends outlined above.[17] It was partly through the informal system of private teaching that progressive young college fellows such as George Peacock and Richard Gwatkin had introduced Continental mathematics to the most able young undergraduates. Mastering analytical mathematics required years of tough progressive study, and the Cambridge system, developed piecemeal over the previous half century, was easily and advantageously adapted to provide such a training: thrice-weekly meetings with an experienced private tutor in groups of between two and ten pupils, notes carefully prepared by the tutor for students to copy out in their own time, weekly example sheets to develop technical proficiency and monitor progress, and a system of competitive written examinations that required ambitious students to work as hard as they possibly could. These resources enabled the rapid uptake of analytical mathematics in Cambridge, placed that mathematics at the heart of the competitive training process, and made Cambridge the center of British mathematics throughout the nineteenth century.

For present purposes, the most important aspect of these developments is the transformation they wrought in the aspirations, attributes, and lifestyle of the ideal undergraduate. The respecification of university education as a reliable

---

17. Hopkins, *Remarks on the Mathematical Teaching,* discusses analytical mathematics and coaching. On the increase in student numbers after 1815 see Venn, *A Biographical History,* vol. 3.

mechanism of self-improvement for bright students from the middle classes proved a powerful incentive to hard study. Those who came to Cambridge in search of meritocratic advancement were obliged to submit themselves to its tough regime of disciplined learning. Indeed, in a system based upon competitive study and open-ended examination papers (candidates solved as many problems as they could in the time allowed), the academic standard was defined, and continually inflated, by the combined efforts of the most able students and their private tutors. By the 1830s, working under a professional tutor was a virtual prerequisite for Tripos success, and undergraduates coined the term "coach"— from mail coach or stagecoach—to capture the way a tutor drove his "team" along a carefully contrived course of ordered topics and graded examples.[18] It is also important to appreciate that, prior to the establishment of the Classical Tripos in 1824, the Senate House Examination (in mathematics) was the *only* means by which a student could graduate. With the establishment of the Classical Tripos, the Senate House examination became known formally as the Mathematical Tripos but, until 1850, students were still *required* to attain honors in the latter before they could take the Classical Tripos. In practice, the Mathematical Tripos remained the elite Cambridge degree to which the most able and ambitious students aspired until the 1890s.

This remarkable emphasis on mathematics in what was regarded in early Victorian Cambridge as a "liberal education" requires comment. The chief architect of the Mathematical Tripos during the 1830s and 1840s, William Whewell, argued forcefully that the prime purpose of a liberal education was to "develop the whole mental system of man" such that his "speculative inferences coincide with his practical convictions." Whewell believed that this coincidence was best nurtured through practical examples that taught students how to "proceed with certainty and facility from fundamental principles to their consequences."[19] One might imagine logic to be the obvious pedagogical means to this end, but, Whewell pointed out, although logic enabled one to draw proper conclusions from given premises, it could not guarantee the truth of the premises themselves. A premise could only become a reliable "fundamental principle" if it contained an intuitive dimension that rendered it self-evident to the student. Whewell believed that mathematics—especially Euclidian geometry and Newtonian mechanics—provided both the soundest fundamental principles and the most reliable means of proceeding through "strict reasoning" to reliable conclusions. In the case of mechanics, for example, it was the mathematical definition

18. The development of the "coaching" system is discussed in Warwick, *Masters of Theory*.
19. Whewell, *Thoughts on the Study of Mathematics*, 5.

of "force" that enabled one to use the calculus to solve difficult mechanical problems. It was, however, the familiar sensation of muscular effort that enabled the student literally to feel the concept of force and thereby to shift its status from a mere algebraic variable to a practical conviction.[20] In passing correctly through the steps of the solution to a difficult problem in mechanics, a student was therefore able to display the ability to make his speculative inferences coincide with his practical convictions.

The place of mathematics at the heart of a Cambridge liberal education was justified, then, not as a vehicle for training professional mathematicians, but as the best discipline through which to teach students, via practical examples, to reason properly and reliably. It was for this reason that Whewell and many of his contemporaries were hostile to the wholesale introduction of Continental analysis into the undergraduate curriculum. They believed that some analytical methods obscured the proper logical steps to be followed in solving a physical problem, and preferred to retain a "mixed mathematics" in which the student never lost sight of the mechanical or geometrical principles behind a problem. In this sense, mixed mathematics was a language through which rational students could be identified and ranked relative to their peers. Furthermore, those who excelled in the Mathematical Tripos revealed themselves as potential leaders who would occupy positions of responsibility as teachers, clergymen, lawyers, statesmen, Imperial administrators, and men of business. This role of mathematics, to prepare the nation's intellectual elite, lent an extremely important moral dimension to the study of the discipline. It was the duty of every student to display his ability through mathematical attainment and to strive, without regard to his own suffering, for high placing in the order of merit. The pressure generated in this ruthlessly competitive environment led ambitious students to seek ways of improving their performance. An obvious strategy was to work extremely long hours, but students soon came to believe that prolonged periods of intense mental activity could produce unpleasant and even dangerous side effects. Promising undergraduates, and their tutors, gradually agreed that hard study was most efficiently and safely accomplished when interspersed with periods of more leisurely activity and recreation. Some students relaxed by socializing or playing musical instruments, but, for reasons not entirely clear, the most ambitious undergraduates gradually transformed the traditional afternoon ramble or promenade into a daily regimen of measured physical exercise.[21] This exercise became the recognized complement of hard study, and students experimented with dif-

20. Ibid., 24–25.
21. On the leisurely activity of regular walking in eighteenth-century Cambridge see Rothblatt, "The Student Sub-culture," 247.

ferent regimes of working, exercising, and sleeping until they found what they believed to be the most productive combination.[22] As the Mathematical Tripos became yet more demanding and competitive through the 1820s and 1830s, this nascent athleticism embraced competitive sport so that, by the mid-Victorian period, the training of mind and body had become equally important factors in shaping the ideal manly character.

■ *Mathematical Rigor and the Wrangler-Making Process* ■

[A] man must be healthy as well as strong—"in condition" altogether to stand the work. For in the eight hours a-day which form the ordinary amount of a reading man's study, he gets through as much work as a German does in twelve; and nothing that [American] students go through can compare with the fatigue of a Cambridge examination. If a man's health is seriously affected he must give up honours at once.

Charles Bristed, 1852.[23]

HAVING DISCUSSED THE emergence of progressive and competitive study in Cambridge in the early nineteenth century, I turn to a more detailed discussion of the way these developments altered the undergraduate learning experience. Most historical studies of Cambridge mathematics in this period have focused either on what is seen as the belated arrival of Continental methods at the university in the 1810s and 1820s or on the subsequent struggles between radicals and conservatives to control the precise content of the undergraduate syllabus. Moreover, the successful assimilation of the new mathematics, especially analysis, by the university is generally portrayed as a wholly positive event that enabled Cambridge to emerge as an important center of mathematical research in the Victorian era. But if we are to understand how the student's body became explicitly implicated in mathematics pedagogy during the same period, it is important to appreciate that the new regimes of teaching and examining through which Continental methods were effectively introduced were seen to have *destructive* as well as *productive* dimensions. From the perspective of those who found Cambridge pedagogy productive, the Mathematical Tripos offered the most able students a thorough training in a range of mathematical techniques and provided them with the motivation to work hard. Some students thrived in this competitive environment, at least for a while, finding pleasure in the rapid mastery of new skills and the timed solution of difficult problems. From the perspective of those who found it destructive, however, students lacking the requisite qualities were

22. For an example of a senior wrangler from around 1815 who relaxed by playing the flute see Atkinson, "Struggles of a Poor Student," 510.
23. Bristed, *Five Years in an English University,* 1:331.

alienated, dispirited, and sometimes damaged by the system. Even those who were successful in the early years often found it hard to live up to the promise they had shown and, in their final year, invariably found themselves being driven at an exhausting pace.[24]

An informative contemporary account of both dimensions of competitive study occurs in the undergraduate correspondence of Francis Galton. In 1840 Galton went to up to Cambridge, where he worked with the first of the outstanding mathematical coaches, William Hopkins. Selected by Hopkins as a potential high wrangler, Galton initially enjoyed the coaching experience. Writing home to his father at the beginning of his second year, Galton praised Hopkins's technique: "Hopkins, to use a Cantab expression, is a regular brick; tells funny stories connected with different problems and is in no way Donnish; he rattles us on at a splendid pace and makes mathematics anything but a dry subject by entering thoroughly into its metaphysics. I never enjoyed anything so much before."[25]

The letter reveals how the successful coach developed an avuncular intimacy with his pupils that would have been quite alien to most college lecturers and university professors in early Victorian Cambridge. In the friendly atmosphere of his coaching room, Hopkins combined the admiration of his "team" with his own infectious enthusiasm for mathematics to promote the competitive ethos and a dedication to hard work. Galton claimed to "love and revere" Hopkins like no other teacher, but submission to his tutelage was not enough to guarantee success.[26] In the middle of his second year, Galton informed his father that two of Hopkins's pupils were leaving because they could not take the pace and that his own health was suffering. A few weeks into his third year, Galton's health began to fail completely and he informed his father "my head very uncertain so that I can scarcely read at all." This letter also reported that the three best mathematicians in the college in the year above him were all graduating as poll men because their health had broken down under the pressure of hard study. Galton concluded that the unremitting emphasis on competition in Cambridge undergraduate studies was in desperate need of reform because the "satisfaction enjoyed by the gainers is very far from counter-balancing the pain it produces among the others."[27] He subsequently suffered a complete breakdown and had to leave Cambridge for a term. At home Galton experienced an "intermittent

24. On the rise of the fear of failure that accompanied the development of meritocratic examination see Rothblatt, "Failure in Early Nineteenth-Century Oxford and Cambridge."

25. Pearson, *Life, Letters and Labours of Galton,* 163.

26. Galton, *Memories of My Life,* 65.

27. Pearson, *Life, Letters and Labours of Galton,* 166.

pulse and a variety of brain symptoms of an alarming kind"; a "mill seemed to be working inside my head," he complained, and he could not "banish obsessive ideas."[28]

The American Charles Bristed, who came to study for the Mathematical Tripos in 1840, was also surprised by the extraordinary toll the work took on the health of many students. As the epigraph to this section reveals, Bristed soon realized that in order to have a reasonable chance of withstanding the pressure of intense coaching, undergraduates needed to possess and retain strong and healthy constitutions. His remarks are the more poignant as he too suffered a mild mental breakdown shortly before the Tripos examination of 1845:

> About ten days before the examination, just as I was . . . expecting not merely to pass, but to pass *high* among the Junior Optimes . . . there came upon me a feeling of utter disgust and weariness, muddleheadedness and want of mental elasticity. I fell to playing billiards and whilst in very desperation gave myself up to what might happen. At the same time, one of our scholars who stood a much better chance than myself, gave up from mere "funk," and resolved to go out in the poll [become a poll man].[29]

These rare personal comments by Bristed and, especially, Galton, begin to reveal the extent to which the training process at Cambridge made very tough demands on the students. The coaches and examiners had clearly been successful in establishing what was widely regarded as an objective scale of intellectual merit. Neither Galton nor any of his similarly disaffected contemporaries challenged the fairness of the Tripos examination, but they learned, by painful experience, that this was a system that celebrated a handful of winners—especially the "upper ten" wranglers—while remaining largely indifferent to the fate of the rest.

The suffering produced by hard study was by no means confined to the mediocre. Archibald Smith, senior wrangler in 1836, informed his sister at the beginning of his final term's coaching with Hopkins that he was "getting heartily sick of mathematics—and the pleasure I anticipate from being again at home is much increased by the thought that I shall by that time have done for ever with the drudgery of mathematics and be able to apply myself to more pleasant and more profitable studies." Three months later, having been through the grueling examination, he further confided to his sister that he was "quite tired of, I might almost say disgusted with, mathematics."[30] The celebrated senior wrangler of 1840, Leslie Ellis, wrote even more damning comments on his undergraduate

28. Galton, *Memories of My Life,* 79.
29. Bristed, *Five Years in an English University,* 1:229.
30. Both quotations from Smith and Wise, *Energy and Empire,* 56.

experience in a private journal kept daily during 1839–40, his final undergraduate year. Although exceptionally well prepared in mathematics before entering Cambridge, Ellis had been warned by J. D. Forbes that his delicate health would make the Mathematical Tripos a risky venture.[31] And despite his head start, Ellis did indeed suffer appallingly as an undergraduate, especially during his final year of hard coaching with Hopkins. He recorded in his journal how early success in college examinations had marked him out as a potential senior wrangler and how his subsequent attempts to live up to these expectations had gradually replaced his "freshness and purity of mind" with "vulgar and trivial ambition."[32] Ellis privately expressed his sense of apprehension and despondency upon returning to Cambridge in February of his final undergraduate year: "And so here I am again, with a little of that sickening feeling, which comes over me from time to time, and which I can but ill describe, and with some degree of [,] harness bitter dislike of Cambridge and of my own repugnance to the wrangler making process."[33] Like Smith before him, Ellis longed for the Tripos to be over. He confided despairingly to his journal that "this must and will pass away—if not before—when I leave this [place] and shake the dust off my shoes for a testament against the system."[34] Ellis's private comments are especially interesting because they reveal that where Galton wondered whether the wrangler-making process was worth the suffering it produced in the losers, Ellis clearly doubted that it was worth the suffering it produced in the winners.

The two most outstanding mathematical physicists produced by Cambridge in the mid–nineteenth century, William Thomson (later Lord Kelvin) and James Clerk Maxwell, were similarly disaffected by their undergraduate experience. More than nine months before he took the Tripos examination of 1845, the extraordinarily able and energetic Thomson informed his father that "three years of Cambridge drilling is quite enough for anybody."[35] During the equivalent year of Maxwell's undergraduate career, 1853, he was taken ill while working "at high pressure" for the Trinity College summer-term examinations.[36] Struck down by what was described as a "brain fever," Maxwell was initially left unable to "sit up without fainting" and remained disabled for more than a month. During his first two years at Cambridge, Maxwell had resisted pressure to concen-

31. Forbes wrote, "I hope you will not go to Cambridge unless you are equal to the fatigue of such a career": Forbes to Ellis, 14 February 1836, J. D. Forbes Papers, University of Saint Andrews Library (hereafter cited as JDF-UAL).

32. Ellis Journal, 1 June 1839, Leslie Ellis Collection, Wren Library, Trinity College Cambridge (hereafter cited as LE-TCC).

33. Ibid., 8 February 1839.

34. Ibid.

35. Smith and Wise, *Energy and Empire*, 56.

36. Campbell and Garnett, *Life of James Clerk Maxwell*, 167, 170.

trate solely on preparations for the Mathematical Tripos and had continued to read and discuss literature and philosophy.[37] Upon his return to Cambridge, the still-weakened Maxwell abandoned all but his mathematical studies, doing only what "Hopkins prescribe[d] to be done, and avoiding anything more."[38]

It was not only the years of hard study that tested the mettle of ambitious students. The examinations themselves were intended partly as tests of endurance, taking place on consecutive mornings and afternoons for four and five days together. Bristed experienced the communal sense of mounting apprehension as examinations approached, and noted that students were especially prone to "nervous attacks from over work just before, or excitement at an examination."[39] The earliest references to the frantic nature of written examinations in Cambridge occur in the mid-1810s, just at the time the analytical revolution was taking place. James Wright, for example, a mathematics student at Trinity College during this period, described the beginning of a college examination in 1816: "The utmost anxiety is depicted upon the countenances of the Reading-men. Some you see entering the hall with a handful of the *very best pens*, although there is an ample supply upon every table, so fearful are they lest a moment should be lost in mending the same."[40]

Wright also reported how the examinations heaped further stress upon students already exhausted by months of hard study. After several days of consecutive examinations from dawn to dusk in 1816, he described how the "martyrs of learning," already "pale and death-like . . . from excessive reading before the Examination commenced," grew "paler and paler" as the examination proceeded.[41] For some students the examinations proved the final straw and they physically collapsed in the examination room. Wright also gives a rare firsthand account of such a happening—known as a "funking fit"—in 1817. Passing by the open door of Trinity College Hall he chanced to see a student collapse to floor "as lifeless as a corpse." The student was "carried off to his rooms by his fellow-candidates," who immediately rushed back to the examination hall leaving their "inanimate competitor" to be revived with ether.[42]

Yet another affliction feared by would-be high wranglers during examinations was insomnia. Henry Fawcett, for example, another of Hopkins's subsequently famous pupils, was tipped to be senior wrangler in the Tripos of 1856, but his chance slipped away due to "over excitement." As his contemporary and

37. Ibid., 190.
38. Ibid., 193.
39. Bristed, *Five Years in an English University,* 1:331.
40. Wright, *Alma Mater,* 1:218.
41. Ibid., 1:249.
42. Ibid., 2:17.

biographer, Leslie Stephen, records, "In the Tripos, for, as I imagine, the first and last time in his life, Fawcett's nerve failed him. He could not sleep, though he got out of bed and ran round the college quadrangle to exhaust himself."[43] Fawcett slipped to sixth place, a shame Stephen still felt it necessary to mitigate thirty years later by analyzing his examination marks day by day to show how lack of sleep had gradually undermined his technical skill.[44] Fawcett was by no means the only wrangler to experience difficulty sleeping during the Tripos examinations. John Hopkinson, senior wrangler 1871, attributed an attack of insomnia in the examinations to the fact that "having had one's brain in vigorous activity it is hard to get the blood out of it." Following a night in which he lay awake until five in the morning, he acquired some chloral, which, the following night, "secure[d] immediate sleep."[45] J. J. Thomson, second wrangler in 1880, barely slept at all during the last five days of the examination and pointed out how the insomniac's misery is exacerbated in Cambridge by the several clocks that audibly chime each quarter of an hour. Thomson's exhaustion was such that he actually fell asleep during one paper of the subsequent Smiths Prize examinations.[46] Lord Rayleigh, senior wrangler in 1865, tells the other side of the story. He partly attributed his success to his ability to remain relaxed throughout the examination and, especially, to the possibly "unique feat" of taking a short nap before the afternoon examination papers.[47] The contrast between Fawcett's failure of nerve and Rayleigh's cool head under pressure is an important one to which we return in the section on athleticism.

The training process developed in Cambridge during the early decades of the nineteenth century clearly worked ambitious undergraduates to the limits of their emotional and intellectual tolerance. To what extent the system was consciously designed or developed this way is difficult to assess, but visitors to the university saw skilled coaching and competitive examinations as an effective industrialization of the learning process that reflected the wider manufacturing ideals of British culture. Bristed, for example, viewed the strong emphasis on the mastery of advanced mathematical technique as characteristic of a country "where the division of mental labour, like that of mechanical labour, is carried out to a degree which must be witnessed and experienced to be conceived."[48]

43. Stephen, *Life of Henry Fawcett*, 32. Exercise was widely used to induce sleep at examination time; see Bristed, *Five Years in an English University*, 1:314.
44. Stephen, *Life of Henry Fawcett*, 32.
45. Hopkinson, *Original Papers by the Late John Hopkinson*, 1:xxiii.
46. Thomson, *Recollections and Reflections*, 63. The Smiths Prize examination was less stressful than the Mathematical Tripos, the candidates having as much time as they needed to complete the problems.
47. Strutt, *Life of John William Strutt*, 34.
48. Bristed, *Five Years in an English University*, 1:88.

He also used explicitly industrial terms to describe the activities of the students: those making special preparation for an examination were "getting up steam," while the perfectly trained high wrangler could solve mathematical problems with the "regularity and velocity of a machine."[49] Another American visitor to both Oxford and Cambridge during the late 1840s, Ralph Waldo Emerson, was also struck by the distinctly industrial flavor of undergraduate teaching. The English "train a scholar as they train an engineer" he wrote, and he emphasized the importance of selecting good raw materials for the manufacturing process; when born with "bottom, endurance, wind" and "good constitution," he claimed, Cambridge students made "those eupeptic studying-mills, the cast iron men, the *dura ilia* [men of guts], whose powers of performance compare with [an American's] as the steam-hammer with the music-box."[50] Cambridge students often choose similarly industrial terms to describe their experiences. When William Thomson arrived in Cambridge from industrial Glasgow, for example, he described how Hopkins had invited the freshmen to an informal party at the beginning of term at which he could assess their potential as "raw materials for manufacture."[51] Likewise Galton, reflecting upon his breakdown in 1842, felt that it was as if he had "tried to make a steam-engine perform more work than it was constructed for, by tampering with its safety valve and thereby straining its mechanism."[52]

Galton's evocative notion of the Cambridge training process tampering dangerously with the human safety valve is in some respects very appropriate. The system of financial inducements in the form of scholarships and fellowships, the pervasive atmosphere of competition regulated by examination, and the disciplined training mechanism provided by the best coaches incited and enabled students to participate willingly in the manufacturing process. Furthermore, even the most able young mathematicians knew that high placing in the order of merit depended crucially upon a man's ability to judge how much sustained hard work he could tolerate without suffering a debilitating mental breakdown. Bristed certainly believed the "excessive devotion" demanded of undergraduate mathematicians to be responsible for the "disgust and satiety" that they frequently felt for their "unattractive studies," a comment he illustrated with the example of a "high wrangler" who, just before his Tripos examination, was so sick of mathematics that he wished he had never opened a mathematical book and "never wanted to see the inside of one again."[53] Submission to the training

49. Ibid., 1:13, 319.
50. Emerson, *English Traits*, 131, 135.
51. Thompson, *Life of William Thomson*, 1:32.
52. Galton, *Memories of My Life*, 79.
53. Bristed, *Five Years in an English University*, 1:307.

process and stoicism under its rigors were, as we have seen, qualities of character wranglers were expected to display. For some, the drawn features of an exhausted scholar even became an aesthetic hallmark of scholarly piety and intellectual strength. As Leslie Ellis walked the length of a packed Senate House to tumultuous applause to receive his degree in 1840, William Walton was awed by the way his "pale and ill" countenance enhanced his "intellectual beauty." Even more strikingly, another onlooker remarked to Walton "pithily" that had he seen Ellis before the examination he would have known him to be unbeatable.[54]

The code of stoicism, discussed further below, also prohibited most students from speaking publicly of their undergraduate suffering. After winning the coveted senior wranglership, Ellis became a coach, examiner and staunch advocate of the very "wrangler making process" he had privately despised. Likewise Galton, who eventually left Cambridge as one of hoi polloi, later cited the order of merit as objective proof of the uneven distribution of natural intellectual ability among men, without mentioning his own miserable experience as an undergraduate.[55] Only rare contemporary comments—generally made under duress and in private journals or correspondence—and the ethnographical observations of strangers such as Bristed and Emerson (and several others quoted below) give personal insight into actual undergraduate experience. That experience produced remarkable technical proficiency in those who survived the wrangler-making process while, necessarily, placing enormous strain on all students and making a strong constitution a virtual prerequisite to successful study. But just as hard reading was prone to push students beyond their natural capacities, so they found ways to increase that capacity and to place an effective physical governor on the manufacturing machine.

■ Exercising the Student Body ■

In short, it is a safe rule to lay down, that, to keep a student in good working order for a length of time, the harder he applies himself to his studies while studying, the more diversion he requires when taking exercise.

Charles Bristed, 1852.[56]

THE MOST POPULAR method employed by hard-reading students to manage their working day was the taking of regular physical exercise, a practice that emerged in the late 1810s. At the beginning of the nineteenth century, sporting activity in Cambridge was largely confined to such aristocratic outdoor pastimes as hunting, shooting, and angling.[57] In only a handful of English public schools

54. Walton, *Mathematical and Other Writings of Robert Leslie Ellis*, xvii.
55. Galton, *Hereditary Genius*, 16–22.
56. Bristed, *Five Years in an English University*, 2:30.
57. Haley, *The Healthy Body*, 123.

were competitive games encouraged, mainly for amusement and to develop physical fitness and a manly animality. These qualities were not highly regarded in the universities, however, where the cultivation of gentlemanly manners remained an important component of the education provided.[58] "What use is the body of an athlete," wrote Sydney Smith in the *Edinburgh Review* in 1810, "when a pistol, a postchaise, or a porter, can be hired for a few shillings? A gentleman does nothing but ride or walk."[59] Smith also considered it most inappropriate that young gentlemen, even at school, should spend time mastering the technical skills of what were seen as subservient, laboring activities. A gentleman had no need to "row a boat with the skill and precision of a waterman" when, as Wordsworth confirms, rowing was viewed solely as a means of transport in early nineteenth-century Cambridge.[60]

But as the analytical revolution intensified progressive study during the late 1810s, regular physical exercise began to be regarded as a necessary and appropriate companion to hard study. The first and most enduring form of physical exercise used for this purpose was daily long-distance walking (fig. 8.1). When, for example, the eighteen-year-old George Airy arrived in Cambridge in October 1819, one of the chief advocates of the analytical revolution, George Peacock, not only offered to oversee his mathematical studies but "warned" Airy to "arrange for taking regular exercise, and prescribed a walk of two hours every day before dinner."[61] Airy thereafter followed Peacock's advice religiously, later ascribing his lifelong good health to regular exercise. Peacock's warning to the ambitious young mathematician was also novel in 1819, as Wordsworth records that the use of regular walking as a means of exercise and relaxation was a nineteenth-century "refinement."[62] Walking remained the main form of daily exercise in the 1820s and 1830s, the roads around Cambridge being thick with students and college fellows between two and four o'clock in the afternoon as they took their daily "constitutionals." Several walks became so well established that they were given names such as the "Granchester grind" and "wranglers' walk."[63] Another

58. Rothblatt notes that in the eighteenth century undergraduate sports and games were not associated with "moral or character formation" or "physical exercise": Rothblatt, "The Student subCulture," 259.

59. Smith, "Remarks of the System of Education," 328–29.

60. Wordsworth, *Social Life at the English Universities,* 175.

61. Airy, *Autobiography of Sir George Biddell Airy,* 22–23.

62. Wordsworth, *Social Life at the English Universities,* 170; see also Atkinson, "Struggles of a Poor Student through Cambridge," who claimed to be unusual around 1815 in using afternoon rambles to regulate his studies.

63. The "Granchester grind" was a walk that included the village of Granchester, about three miles south of Cambridge. For an interesting account of Thomson and Hopkins pacing "wranglers' walk" together see Thompson, *Life of William Thomson,* 1:113. "Senior wrangler's walk" retained the air of a promenade along which final-year students displayed their ambition for the senior wrangler-ship; see Wright, *Alma Mater,* 1:57.

Keeping an exercise.

FIGURE 8.1. Roget's humorous sketch puns on the ambiguity already inherent in the word "exercise" in Cambridge in the 1840s. Roget was second senior optime in 1850. From J. L. Roget, *Cambridge Customs and Costumes* (Cambridge, 1851), pl. 1. Reproduced by permission of the Syndics of Cambridge University Library.

outstanding mathematician who recorded his walking habits is George Stokes. Arriving in Cambridge in 1837, Stokes made sure to read for only eight hours each day and, like Airy, always took his constitutional walk in the afternoon. As his son later wrote, "This habit [Stokes] maintained in youth, and until long past middle life long walks were the custom, both summer and winter, at a pace of nearly four miles per hour. At eighty-three years of age he still went the Gran-chester "Grind," of three or four miles, and other equally long walks as his after-noon exercise."[64]

By the 1840s, regular physical exercise had become such an established as-pect of wrangler life that Bristed believed it to be the "great secret" of scholarly success.[65] For eight or nine months of the year the Cambridge undergraduate was now in a "regular state of training," his exercise being "as much a daily ne-cessity to him as his food."[66] Bristed also emphasized that it was the hardest reading men who took the hardest exercise, a comment supported by contem-porary student satire that ridiculed the "studious" rather than the idle freshman as one who "taketh furious constitutional walks."[67] Emerson too noticed that it was not the lazy or unintelligent students but the "reading men" who were kept "at the top of their condition" by "hard walking, hard riding and measured eat-ing and drinking."[68] The range of physical activities engaged in by reading men had also expanded to encompass a wide range of competitive sports. The serious scholar now took

> "Constitutionals" of eight miles in less than two hours, varied with jump-ing hedges, ditches, and gates; "pulling" on the river, cricket, football, rid-ing twelve miles without drawing bridle; all combinations of muscular ex-ertion and fresh air which shake a man well up and bring big drops from all his pores; that's what he understands by his two hours exercise.[69]

Students were also discouraged from exercising alone. The solitary walker or rider might find his mind drifting back to his Tripos problems, and to ward off this possibility men would exercise in twos and threes or participate in team games. According to Bristed, talking "shop" on long walks was restricted to dis-cussions about the relative academic abilities of one's peers.[70]

It was during the 1830s that rowing began to rival walking as method of tak-ing regular exercise. As we have seen, in the early nineteenth century the sports pursued in a few public schools had not penetrated the universities, but the in-creasing enthusiasm for healthy exercise altered the situation. During the mid-1820s, occasional recreational rowing on the Cam was transformed when several ex–public school boys formed a boat club at Trinity College for the purpose of

64. Stokes, *Memoirs and Scientific Correspondence of Sir George Stokes,* 1:7.
65. Bristed, *Five Years in an English University,* 2:27.
66. Ibid., 2:327–28.
67. Ibid., 2:57; and "Characters of Freshman," 176.
68. Emerson, *English Traits,* 131. Curthoys and Jones, "Oxford Athleticism," emphasize that hard reading and hard exercise were similarly linked in Victorian Oxford.
69. Bristed, *Five Years in an English University,* 1:328.
70. Ibid., 1:331–32.

Boat Races on the Cam, 1837.

FIGURE 8.2. This scene, about a mile down the Cam from Cambridge (seen in the background), depicts intercollegiate rowing between "eights" as a popular spectator sport in the late 1830s. Walter W. Rouse Ball, *A History of the First Trinity Boat Club* (Cambridge, 1908). Reproduced by permission of the Syndics of Cambridge University Library.

competitive rowing.[71] Saint John's responded within a few months by setting up the Lady Margaret boat club and importing an "eight" (with which to race against Trinity) directly from Eton school.[72] Through the 1830s many other colleges set up boat clubs, and intercollegiate and intervarsity races became formalized and commonplace (fig. 8.2). Unlike walking, moreover, rowing combined extreme physical exertion with keen intercollegiate competition. An example of an outstanding Cambridge mathematician who balanced hard reading with hard rowing is William Thomson. Already an accomplished mathematician when he arrived in Cambridge in 1841, Thomson was recognized from the start as a potentially outstanding scholar. His father, a mathematics professor in Glasgow, was soon troubled to learn that his son had bought a boat to row on the River Cam and was contemplating joining the college boat club. William quelled his father's fears by pointing out that his coach, Hopkins, not only approved of rowing but actively recommended it as a means of exercise and diversion from study. By the end of his second term, Thomson claimed that his general health had been greatly improved by rowing, and that he could "read with much greater

71. For a concise critical discussion of the origins of competitive rowing in Oxbridge see Curthoys and Jones, "Oxford Athleticism," 306; Rothblatt, "The Student Sub-culture."

72. Ball, *History of the Study of Mathematics at Cambridge,* 157–58; on the origins of competitive rowing in England see Halladay, *Rowing in England.*

FIGURE 8.3. This sketch, from a plate entitled "The Boats," puns on the word "training" and captures the physical exertion (compare with fig. 8.1) associated with competitive rowing. From J. L. Roget, *Cambridge Customs and Costumes* (Cambridge, 1851), pl. 6. Reproduced by permission of the Syndics of Cambridge University Library.

vigour than [he] could when he had no exercise but walking."[73] In his second year, Thomson went into a period of intense physical training, ending up number-one oarsman in the first Trinity boat (fig. 8.3). By the 1860s, rowing was described as a "mania" in Cambridge that, together with cricket, shared the "honour" of being the "finest physical exercise that a *hard reading* undergraduate can regularly take."[74]

The outstanding mathematical physicist produced by Cambridge during the 1850s, James Clerk Maxwell, also used walking, swimming, and rowing to regulate his studies. Second wrangler in 1854, Maxwell experimented on his own mind and body by trying different regimens of work, exercise, and rest. One daily routine involved working late at night and then taking half an hour's vigor-

73. Quoted in Smith and Wise, *Energy and Empire*, 73.
74. "The Boating Mania," 42 (my italics).

ous physical exercise. A fellow student who shared his lodgings in King's Parade recorded the downfall of this system: "From 2 to 2:30 A.M. he took exercise by running along the upper corridor, *down* the stairs, along the lower corridor, then *up* the stairs, and so on, until the inhabitants of the rooms along his track got up and lay *purdus* behind their sporting-doors to have shots at him with boots, hairbrushes, etc as he passed."[75] Maxwell's father was partially successful in dissuading his son from engaging in such unsociable practices, but Maxwell continued his experiments after moving into rooms in Trinity College. In the summer he exercised in the River Cam: Maxwell would run up to the river and "take a running header from the bank, turning a complete somersault before touching the water."[76] Alternatively, as P. G. Tait records, Maxwell would "go up on the Pollard on the bathing-shed, throw himself *flat on his face* in the water, dive and cross, then ascend the Pollard on the other side, [and] project himself *flat on his back* in the water."[77] Maxwell claimed, humorously, that this method of exercise improved his circulation.

Yet another outstanding wrangler who made use of physical exercise is W. K. Clifford, second wrangler in 1867. Clifford took a "boyish pride in his gymnastic prowess," drawing upon his "great nervous energy" to perform "remarkable feats": he could "pull up on a bar" with one hand and, on one occasion, "hung by his toes from the cross-bar of a weathercock on a church tower."[78] Praise of Clifford's athletic excellence apparently gratified him "even more than official recognition of his intellectual achievements"; he once described himself as in "a very heaven of joy" because an athletic exercise he had invented—the "corkscrew"—was publicly applauded.[79] These antics are more than mere curiosities because, as his biographer pointed out, Clifford undertook athletic exercises as "experiments on himself" intended to find the best way of "training his body to versatility and disregard of circumstances."[80] Clifford shared the psychophysiological belief popular among mid-Victorian intellectuals that the health and strength of mind and body were intimately related and reliant one upon the other.[81] This belief enabled Clifford to rationalize the Cambridge athletic tradition as a method of "making investments" in nervous energy that, in his case, sustained such habits as studying through the night without reduction

75. Campbell and Garnett, *Life of James Clerk Maxwell,* 153.
76. Ibid., 164.
77. Ibid., 165.
78. *Dictionary of National Biography,* s.v. Clifford, William K.
79. Clifford, *Lectures and Essays,* 1:7. The "corkscrew" consisted in "running at a fixed upright pole which you seize with both hands and spin round and round descending in a corkscrew fashion."
80. Ibid., 1:25.
81. Haley, *The Healthy Body,* chap. 2.

in daytime activities. His exploits turned out to have been sadly ill advised as he suffered with "pulmonary disease" that, exacerbated by extreme physical exertion and long hours of work, led to his early death at the age of thirty-four.[82]

These examples could easily be multiplied, as virtually every high wrangler (for whom records exist) participated in some form of regular physical exercise to preserve his strength and stamina. As a final example, however, it is perhaps more instructive to look in greater depth at the sporting activities of the honors graduates of the Mathematical Tripos for a typical year. The Tripos of 1873 was of special academic significance, as a new set of regulations came into force that year expanding the range of topics studied.[83] These changes heightened both local and national interest in the order of merit, prompting one Cambridge newspaper to provide in-depth coverage of the successful candidates. Almost two-thirds of the article was devoted to coverage of the sporting activities of the honors candidates, thereby providing unique insight into the variety and popularity of undergraduate sport. Both the senior and second wranglers, Thomas Harding and Edward Nanson, were keen oarsmen and swimmers (Nanson being a university champion), while the third wrangler, Theodore Gurney, was president of the Athletic Club, an outstanding athlete and a first-class racquets and tennis player.[84] The article also provides a useful indication of the popularity of rowing. Of the undergraduate mathematicians interviewed, there were more than "three dozen" notable oarsmen, all of whom had rowed since first coming into residence in Cambridge. Indeed, of those who had passed the Mathematical Tripos with honors in 1873, more than 40 percent were identified as "notable" practitioners of at least one sport. Bearing in mind that the reporter did not interview all the students, identified only those who were especially notable sportsmen, and had not included those with interests in football, cricket, and canoeing (or walking), it is quite reasonable to infer that a very high percentage of undergraduate mathematicians, certainly a substantial majority, participated in some form of regular physical activity.

This conclusion is supported by the comments of visitors to Cambridge in the late 1870s and 1880s. Samuel Satthianadhan, an Indian mathematics student at Corpus Christi College from 1878 to 1882, observed with astonishment that his fellow students paid "as much attention to their bodily as to their mental development."[85] Satthianadhan reckoned that nothing would more surprise a visi-

82. Clifford, *Lectures and Essays*, 1:25.
83. Wilson, "Experimentalists among the Mathematicians," 335.
84. The details in this paragraph are taken from *The Cambridge Chronicle*, 25 January 1873, 8. Harding subsequently became a mathematics master at Marlborough School, while Nanson and Gurney both became professors of mathematics in Australia.
85. Satthianadhan, *Four Years in an English University*, 38.

tor to Cambridge than the "fine, stalwart, muscular figure of English students" who so loved sport that any of their number who did not take two hours' exercise each afternoon would be looked upon as "an abnormal character and snubbed by other members of the College."[86] Likewise Michael Pupin, an American immigrant from Europe who came to Cambridge in 1883, echoed Bristed's comments, noting that every student "took his daily exercise just as regularly as he took his daily bath and food."[87] Upon commencing his studies with the greatest of the mathematics coaches, Edward Routh, Pupin was initially "bewildered" to find that the same students who studied hard in the mornings "like sombre monks" came out cheerfully in the afternoons to take their "athletic recreation," an apparent ambiguity that Pupin resolved by referring to his peers as "mathematical athletes."[88] On the advice of his King's College tutor, Oscar Browning, Pupin too adopted this "universal custom" and soon won a place in the college boat. Mastering the skills of the mathematician in the morning and those of the oarsman in the afternoon, Pupin described the daily events of his academic life in Cambridge as shaped by "Routh and rowing."[89]

The practice of combining hard study with regular, often competitive physical exercise was a tradition deeply entrenched in undergraduate life in early and mid-Victorian Cambridge. Physical exercise was believed to develop all-around stamina, provide a forced and wholesome break from study, and leave students sufficiently tired to get a good night's sleep. Students reckoned that, with the aid of this daily regimen, they would be able to withstand ten terms of highly competitive learning without succumbing to the mental anguish or "funk" described by Smith, Ellis, Galton, and Bristed. Furthermore, student participation in daily physical activity was, as we have just seen, effectively policed through peer pressure; any student who did not participate was treated as antisocial and abnormal. Students and tutors also became familiar with the symptoms of overwork so that serious illness could be averted or else attributed to incorrect habits of study. Maxwell's difficulties of 1853, for example, were made public by his biographers in 1882, but attributed to his having studied topics not prescribed by his coach—a practice warned against by "grave and hard-reading students [who] shook their heads at [his] discursive talk and reading."[90] The importance of student participation in an increasing range of competitive sporting activities in mid-Victorian Cambridge was not, however, confined to balancing the effects of hard

86. Ibid., 38, 90.
87. Pupin, *From Immigrant to Inventor*, 173.
88. Ibid., 170, 177.
89. Ibid., 173.
90. Campbell and Garnett, *Life of James Clerk Maxwell*, 173.

study. During the 1850s and 1860s, athleticism in Cambridge began to assume a more complex series of meanings that cannot properly be understood in isolation from broader developments taking place in the English education system.

■ *Mathematics, Athleticism, and Manliness* ■

[S]uch manly exercises as cricket, boating, &c . . . are among the most important means of bringing into existence and fostering those grand moral qualities patience, perseverance, pluck and self-denial, without which a nation can never excel in anything.

Samuel Satthianadhan, 1890.[91]

I HAVE ALREADY noted that at the beginning of the nineteenth century the encouragement of competitive games in education was largely confined to a few public schools. As Mangan has shown, however, during the 1850s and 1860s a number of reforming headmasters used such games to transform student life in the public schools and to foster a more liberal notion of Christian manliness.[92] Competitive sport was employed to improve discipline, health, and appetite while keeping students away from illicit activities when not in the classroom. Sport was also used to build bonds of respect and admiration between masters and pupils and, crucially, as a medium through which to reconcile qualities of manliness that, to an earlier generation, would have seemed incompatible: personal strength with compassion for others, self-interest with team loyalty, and a determination to win with respect for rules and established authority. These were the virtues of liberal manliness captured pithily by Sandars in 1857 in the phrase "muscular Christianity."[93] These public school ideals, which made the river and the cricket field as important as the classroom and chapel as sites for moral education, were soon felt in Cambridge. The public schools provided students for the ancient universities, which in turn provided masters for the public schools, a circulation that legitimated and mutually reinforced such values in both institutions. Many Cambridge dons recognized that students coming to the university from around 1860 were "morally superior" to their predecessors.[94] Traditional pastimes of gambling, drinking, horse racing, and womanizing were being replaced by a love of exercising and training bodies to physical endurance. The new students also brought new sports to Cambridge, and the mid-Victorian

91. Satthianadhan, *Four Years in an English University,* 95.
92. Mangan, *Athleticism in the Victorian and Edwardian Public School,* chap. 1. On changing notions of manliness in mid-Victorian England see Hilton, "Manliness, Masculinity and the Mid-Victorian Temperament."
93. Thomas Collett Sandars coined the term "musclar Christianity" in a review of Hughes's *Tom Brown's Schooldays* in the *Saturday Review,* 21 February 1857, 176. See Haley, *The Healthy Body,* 108.
94. Mangan, *Athleticism in the Victorian and Edwardian Public School,* 122.

period witnessed the rapid formation of new sports clubs through which intercollegiate and intervarsity competitions could be organized: first cricket and rowing, then racquets, tennis, athletics, rugby, soccer, cycling, skating, boxing, swimming, and fencing; by the end of the century, twenty-three recognized sports were
contested annually between the ancient universities.[95] The sporting boom in
Oxbridge also played an important role in popularizing athleticism throughout
Victorian society. As Haley has shown, during the 1850s the Oxford-Cambridge
boat race became the first mass spectator sport while its coverage in the national
press marked the beginning of modern sports journalism.[96]

The clean living, discipline, and competitive ideals of muscular Christianity
found obvious resonances in wrangler life. The fashioning of adolescent sexuality away from heterosexual desire toward an ideal of manly love between master
and student provided a means of managing an obvious distraction from hard
work while enhancing student loyalty to coaches and tutors. Bristed's account
of Cambridge undergraduate life reveals that, as early as the 1840s, sexual abstinence was becoming a factor in Tripos preparation. In and around a town containing a disproportionately large number of young men, many officially sworn
to celibacy, several brothels had appeared, most notoriously in the nearby village
of Barnwell.[97] Bristed was appalled to find that many students thought nothing
of visiting prostitutes after hall in the evening, and he recounted with disgust a
conversation in which a don dismissed a group of poor young Cambridge girls
as nothing more than "prostitutes for the next generation."[98] Students similarly
talked about "gross vice" in a "careless and undisguised" way, considering abstinence from "fornication" an oddity practiced only by those of "extraordinary
frigidity of temperament or high religious scruple."[99] But for the would-be
wrangler, abstinence from fornication, indeed all sexual activity, was already being practiced as a means of "training with reference to the physical consequences
alone."[100] It was a striking example to Bristed of how "physical considerations"
were apt to overcome all others in Cambridge, that students would frequently
remained chaste because common medical opinion held that sexual activity affected their working condition. Late-night hours spent in brothels deprived a
man of valuable sleep and exposed him to the risk of disease or arrest by the university proctors. Moreover, masturbation and other irregular sexual activity (as-

95. Ibid., 125.
96. Haley, *The Healthy Body*, 127.
97. Very little has been written about the Cambridge underworld, but see Desmond and Moore,
Darwin, 54; and, esp., Mansfield, "Grads and Snobs," 187–91.
98. Bristed, *Five Years in an English University*, 2:48–49.
99. Ibid., 40.
100. Ibid.

sociated with extramarital sex) were widely believed to induce precisely the kind of nervous debility most feared by ambitious students.[101] Bristed recounts the case of George Hemming, fellow oarsman of William Thomson and senior wrangler in 1844, who "preserved his bodily purity *solely and avowedly* because he wanted to put himself head of the Tripos and keep his boat head of the river."[102] In this climate, team games provided a medium through which to develop the ideal of manly love untainted by eroticism: on the playing field, physical contact was channeled into the maul, emotion into self-sacrifice and comradeship while, in the classroom, the coach or tutor took the place of the heroic schoolmaster.[103]

As parliamentary pressure for Oxbridge reform increased during the 1850s and 1860s, some young college fellows began to employ both the leadership qualities pioneered by reforming headmasters and the teaching methods long used by successful coaches. These men eschewed the aloof donnishness of traditional college fellows to become the manly and avuncular student leaders who promoted both competitive sports and intellectual endeavor. An exemplary life in this respect was that of the wrangler and reformer Leslie Stephen. As a child, Stephen was believed to have such a "precociously active brain" that it affected his physical health and hindered his academic progress. Only when he entered Cambridge and began to combine the study of mathematics with a regimen of hard physical exercise did his condition improve. While acquiring a reputation as a scholar in mathematics and literature, Stephen distinguished himself as a long-distance runner, a walker of unusual endurance, and a fanatical oarsman. Stephen was twentieth wrangler in 1854 and, following his success, became tutor at Trinity Hall. Like many of his contemporaries, he reckoned that hard physical exercise had restored the inner balance between his mind and body while increasing the stamina of both. As a college tutor, Stephen was in the vanguard of introducing team sports and encouraging intercollegiate and intervarsity competition; he joined enthusiastically in student games, eventually earning himself a reputation as first of the muscular Christians in Cambridge.[104]

This combination of intellectual and athletic activity, together with the increasing emphasis on sport as a means of moral education, gradually made competitive sport and academic examination complementary tests of character. From an intellectual perspective, the Mathematical Tripos was reckoned the most

101. Mason, *The Making of Victorian Sexuality*, 205–27.
102. Bristed, *Five Years in an English University*, 2:41 (my italics). Bristed does not name Hemming in this passage, but his identity is easily inferred from other references such as Bristed 2:24; and Thompson, *Life of William Thomson*, 1:36.
103. Mangan, *Athleticism in the Victorian and Edwardian Public School*, 186.
104. *Dictionary of National Biography*, s.v. Stephen, Leslie; Mangan, *Athleticism in the Victorian and Edwardian Public School*, 124.

severe and accurate test of the rational mind ever devised. Successful candidates
had to learn rapidly in a highly competitive environment and accurately reproduce what they had learned under stressful examination conditions. Likewise on
the river or playing field a man had to display a rational control of the body that
transcended adversity, exhaustion, or pain. Just as Bristed described the ideal
wrangler as solving problems hour after hour with the "regularity and velocity of
a machine," so the ideal rowing crew achieved and maintained an "entire uni-
formity and machine-like regularity of performance." [105] But in order to ensure
that the successful candidates were really made of the right stuff, competition in
the examination room, as on the playing field, had to push some competitors to
the breaking point. Contemporary observers noted that it was not the primary
purpose of the Tripos to produce mathematicians, nor even to test for a sound
knowledge of mathematics; it was, rather, a test of the rational mind under duress. As Robert Seeley wrote revealingly in 1868, the Tripos was as much intended
to detect weakness as it was to detect ability.[106] Seeley also made explicit the link
between examination stress and good character. For him, the "supreme quality
of great men is the power of resting. Anxiety, restlessness, fretting are marks of
weakness." [107] These comments explicate the moral dimension to Rayleigh's claim
to have slept between examination papers in 1865 and, by contrast, the shame
felt by Fawcett who, despite becoming a Cambridge professor and successful
member of Parliament, could never forget that his nerve had failed him in the
Tripos of 1856. The ideal product of the Mathematical Tripos displayed a calm-
ness of spirit together with a robust and unshakable rationality of mind and body.
Characterizing Cambridge in the 1860s, Henry Jackson recalled succinctly that it
was "the man who read hard, the man who rowed hard, and let me add the man
who did both, whom I and my contemporaries respected and admired." [108] As
these comments imply, intellectual toughness had become as much a "manly"
virtue as physical strength. Bristed observed in the 1840s that coaching developed
"manly habits of thinking and reading," so that students acquired a "fondness for
hard mental work." [109] Subsequent commentators, such as Isaac Todhunter, similarly defended the substantial rewards offered to those who excelled in mathematics at Cambridge on the grounds that they provided "incentives to manly exertion." [110] While it was therefore not the prime purpose of the Mathematical
Tripos to train mathematicians, the hard reading and competitive examinations

105. Halladay, *Rowing in England*, 46.
106. Seeley, "A Plea for More Universities."
107. Seeley, "Recreation," 207.
108. Quoted in Rothblatt, *Revolution of the Dons*, 228.
109. Bristed, *Five Years in an English University*, 2:13.
110. Todhunter, *Conflict of Studies*, 242.

in mathematics through which intellectual success was measured made proper mastery of the discipline a manly accomplishment.

The educational and moral economies described above remained intact in Cambridge throughout the mid-Victorian period. As the epigraph at the beginning of this section shows, the visitor to Cambridge in the late 1870s and early 1880s soon recognized both the combination of mental and physical endeavor that defined the ideal undergraduate and the national and imperial ends it was supposed to serve. When he returned to Madras, Satthianadhan sought to inspire the same qualities in his students so that they too would be ideally suited to serve their country. Like Bristed forty years earlier, Satthianadahan learned in his four years at the intellectual hub of the British Empire that physical exercise was "the secret of success in mental advancement." He was now convinced that if there was one thing Indian students could learn from their Cambridge peers, it was "to pay as great an attention to their bodily as to their mental development."[111] The qualities of manliness associated with hard mathematical study were also made starkly visible when women sought to study the discipline. As Winter shows in this volume (chapter 6), wranglers such as De Morgan saw a man's physical strength and constitution as a prerequisite to a thorough mastery of higher mathematics. These assumptions were made even more explicit during the last quarter of the century, when women sought to study mathematics at Cambridge. Apart from the fact that women were believed to lack the emotional stability and intellectual power to compete with men, it was considered highly inappropriate for women to participate in the competitive physical games that necessarily accompanied hard study. The new women's colleges nevertheless provided for competitive sport and adopted the coaching methods employed by the men, a strategy that paid off spectacularly in 1890 when Phillipa Fawcett caused a national sensation by being placed "above the senior wrangler."[112]

■ Mathematics and Athleticism after 1875 ■

During the last quarter of the nineteenth century, the Mathematical Tripos began to lose its elite status. New Triposes, especially the Natural Sciences Tripos, rivaled the Mathematical Tripos in popularity, and high placing in the order of merit ceased to be a unique passport to college or professional success. Wranglers also began to view their training more vocationally. In the 1880s and 1890s they sought work as mathematicians and mathematical physicists in the expanding secondary- and tertiary-education systems throughout Britain and

111. Satthianadhan, *Four Years in an English University*, 39.
112. On the success of Phillipa Fawcett see *The Times*, 9 June 1890, 7. On women in physics in late Victorian Cambridge see Gould, "'A Thing Most Inexpedient and Immodest.'"

the empire. These changes gradually undermined the notion of the wrangler as the ideal intellectual who studied mathematics merely as the most appropriate discipline to hone the rational faculties of his mind. As mathematics was studied for its own sake and for vocational purposes, healthy exercise became more explicitly associated with the regulation of the working day and the maintenance of a healthy body. Ambitious students sought fitness, amusement, and mental stability rather than a physical manliness and outstanding athletic achievement.

Striking an appropriate balance between physical and intellectual endeavor had long been a problem for wranglers. From the late 1830s, when rowing became the sport that best promoted extreme physical fitness and the competitive spirit, ambitious scholars were aware that physical exercise, although crucial, could easily absorb too much of their precious time and energy. Keen oarsman George Hemming, for example, raced in a private boat because he preferred to work to his own training schedule and did not want to be associated with the dissolute ways of some boat club members.[113] His fellow oarsman William Thomson, by contrast, did join the boat club but kept clear of the "dissipated men" and avoided competition during his final year.[114] By the 1880s ambitious wranglers were taking advantage of the wide range of undergraduate sporting activities now available to preserve physical fitness with less concern for competitive sporting success. J. J. Thomson contrasted the closing years of the century with his own experience in the late 1870s, when wranglers often found it difficult to find an appropriate form of exercise. He recalled that those who did not row hard enough to win a place in a college boat were probably not getting enough exercise, while those who did win a place were required to train so hard that they would not be able to manage a second stint of reading in the early evening.[115] It was for this reason that some scholars, himself included, walked rather than rowed. But the rise in popularity of new games such as tennis, golf, and squash, together with the proliferation of mediocre teams for football, hockey, and cricket, meant that scholars could find opponents nearly as bad as themselves, thereby obtaining competitive exercise without having to train too hard.[116]

This sporting ethos remained popular among mathematics undergraduates in Cambridge through the 1890s and well into the Edwardian period. G. H. Hardy, for example, described by his Trinity colleague, C. P. Snow, as "the purest of the pure" mathematicians, acquired a very Cambridge attitude toward mathematics at Winchester. By the 1890s, many public schools had prominent math-

113. Thompson, *Life of William Thomson*, 1:36; Smith and Wise, *Energy and Empire*, 71–78.
114. Thompson, *Life of William Thomson*, 1:61.
115. Thomson, *Recollections and Reflections*, 67.
116. Ibid.

ematics departments run by Cambridge graduates and, even as a schoolboy, Hardy learned to think of mathematics "in terms of examinations and scholarships" and studied hard because he "wanted to beat other boys, and this seemed to be the way in which [he] could do it most decisively."[117] On his way to becoming fourth wrangler in 1898, Hardy played real tennis and cricket, games that, according to Snow, he continued to play competitively into his fifties.[118] Hardy's famous collaborator, J. E. Littlewood, likewise acquired the Cambridge style at Saint Paul's, a school that produced a crop of high wranglers around 1904. Mentally exhausted by his first term's hard coaching in Cambridge, Littlewood became a keen oarsman in his second term; he later became an accomplished rock climber, practicing difficult maneuvers around the masonry of Trinity College.[119] Other famous wranglers from this period, Bertrand Russell and J. M. Keynes for example, were also sports enthusiasts: Russell played tennis, walked, and swam as well as coxswaining the Trinity College first boat; Keynes rowed and played tennis, racquets, and football.[120]

During the same period, Cambridge mathematics itself became increasingly fragmented. Coaches such as Edward Routh and analysts such as Joseph Larmor (who was coached by Routh) strongly defended the tradition of "mixed mathematics," in which mechanics provided the foundation for a comprehensive account of the material universe. These men also maintained Whewell's notion that fundamental principles should be intuitive, Larmor arguing as late as 1900 that "muscular effort" remained the "chief concept" of mechanics.[121] Among the younger generation of mathematicians, however, men like Russell, Hardy, and Littlewood sought to bring the new Continental tradition of "pure mathematics" to Cambridge. They envisaged a self-contained discipline that did not look to mechanics for foundational security. This was a more aesthetic, even narcissistic, vision of mathematics that distanced itself from the physical and athletic sensibilities of muscular exertion. Pure mathematics was characterized by an obsession with proof, rigor, beauty, and elegance and sought its foundations in the disembodied worlds of logic or intuition. Far from being coextensive with physics, pure mathematics could be "applied" only after it had been made foundationally secure by the purists, a shift exemplified in Edmund Whittaker's famous textbook on modern analysis of 1902, in which analytical methods were founded on number theory and only briefly "applied" to the "equations of mathematical

117. Hardy, *A Mathematician's Apology*, 9, 144.
118. Ibid., 18.
119. Littlewood, *Littlewood's Miscellany*, 80–84; Wiener, *I Am a Mathematician*, 152.
120. Moggridge, *Maynard Keynes*, chap. 3; Moorehead, *Bertrand Russell*, 37.
121. Larmor, *Aether and Matter*, 271.

physics."[122] In the words of H. R. Hasse, seventh wrangler 1905, the reformers replaced the "problem age," in which students were drilled by coaches to solve problems by the proper application of advanced mathematical methods, with the "examples age," in which students attended lectures and were required to display an understanding of the foundations and limitations of the methods themselves.[123] But despite this shift in intellectual orientation, the reformers retained other deeply embodied sensibilities of wrangler culture. The belief that a healthy mind was the complement of a healthy body remained commonplace, and students were still expected to work hard, display aggressive individualism, submit themselves to regular competitive examination, and practice sexual restraint. J. J. Thomson's comments, that the new and more genteel sports of the 1880s were very appealing to reading men, also suggest a gradual separation between college hearties who pursued sport for its own sake—the "athletocracy" as Mangan has dubbed them—and those who saw themselves as the intellectual elite. In the late nineteenth century the mid-Victorian ideal of manliness, with its emphasis on earnestness, selflessness, and integrity, gave way to a new cult of neo-Spartan virility with explicit overtones of patriotism, militarism, and anti-intellectualism.[124] This version of manliness, also cultivated in the public schools, undermined the holistic ideal by celebrating the playing field as the primary site of character formation while marginalizing the "highbrow" scholar who showed no special talent for athletics.[125] But although competitive contact sport at the highest level began to acquire unwelcome associations for some aspiring wranglers, regular participation in tennis, cricket, golf, and walking were still regarded by the majority as proper and necessary diversions from hard study.

The robust and competitive qualities nurtured in wranglers also played a role in shaping the research style of those who did go on to become mathematicians. William Thomson, for example, always carried a mathematical notebook about with him and would begin calculating as soon as he had an idea. When the great German physicist, Hermann von Helmholtz, was being entertained on Thomson's private yacht in 1871, he recorded how Thomson astonished his distinguished guests by calculating in their company or simply disappearing below deck to get on with his work. While at sea, Thomson and Helmholtz studied the theory of waves together, the latter being surprised that his host treated the work

122. Whittaker, *A Course of Modern Analysis,* chap. 13.
123. Hasse, "My Fifty Years of Mathematics," 156.
124. Mangan, *Athleticism in the Victorian and Edwardian Public School,* chap. 5 and p. 127; Mangan and Walvin, *Manliness and Morality,* 1.
125. Mangan, *Athleticism in the Victorian and Edwardian Public School,* 107. Curthoys and Jones, "Oxford Athleticism," detect a similar weakening of the relationship between high academic achievement and sports in Oxford around 1900.

as a "kind of race" to such an extent that on one occasion Helmholtz was forbidden to work on the problem while Thomson went ashore.[126] The deeply embedded sensibilities of competitiveness and fair play also appear to have survived the transition from the problem age to the examples age and to have persisted well into the twentieth century. The American mathematician Norbert Wiener, who first visited Cambridge just before the Great War, found a style of mathematics very different the one he had learned in Cambridge, Massachusetts. Surprised by Hardy's skill on the tennis court and Littlewood's prowess as a rock climber, Wiener noted that the young Cambridge mathematician "carried into his valuation of mathematical work a great deal of the adolescent 'play-the-game' attitude which he had learned on the cricket field."[127] Wiener was even more explicit about this Cambridge style in discussing his subsequent collaboration with Raymond Paley, a student of Littlewood's, who worked with him at MIT during the early 1930s. Wiener observed that Paley approached a piece of research mathematics as if it were a "beautiful and difficult chess problem, completely contained within itself." He also offered a telling field-sport analogy to capture the difference between his and Paley's style of mathematics; Paley did not regard Wiener's pragmatic approach to problem solving as "fully sportsmanlike" and was "shocked" by Wiener's "willingness to shoot a mathematical fox if [he] could not follow it with the hunt."[128] Paley's notion that a properly formulated problem had a proper elegant solution—as if set by a cunning examiner—and that resorting to short-cut solutions was unsportsmanlike was fully consistent with both the Cambridge approach to applied mathematics and his broader athletic view of life. He was a keen skier for whom "any concession to danger and to self-preservation was a confession of weakness which he dared not make in view of his desire for the integrity of a sportsman." Paley's collaboration with Wiener was cut short, tragically but appropriately, when the former was killed in an avalanche while deliberately skiing across "forbidden slopes" in the Canadian Rockies.[129]

The Victorian tradition of taking long walks for relaxation was also preserved in the interwar period in an interesting and modified form. Paul Dirac, Cambridge's finest theoretical physicist of this period, famously claimed to have hit upon his analogy between Heisenberg's matrix mechanics and the Poisson's brackets formulation of Hamiltonian mechanics while taking his Sunday constitutional in the Cambridgeshire countryside. Although not an undergraduate at Cambridge, Dirac adopted the local habit of taking long walks on Sundays to

126. Thompson, *Life of William Thomson*, 2:614.
127. Wiener, *I Am a Mathematician*, 152; see also idem, *Ex-Prodigy*, 189.
128. Wiener, *I Am a Mathematician*, 168.
129. Ibid., 170.

recuperate from the "intense studies of the week."[130] Another outstanding Cambridge mathematician and sportsman from the 1930s is Alan Turing. A keen oarsman and runner, Turing had the habit as a young research fellow at King's College to take long runs in the Cambridge countryside in the afternoons. He later described how an important insight for his most famous contribution to mathematics, the Turing machine, came to him while resting in a meadow during such a run through Granchester, the famous "grind" followed by wranglers for more than a century.[131] These examples, which certainly warrant further study, suggest that habits of exercise acquired in undergraduate years assumed additional meaning for the modern researcher. An afternoon walk or run still made a healthy escape from the places and pressures of work, but where talking shop was forbidden in undergraduate sport, the lone walker or runner could reflect, without recourse to books, pen, and paper, upon a research problem. It is, of course, extremely difficult to assess the contribution of such periods of quiet reflection to the creative process but as the stories told by Dirac and Turing make clear, it was solitude in the Cambridge countryside, rather than the library or college room, that provided a memorable and romantic discovery moment for the Cambridge theoretician.

■ *Conclusion* ■

ILIFFE'S ESSAY IN this volume (chapter 4) shows that concerns about both the presentation of the scholarly self and what lifestyle best promoted intellectual endeavor were commonplace in Cambridge well before the nineteenth century. There are, nevertheless, several senses in which the athletic tradition discussed above represents a new, modern economy of the scholarly mind and body. First, there is no evidence that undergraduates—as opposed to college fellows—in early modern Cambridge were especially concerned with the care of the scholarly self, nor that they engaged collectively in daily physical exercise to promote hard study. In the early nineteenth century, by contrast, athleticism became a widespread student activity—continued by many scholars into later life—and one that was preserved and passed on within student culture. Second, the novel regimens of daily exercise followed by students appear to have originated not as a new twist in fashionable self-presentation but, at least in part, in response to the emergence in the 1810s and 1820s of what we would now recognize as modern practices of training and examining students in technical disciplines. Only in the early Victorian period did the competitive study of mathematics become a hallmark of cerebral manliness. Third, such activities as daily walking

130. Kragh, *Dirac*, 17.
131. Hodges, *Alan Turing*, 96.

and rowing, together with clean living and sexual restraint, gradually implicated the scholar's body in the learning process itself: athleticism preserved a robust constitution, regulated the working day, precluded what were considered illicit activities, and fueled the competitive ethos. But despite the complementary nature of physical and intellectual endeavor in Cambridge, it is important to remember that both enterprises retained a large measure of autonomy. Wiener, for example, saw the competitive spirit and sense of mathematical fair play exhibited by the young Paley as derivative of an English playing-field mentality; yet it was, as we have seen, an accomplishment of the early Victorian era to make physical exercise an activity of moral education and the competitive study of mathematics a manly pursuit. What Wiener did not appreciate, therefore, was that Paley's competitive spirit and well-defined sense of how a mathematical problem should be formulated and solved was as much derivative of the Cambridge style of teaching mathematics as it was of the ideals of competitive sport. In this sense, Paley's attitudes to mathematics and skiing should be seen as mutually supportive, rather than as causal in either direction.[132]

The intimate relationship between hard training both in the coaching room and on the playing field also indicates that we should treat the traditional distinction between "reading" and "rowing" men—between aesthetes and athletes—with caution.[133] While it is certainly the case that rowing men were not scholars, most reading men were sportsmen. Indeed, from the 1840s until at least the 1870s, it was the hardest-reading men who took the hardest exercise. Only in the climate of anti-intellectual athleticism toward the end of the century did the distinction between the elite mathematician and the elite sportsman begin to become a distinction between the serious scholar and the college hearty. The athletic accomplishments of men like Hardy, Littlewood, and Paley remind us, nevertheless, that many successful mathematicians did continue to participate in competitive college sport to the highest level. The combination of physical and intellectual endeavor that typified wrangler studies requires us also to exercise caution in assessing the practical skills of the most able graduates. Conservative ideologues such as Whewell defended a "liberal education" as a means of preserving an elite Cambridge intellectualism against the radical, materialist, and commercial forces that they felt were displacing traditional aristocratic and Anglican values in England. In practice, however, the regimens of hard training

132. For a contemporary discussion of the importance of seeing all mathematical research as "problems" see Pearson, *Karl Pearson,* 27.

133. The terms *reading man* and *rowing man* were current in early Victorian Cambridge. The ambiguity in the latter term (where *row* is pronounced to rhyme with *cow*) captured the stereotypical image of the boisterous, quarrelsome, nonreading man; see Bristed, *Five Years in an English University,* 1:23; and Smith and Wise, *Energy and Empire,* 71.

through which men prepared for the competitive ordeal of the Mathematical Tripos produced graduates with precisely the narrow technical skills against which Whewell and Sydney Smith had cautioned. The claim by observers in the 1840s that Cambridge had effectively industrialized the learning process confirms that ranking students through highly competitive written examinations in mathematics had enabled the very forces of secular, technical meritocracy, which Whewell had resisted, to undermine the ideal of a liberal education by another route. What this system of tough technical training did produce in the later decades of the century was a remarkable school of outstanding mathematical physicists whose mastery of mathematical technique and problem solving was unrivaled elsewhere in Britain and the envy of other major European centers of mathematical physics.[134]

Finally, the disciplining of the student body in ways generally conducive to a new social order in the academy, and to progressive study and competitive examination in particular, became a distinctive feature of technical education in many elite Anglo-American institutions in the late nineteenth and early twentieth centuries. Curthoys and Jones, for example, have shown that undergraduate athleticism developed in Victorian Oxford in broadly similar ways to those discussed above, though apparently with a less direct link between intellectual and physical endeavor.[135] Likewise, Owens has shown that athleticism began to be deployed in American colleges in the 1860s in ways that parallel earlier developments in Oxbridge. Harvard and Yale both encouraged competitive sport in order to improve student discipline, preserve physical fitness, balance intellectual endeavor, and nurture competitiveness, perseverance, individualism, and respect for established authority.[136] Owens does not discuss the origins of these developments in depth but, bearing in mind that Emerson and Bristed wrote about British universities with educational reform in mind, it seems likely that Cambridge played some role as a model for comparable American institutions. Recent ethnographic studies of contemporary student life also suggest that the values of competitive technical training consolidated in early Victorian Cambridge remain embodied in student learning practices in the elite scientific institutions of the late twentieth century.[137] Traweek shows, for example, that would-be high-energy physicists submit themselves to years of progressive training, are extremely com-

134. For the complimentary remarks on mathematical training in Cambridge by Felix Klein (head of the Göttingen school) see Routh, *Die Dynamik der Systeme starrer Körper*, foreword.
135. Curthoys and Jones, "Oxford Athleticism." Oxford developed written examinations around classical rather than mathematical studies and placed much less emphasis on the ranking of students by marks; see Rothblatt, "The Student Sub-culture," 261, 280.
136. Owens, "Pure and Sound Government."
137. Traweek, *Beamtimes and Lifetimes*, 74–105; White, *The Idea Factory*.

petitive, hardworking, and contemptuous of mediocrity, and discountenance extramarital liaisons as "an unworthy distraction of vital energies."[138] There are, furthermore, few modern higher-educational establishments that do not provide extensive facilities for competitive sport and the physical development of the individual. These embodied values of modern academic life are now so commonplace as to be virtually invisible, yet their origin has both historical and epistemological significance. It is tempting to see the emergence of progressive teaching and competitive written examination in the early nineteenth century as an inevitable process in which the mathematical sciences finally shed all cultural baggage to become purely technical disciplines. But as we saw in the section on mathematics, competitive study, and liberal education, these historical events cannot be understood in isolation from the changing role of the ancient universities in Georgian England. From an epistemological perspective, moreover, the cultural values associated with the establishment of a large community of technically expert mathematicians and mathematical physicists can also be explored at the level of everyday undergraduate experience. The private anguish expressed by Ellis, Smith, and Galton makes visible the painful process by which student sensibilities were slowly fashioned to the needs of industrialized learning. In the process, athleticism became an adjunct to competitive study and mathematics became a manly pursuit that selected and shaped the minds of the intellectual elite. Conversely, the mathematics prized in Cambridge came to reflect the ideals of competition, examination, and fair play, prevalent in the coaching room and on the playing field. In exploring the minutiae of everyday student activity we are, therefore, dealing simultaneously with historical questions of local and national significance, a fact acknowledged by Pearson, who, viewing the training of Cambridge mathematicians in the context of Anglo-German competition, saw the "aesthetic ideal, mind and body combined" as a weapon for the British national cause.

138. Traweek, *Beamtimes and Lifetimes*, 84.

■ ACKNOWLEDGMENTS ■

I would like to thank David Edgerton, Rob Iliffe, and Alison Winter for making a number of helpful comments on an earlier draft of this essay. I would also like to thank Joan Richards and Eileen Magnello for advice on the Ellis and Pearson papers respectively. Materials from the Ellis, Browning, Pearson, and Forbes papers are reproduced with kind permission of Trinity College Cambridge Library, King's College Cambridge Library, University College London Library, and University of Saint Andrews Library respectively.

▪ REFERENCES ▪

Airy, George B. *Autobiography of Sir George Biddell Airy.* Ed. Wilfred Airy. Cambridge: Cambridge University Press, 1896.

Atkinson, Solomon. "Struggles of a Poor Student Through Cambridge." *London Magazine,* April 1825, 491–510.

Ball, Walter W. Rouse. *A History of the Study of Mathematics at Cambridge.* Cambridge: Cambridge University Press, 1889.

———. *Notes on the History of Trinity College Cambridge.* London: Macmillan, 1899.

Becher, Harvey. "The Social Origins and Post-graduate Careers of Cambridge Intellectual Elite, 1830–1860." *Victorian Studies* 28 (1984): 97–127.

———. "William Whewell and Cambridge Mathematics." *Historical Studies in the Physical Sciences* 11 (1980): 1–48.

"The Boating Mania." *Eagle* 2 (1861): 41–44.

Bristed, Charles. *Five Years in an English University.* 2 vols. New York: Putman and Co., 1852.

Campbell, Lewis, and William Garnett. *The Life of James Clerk Maxwell.* London: Macmillan, 1882.

"Characters of Freshmen." *Cambridge University Magazine* 1 (1840): 176–79.

Clifford, William K. *Lectures and Essays.* Ed. L. Stephen and F. Pollock. 2 vols. London: Macmillan, 1879.

Curthoys, M. C., and H. S. Jones. "Oxford Athleticism, 1850–1914: A Reappraisal." *History of Education* 24 (1995): 305–17.

Desmond, Adrian, and James Moore. *Darwin.* London: Michael Joseph, 1992.

Emerson, Ralph W. *English Traits.* Ed. H. M. Jones. 1856. Reprint, Cambridge: Harvard University Press, 1966.

Fisch, Menachem, and Simon Schaffer, eds. *William Whewell: A Composite Portrait.* Oxford: Clarendon Press, 1991.

Galton, Francis. *Hereditary Genius: An Inquiry into Its Laws and Consequences.* London: Macmillan, 1869.

———. *Memories of My Life.* London: Methuen and Co., 1908.

Gascoigne, John. "Mathematics and Meritocracy: The Emergence of the Cambridge Mathematical Tripos." *Social Studies of Science* 14 (1984): 547–84.

Goldman, Lawrence, ed. *The Blind Victorian: Henry Fawcett and British Liberalism.* Cambridge: Cambridge University Press, 1989.

Gould, Paula. "'A Thing Inexpedient and Immodest': Women and the Culture of University Physics in Late Nineteenth Century Cambridge." Unpublished undergraduate dissertation, Whipple Museum Library, Cambridge, 1993.

Haley, Bruce. *The Healthy Body and Victorian Culture.* Cambridge: Harvard University Press, 1978.

Halladay, Eric. *Rowing in England: A Social History.* Manchester: Manchester University Press, 1990.

Hardy, G. H. *A Mathematician's Apology.* 1940. Reprint, Cambridge: Cambridge University Press, 1992.

Hasse, Henry R. "My Fifty Years of Mathematics." *Mathematical Gazette* 35 (1951): 153–64.

Hilton, Boyd. "Manliness, Masculinity and the Mid-Victorian Temperament." In Goldman, 60–70.

Hodges, Andrew. *Alan Turing: The Enigma.* London: Vintage, 1983.

Hopkins, William. *Remarks on the Mathematical Teaching of the University of Cambridge.* Cambridge, 1854.

Hopkinson, Bertrand, ed. *Original Papers by the Late John Hopkinson.* 2 vols. Cambridge: Cambridge University Press, 1901.

Kragh, Helge. *Dirac: A Scientific Biography.* Cambridge: Cambridge University Press, 1990.

Larmor, Joseph. *Aether and Matter.* Cambridge: Cambridge University Press, 1900.

Littlewood, J. E. *Littlewood's Miscellany*. Ed. B. Bollobas. Cambridge: Cambridge University Press, 1986.

MacLeod, Roy, ed. *Days of Judgement: Science, Examination and the Organisation of Knowledge in Late Victorian England*. Driffield: Nafferton, 1982.

Mangan, James A. *Athleticism in the Victorian and Edwardian Public School: The Emergence and Consolidation of an Educational Ideology*. Cambridge: Cambridge University Press, 1981.

Mangan, James A., and James Walvin, eds. *Manliness and Morality: Middle-Class Masculinity in Britain and America 1800–1940*. Manchester: Manchester University Press, 1987.

Mansfield, N. "Grads and Snobs: John Brown, Town and Gown in Early Nineteenth-Century Cambridge." *History Workshop Journal* 35 (1993): 184–98.

Mason, Michael. *The Making of Victorian Sexuality*. Oxford: Oxford University Press, 1995.

Moggridge, D. E. *Maynard Keynes: An Economist's Biography*. London: Routledge, 1992.

Moorehead, Caroline. *Bertrand Russell: A Life*. London: Sinclair Stevenson, 1992.

Owens, Larry. "Pure and Sound Government: Laboratories, Playing Fields, and Gymnasia in the Nineteenth-Century Search for Order." *Isis* 76 (1985): 182–94.

Pearson, E. S. *Karl Pearson: An Appreciation of Some Aspects of His Life and Work*. Cambridge: Cambridge University Press, 1938.

Pearson, Karl. *The Life, Letters and Labours of Francis Galton*. Cambridge: Cambridge University Press, 1914.

Pupin, Michael. *From Immigrant to Inventor*. New York: Scribner's, 1926.

Rothblatt, Sheldon. "Failure in Early Nineteenth-Century Oxford and Cambridge." *History of Education* 11 (1982): 1–21.

———. *The Revolution of the Dons: Cambridge Society in Victorian England*. London: Faber and Faber, 1968.

———. "The Student Sub-culture and the Examination System in Early 19th Century Oxbridge." In *The University in Society*, ed. Stone, 1:247–303.

Routh, E. *Die Dynamik der Systeme starrer Körper*. Trans F. Klein. Leipzig: A. Schepp, 1898.

Satthianadhan, Samuel. *Four Years in an English University*. Madras: Srinivasa, 1890.

Seeley, Robert. "A Plea for More Universities." *London Student*, April 1868, 9.

———. "Recreation." *London Student*, July 1868, 205–8.

Smith, Crosbie, and M. Norton Wise. *Energy and Empire: A Biographical Study of Lord Kelvin*. Cambridge: Cambridge University Press, 1989.

Smith, Sydney. "Remarks of the System of Education in Public Schools." *Edinburgh Review* 16 (1810): 326–34.

Stephen, Leslie. *Life of Henry Fawcett*. London: Smith, Elder and Co., 1885.

Stokes, George G. *Memoirs and Scientific Correspondence of Sir George Stokes*. 2 vols. Cambridge: Cambridge University Press, 1907.

Stone, Lawrence, ed. *The University in Society*. 2 vols. Princeton: Princeton University Press, 1974.

Strutt, R. J. *The Life of John William Strutt, Third Baron Rayleigh*. London: Edward Arnold and Co., 1924.

Thompson, S. P. *The Life of William Thomson*. 2 vols. London: Macmillan, 1910.

Thomson, Joseph J. *Recollections and Reflections*. London: G. Bell and Sons, 1936.

Todhunter, Isaac. *The Conflict of Studies and Other Essays on Subjects Connected with Education*. London: Macmillan, 1873.

Traweek, Sharon. *Beamtimes and Lifetimes: The World of High Energy Physicists*. Cambridge: Harvard University Press, 1988.

Venn, J. *A Biographical History of Gonville and Caius College*. 5 vols. Cambridge: Cambridge University Press, 1901.

Walton, W. *The Mathematical and Other Writings of Robert Leslie Ellis*. Cambridge: Deighton, Bell, and Co., 1863.

Warwick, Andrew C. *Masters of Theory: The Pursuit of Mathematical Physics in Victorian Cambridge*. Cambridge: Cambridge University Press, forthcoming.

Whewell, William. *Thoughts on the Study of Mathematics as Part of a Liberal Education.* Cambridge: Cambridge University Press, 1835.

White, Pepper. *The Idea Factory: Learning to Think at MIT.* New York: Dutton, 1991.

Whittaker, Edmund. *A Course of Modern Analysis.* Cambridge: Cambridge University Press, 1902.

Wiener, Norbert. *Ex-Prodigy: My Childhood and Youth.* New York: Simon and Schuster, 1953.

———. *I Am a Mathematician: The Later Life of a Prodigy.* Cambridge: MIT Press, 1956.

Williams, Perry. "Passing on the Torch: Whewell's Philosophy and the Principles of English Education." In Fisch and Schaffer, 117–47.

Wilson, David B. "Experimentalists among the Mathematicians: Physics in the Cambridge Natural Sciences Tripos, 1851–1900." *Historical Studies in the Physical Sciences* 12 (1982): 325–71.

Wordsworth, Christopher. *Scholae Academicae: Some Account of the Studies at the English Universities in the Eighteenth Century.* Cambridge: Cambridge University Press, 1877.

———. *Social Life at the English Universities in the Eighteenth Century.* Cambridge: Deighton, Bell, 1874.

Wright, James M. F. *Alma Mater: Or Seven Years at the University of Cambridge.* 2 vols. London: Black, Young, and Young, 1827.

CONTRIBUTORS

JANET BROWNE
   The Wellcome Institute for the
      History of Medicine
   183 Euston Road
   London NW1 2BE
   England

PETER DEAR
   Department of History
   McGraw Hall
   Cornell University
   Ithaca, NY 14853

ROB ILIFFE
   Imperial College of Science and
      Technology
   Sherfield Building
   Exhibition Road
   London SW7 2AZ
   England

CHRISTOPHER LAWRENCE
   The Wellcome Institute for the
      History of Medicine
   183 Euston Road
   London NW1 2BE
   England

SIMON SCHAFFER
   Department of History and
      Philosophy of Science
   University of Cambridge
   Free School Lane
   Cambridge CB2 3RH
   England

STEVEN SHAPIN
   Department of Sociology
   University of California, San Diego
   La Jolla, CA 92093-0533

ANDREW WARWICK
   Imperial College of Science and
      Technology
   Sherfield Building
   Exhibition Road
   London SW7 2AZ
   England

ALISON WINTER
   Division of the Humanities and Social
      Sciences 228-77
   California Institute of Technology
   Pasadena, CA 91125

INDEX

Page references to figures are set in italics.

abstinence: definition of, 31–32n. 31; effects of, 34–35, 127–28, 130; and examinations, 312–13; and fasting, 35, 105, 148; versus moderation, 30–33; praise for, 147–48; and Pythagoras, 26; recommendations on, 30–31; significance of, 37–39. *See also* temperance

academics: and embodiment, 4–7; status of, 15. *See also* intellectuality; scholars

Acland, Henry Wentworth, *173,* 216

Adam (biblical): bodily identity of, 11; capabilities of, 93–94; diet of, 28; and knowledge, 83–84

Adams, Charles, 53n. 4

Aguecheek, Andrew (character), 186

Airy, George Biddell, 240–41, 251–52, 303

alchemy, 126, 139–40, 142

alcohol: consumption of, 36; and disease, 135; effects of, 39, 128, 130; recommendations on, 123–24; treatment with, 62. *See also* drugs

Alexander the Great, 25

algebra, development of, 218

alienists, and doctor-patient relations, 225–26

Allen (doctor), 98, 100

Allingham, William, 248–49

analyst, Lovelace as, 203, 218, 227–28

Analytic Engine (Babbage's), 202, 213, 226–30

Anatomy of Melancholy (Burton), 37, 125–26

angels, knowledge of, 83

animals: and automata, 66; experiments on, 95–96; and transfusions, 97–102, *99*

anonymity, and publication, 73

Anthony (Saint), 27, 73n. 81, 146

Antiquity, 28–31, 33, 43

Apollonius, 35

apothecaries, 104, 167–68, 187n. 81

Appleton, Thomas, 271

Arbuthnot, John, 91, 149–51

Archimedes, 125, 134

Aristotle, 30, 37, 58, 124

Arrowsmith, Martin (character), 42

art: and nature, 77; and truth, 14

asceticism: and bodily identity, 10–11; and chastity, 146–48; Christian tropes of, 26–33; classical tropes of, 24–26; critique of, 35–36; demise of notion of, 45–46; Jewish traditions of, 29–30; as model, 56; versus moderation, 30–33; recommendations on, 124; subversiveness of, 31–32

Askew, Anthony, *172*

astronomy, mechanical analogies in, 59

atheism, and cures for disease, 110–11

athleticism: culture of, 289–90; development of, 294–95, 322; and mathematics, 311–20; meanings of, 311–15, 320–23; popularization of, 312; role of, 302–11

Atkinson, Solomon, 303

Aubrey, John, 36n. 44, 141

Augustine (Saint), 29, 65

Austen, Jane, 248

authority, and mind/body issues, 225–26

automata: and animals, 66; and Descartes's philosophy, 58–68; intelligibility of, 77; and language, 66–67; people as, 60, 78; shortcomings of, 74–76; study of, 213–14. *See also* instruments; machines; mechanics

autonomy, and Descartes, 74–78

Averroës (Ibn Rushd), 75

Avicenna, 123

Babbage, Charles: Analytic Engine of, 202, 213, 226–30; and Lovelace's illness, 235; Lovelace's relation with, 202, 213–14, 216, 226–28

Babington, Humphrey, 135n. 21, 138

Bacon, Francis, 35–36, 84

Baillet, Adrien, 54–56, 72, 74

Bakhtin, Mikhail, 43n. 67

Balsam, Lucatello's, 135, 152

Balzac, Guez de, 65

barbers, versus surgeons, 183

passions, 69–72; of physical world, 58–59, 76–78. *See also* automata; instruments

Medicatrix or the Woman-Physician (Trye), 159

medicine: and abstemiousness, 30–31; attitudes toward, 61–62; and body, 156–60; cultural uses of, 242; and dietary excess, 36n. 44; ethics in, 170; and mind-body dualism, 11–12; and moderation, 33; Montaigne's attitudes toward, 34–35; and nakedness, 64; and Newton's concoctions, 134–37, 152; and passions, 69; practice of, 165–68. *See also* diseases; health (physical); invalidism; physicians; surgeons

Meditations (Descartes), 58, 65, 75

melancholia: causes of, 34, 36–39, 125–26, 144–45; conditions associated with, 124–25; countermeasures for, 125–26; and cures for disease, 108; definition of, 37; discourse on, 121; Locke on, 127–28n. 12

memory, mechanization for, 226–30

Menabrea, L. F., 202

Mencke, Otto, 140n. 31, 143

Merret, Christopher, 167

Mersenne, Marin, 56n. 17, 70n. 69

mesmerism, 210–11, 225–29, 234

metaphysics, 61

Micrographia (Hooke), 93

microscopes, 83, 92–94

middle class, and competitive examinations, 291

midwifery, as manual operation, 168

military, as human automaton, 60

De militia Romana (Lipsius), 60

Mill, John Stuart, 210

millenarianism, 108

Millington, John, 140–42

Milton, John, 84, 209

mind: and aesthetic ideal, 288–90; approach to, 4–9; and bodily identity, 10–13; bodily support for, 132; body as antagonistic to, 245–46; care of, 37–40; development and management of, 221–22; and disembodiment, 1–4; and mechanical representation, 226–27; passions controlled by, 68–71; properties of, 75–76; and self-experimentation, 231–33; surroundings of, *279,* 279–80. *See also* body; disembodiment; knowledge; souls

miracles: claims of, 109–13, 115; discourse on, 88; explanations of, 113–14; by monarchy, 85, *87,* 87–88, 108, 111, 116

The Miraculous Conformist (Stubbe), 109, 111

moderation: versus abstemiousness, 30–33; as civilized, 63–64; recommendations for, 33–37

modernism, 6, 15

modernization, and bodily control, 9

Molière, Jean-Baptiste Poquelin, 62

monarchy: celebration of, 85, *86;* maintenance of, 84; and miracles, 85, *87,* 87–88, 108, 111, 116

monasticism: criticism of, 34n. 39, 146, 148; dietetics of, 31–32, 44. *See also* abstinence

Mondeville, Henry de, 186

Monmouth (Lord, Charles Mordaunt), 138–39

Montague, Charles, 138–39, 144

Montaigne, Michel Eyquem de, 34–35

Moore, James, 241

morals: of animals versus humans, 98; conforming to, 63–65; and Descartes's philosophy, 58–68; importance of, 61; and physical exercise, 311–15, 321; and physicians, 169; and transfusions, 99–100. *See also* sanative virtue

Moray, Robert, 91

More, Henry: and bodily identity, 23; and complexion, 39–40; and cures for diseases, 107–13; dietetics of, 38–39, 39n. 53; on melancholia, 121; and temperance, 147–48; and vision, 90

More, L. T., 41

More, Thomas, 186

Morgan, Vance G., 70

Muchembled, Robert, 63

Murray, John, 233–34, 250

Muses, creation of, 24

music, and passions, 70n. 69

Naden, Constance, 210n. 35

Nanson, Edward, 309

Napier, Richard, 124–25

natural philosophy: and Adam, 84; and civilizing process, 63–65; claims of, 88–89; and cures for disease, 112–16; and mechanization, 11, 53; and transfusion, 96–105

nausea, theories of, 245

Nedham, Marchamont, 164, 165–66

neo-Stoicism, 60–61n. 31, 68

nervous system, medical treatment of, 246–47

Netherlands: Descartes in, 55–58, 64–65, 72–73; development of, 60

Newton, Humphrey, 132–33, 134n. 20, 136, 147